Springer Monographs in Mathematics

Editors-in-chief

Isabelle Gallagher, Paris, France
Minhyong Kim, Oxford, UK

Series editors

Sheldon Axler, San Francisco, USA
Mark Braverman, Princeton, USA
Maria Chudnovsky, Princeton, USA
Tadahisa Funaki, Tokyo, Japan
Sinan C. Güntürk, New York, USA
Claude Le Bris, Marne la Vallée, France
Pascal Massart, Orsay, France
Alberto Pinto, Porto, Portugal
Gabriella Pinzari, Napoli, Italy
Ken Ribet, Berkeley, USA
René Schilling, Dresden, Germany
Panagiotis Souganidis, Chicago, USA
Endre Süli, Oxford, UK
Shmuel Weinberger, Chicago, USA
Boris Zilber, Oxford, UK

This series publishes advanced monographs giving well-written presentations of the "state-of-the-art" in fields of mathematical research that have acquired the maturity needed for such a treatment. They are sufficiently self-contained to be accessible to more than just the intimate specialists of the subject, and sufficiently comprehensive to remain valuable references for many years. Besides the current state of knowledge in its field, an SMM volume should ideally describe its relevance to and interaction with neighbouring fields of mathematics, and give pointers to future directions of research.

More information about this series at http://www.springer.com/series/3733

David E. Edmunds · W. Desmond Evans

Elliptic Differential Operators and Spectral Analysis

 Springer

David E. Edmunds
Department of Mathematics
University of Sussex
Brighton, UK

W. Desmond Evans
School of Mathematics
University of Cardiff
Cardiff, UK

ISSN 1439-7382 ISSN 2196-9922 (electronic)
Springer Monographs in Mathematics
ISBN 978-3-030-02124-5 ISBN 978-3-030-02125-2 (eBook)
https://doi.org/10.1007/978-3-030-02125-2

Library of Congress Control Number: 2018960175

Mathematics Subject Classification (2010): 35Jxx, 35Pxx, 35Qxx, 47A10

© Springer Nature Switzerland AG 2018
This work is subject to copyright. All rights are reserved by the Publisher, whether the whole or part of the material is concerned, specifically the rights of translation, reprinting, reuse of illustrations, recitation, broadcasting, reproduction on microfilms or in any other physical way, and transmission or information storage and retrieval, electronic adaptation, computer software, or by similar or dissimilar methodology now known or hereafter developed.
The use of general descriptive names, registered names, trademarks, service marks, etc. in this publication does not imply, even in the absence of a specific statement, that such names are exempt from the relevant protective laws and regulations and therefore free for general use.
The publisher, the authors and the editors are safe to assume that the advice and information in this book are believed to be true and accurate at the date of publication. Neither the publisher nor the authors or the editors give a warranty, express or implied, with respect to the material contained herein or for any errors or omissions that may have been made. The publisher remains neutral with regard to jurisdictional claims in published maps and institutional affiliations.

This Springer imprint is published by the registered company Springer Nature Switzerland AG
The registered company address is: Gewerbestrasse 11, 6330 Cham, Switzerland

Preface

In 1987, our book *Spectral Theory and Differential Operators* was published. It emphasised the symbiotic relationship, as we saw it, between the theory of compact linear operators acting between Banach spaces and the study of boundary value problems for elliptic partial differential equations. Although there have been many advances in the theory since then, the book contained much material of lasting importance that is relatively unaffected by recent progress. Nevertheless, we feel that it is now appropriate to give an account of some topics, both old and new, not covered in the previous book. The central theme is that of elliptic operators: this is a huge subject, well able to support many variations in approach. Those given here reflect our interests and limitations of our knowledge.

The reader is assumed to be familiar with the theory of Lebesgue integration and basic facts concerning functional analysis and the theory of function spaces. A summary of the most fundamental results in these topics is provided in Chap. 1 together with references to more detailed accounts. With an eye to later usage, a Banach space version of the Lax-Milgram lemma is established, as is the determination (due to Pichorides) of the exact value of the norm of the Riesz transform as a map from L_p to itself. The next two chapters cover well-known results involving the Laplace operator, such as maximum principles, Weyl's lemma and the Perron approach to the Dirichlet problem, together with maximum principles for second-order elliptic operators. Chapter 4 studies the classical Dirichlet problem for second-order elliptic operators and presents such familiar matters as Kellogg's theorem and the Schauder boundary estimates. The approach given is that due to M. König in a series of papers and does not seem to have appeared in book form before now: we believe that it has pedagogical advantages over more traditional methods in being simpler and more direct. The next chapter provides various notions of ellipticity for operators of arbitrary order and establishes Gårding's inequality for uniformly strongly elliptic operators. The Dirichlet problem for such operators is discussed: to prove the existence of a classical solution would entail the establishment of a higher-order version of the Schauder estimates, and as we shrink from this formidable task, it is shown that a weak solution exists. Such solutions belong to Sobolev spaces based on the Hilbert space L_2, and regularity theory is needed to show that the weak solution is actually classical: indications of how such arguments go are given for the Poisson equation. The Dirichlet eigenvectors of the Laplacian are next considered: Courant's min-max principle is proved, as are the analyticity of eigenvectors and the Faber–Krahn inequality for the first eigenvalue. The chapter concludes with a brief discussion of semigroup theory and its connection with spectral independence.

Chapter 6 continues the Hilbert space approach of the last chapter and is devoted to self-adjoint extensions of symmetric operators acting in a Hilbert space. After a brief resumé of the von Neumann theory, characterisations of self-adjoint extensions in terms of linear relations and boundary triplets are given, and associated gamma fields and Weyl M-functions are introduced and their main properties noted; a comprehensive treatment may be found in the book by Schmüdgen [190]. The main theme of Chap. 6 is an account of the Krein-Vishik-Birman theory concerning the positive self-adjoint extensions of positive symmetric operators, and Grubb's extension of the theory to adjoint pairs of closed operators. Analogous results of Arlinski and his co-authors on the m-sectorial extensions of sectorial operators are also included. Application of these abstract results is made in the next chapter to realisations of second-order elliptic operators. The first four sections of Chap. 7 deal with symmetric Sturm–Liouville operators satisfying minimal assumptions, followed by a brief description for coercive sectorial operators of Sturm–Liouville type. Then, an outline is given of the work of Grubb in which her abstract theory for determining all the closed realisations of an adjoint pair of operators A, A' is applied to uniformly elliptic operators generated by differential expressions A with smooth coefficients and formal adjoint A', defined on a smooth domain $\Omega \subset \mathbb{R}^n$ ($n \geq 2$). This leads to the identification of all closed realisations of A by means of boundary conditions on $\partial\Omega$ expressed in terms of differential operators acting between function spaces defined on $\partial\Omega$.

Chapter 8 marks a break with the Hilbert space approach: the necessity for this stems from consideration of the Poisson equation with right-hand side f that belongs to L_p for some $p \in (1, 2)$ but does not belong to L_2. The methods of Chap. 5 are then not applicable, but it turns out that the existence of an appropriate type of weak solution in the context of L_p can be established. To do this, some techniques introduced by Simader and Sohr can be used: we give a simplified version of their approach adapted to the particular case we consider. Chapter 9 is devoted to the p–Laplacian, the literature on which is so enormous as to appear overwhelming to those unfamiliar with the topic. Our object here is modest: by concentrating on a small number of problems, we hope to encourage the novice to pluck up enough courage to venture more deeply into the subject. The existence of a solution of the corresponding Dirichlet problem is proved, together with a variety of results concerning eigenvalues, including a version of the Courant nodal domain theorem. The final three chapters are intended to give some idea of current work in which we are interested. Chapter 10 describes some very recent forms of the Rellich inequality; Chap. 11 provides further properties of Sobolev embeddings, such as necessary and sufficient conditions for a Sobolev embedding to be nuclear, and a characterisation of the subspace of a Sobolev space consisting of functions with zero trace; Chap. 12 discusses positive operators which model relativistic properties of the Dirac operator, special attention being paid to the self-adjoint realisations of the Brown–Ravenhall operator defined on a domain $\Omega \subsetneq \mathbb{R}^3$. Notes are given at the end of most chapters to provide the reader with further references and indications of directions taken by current research.

It is hoped that the cocktail of results and techniques presented here with which diverse questions related to the spectral theory of differential operators may be attacked will prove to be of interest, especially since a good deal of the material is hard to find in book form.

Chapters are divided into sections, and some sections are divided into subsections. Theorems, corollaries, lemmas, propositions, remarks and equations are numbered consecutively in each section.

Brighton, UK David E. Edmunds
Cardiff, UK W. Desmond Evans

Contents

1	**Preliminaries**	1
	1.1 Integration	1
	1.2 Functional Analysis	2
	1.3 Function Spaces	9
	1.3.1 Spaces of Continuous Functions	9
	1.3.2 Sobolev Spaces	14
	1.4 The Hilbert and Riesz Transforms	24
2	**The Laplace Operator**	35
	2.1 Mean Value Inequalities	35
	2.2 Representation of Solutions	42
	2.3 Dirichlet Problems: The Method of Perron	51
	2.4 Notes	56
3	**Second-Order Elliptic Equations**	57
	3.1 Basic Notions	57
	3.2 Maximum Principles	58
4	**The Classical Dirichlet Problem for Second-Order Elliptic Operators**	65
	4.1 Preamble	65
	4.2 The Poisson Equation	65
	4.3 More General Elliptic Operators	73
	4.4 Notes	81
5	**Elliptic Operators of Arbitrary Order**	83
	5.1 Preliminaries	83
	5.2 Gårding's Inequality	86
	5.3 The Dirichlet Problem	91
	5.4 A Little Regularity Theory	95
	5.5 Eigenvalues of the Laplacian	99

5.6	Spectral Independence	106
5.7	Notes	113

6 Operators and Quadratic Forms in Hilbert Space ... 115
- 6.1 Self-Adjoint Extensions of Symmetric Operators ... 115
- 6.2 Characterisations of Self-Adjoint Extensions ... 121
  - 6.2.1 Linear Relations ... 121
  - 6.2.2 Boundary Triplets ... 123
  - 6.2.3 Gamma Fields and Weyl Functions ... 128
- 6.3 The Friedrichs Extension ... 130
- 6.4 The Krein-Vishik-Birman (KVB) Theory ... 136
- 6.5 Adjoint Pairs and Closed Extensions ... 144
- 6.6 Sectorial Operators ... 151
  - 6.6.1 The Friedrichs Extension ... 153
  - 6.6.2 The Krein-von Neumann Extension ... 156
- 6.7 Notes ... 158

7 Realisations of Second-Order Linear Elliptic Operators ... 159
- 7.1 Sturm–Liouville Operators: Basic Theory ... 159
  - 7.1.1 The Regular Problem ... 160
  - 7.1.2 One Singular Point ... 163
  - 7.1.3 Two Singular End-Points ... 165
  - 7.1.4 The Titchmarsh–Weyl Function and Spectrum ... 166
- 7.2 KVB Theory for Positive Sturm–Liouville Operators ... 168
  - 7.2.1 Semi-boundedness and Oscillation Theory ... 168
  - 7.2.2 Kalf's Theorem ... 171
- 7.3 Application of the KVB Theory ... 175
  - 7.3.1 The Limit-Point Case at b ... 175
  - 7.3.2 The Case of b Regular or Limit Circle, and $\tau u = 0$ Non-oscillatory at b ... 179
  - 7.3.3 Limit-Point and Limit-Circle Criteria ... 183
- 7.4 Coercive Sectorial Operators ... 184
  - 7.4.1 The Case $dim(ker\ T^*) = 2$ ... 188
- 7.5 Realisations of Second-Order Elliptic Operators on Domains ... 188
- 7.6 Notes ... 201

8 The L_p Approach to the Laplace Operator ... 203
- 8.1 Preamble ... 203
- 8.2 Technical Results ... 204
- 8.3 Existence of a Weak L_p Solution ... 207
- 8.4 Other Procedures ... 210
- 8.5 Notes ... 211

9	**The *p*-Laplacian**		**213**
	9.1	Preamble	213
	9.2	Preliminaries	213
	9.3	The Dirichlet Problem	215
	9.4	An Eigenvalue Problem	218
	9.5	More About the First Eigenvalue	228
	9.6	Notes	233
10	**The Rellich Inequality**		**235**
	10.1	Preamble	235
	10.2	The Mean Distance Function	236
	10.3	Results Involving the Laplace Operator	241
	10.4	The *p*-Laplacian	245
11	**More Properties of Sobolev Embeddings**		**249**
	11.1	The Distance Function	249
	11.2	Nuclear Maps	257
	11.3	Asymptotic Formulae for Approximation Numbers of Sobolev Embeddings	265
	11.4	Spaces with Variable Exponent	266
	11.5	Notes	279
12	**The Dirac Operator**		**281**
	12.1	Preamble	281
	12.2	The Dirac Equation	281
	12.3	The Free Dirac Operator	282
	12.4	The Brown–Ravenhall Operator	286
	12.5	Sums of Operators and Coulomb Potentials	287
		12.5.1 The Case $A = \mathbb{D}$	288
		12.5.2 The Case $A = \mathbb{H}$	289
		12.5.3 The Case $A = \mathbb{B}$	292
	12.6	The Free Dirac Operator on a Bounded Domain	294
	12.7	The Brown–Ravenhall Operator on a Bounded Domain	297

Bibliography 303

Author Index 313

Subject Index 317

Notation Index 321

Basic Notation

$\mathbb{C}$: Complex plane
$\mathbb{C}_\pm = \{z \in \mathbb{C} : im\, z \gtrless 0\}$
$\mathbb{C}^n$: n–dimensional complex space
$\mathbb{N}$: Natural numbers; $\mathbb{N}_0 = \mathbb{N} \cup \{0\}$
$\mathbb{R}$: Real line
re z: Real part of $z \in \mathbb{C}$
im z: Imaginary part of $z \in \mathbb{C}$
$\mathbb{R}^n$: n–dimensional Euclidean space
$\mathbb{Z}$: All integers
$D_i u = \partial u / \partial x_i$
If $\alpha = (\alpha_1, \ldots, \alpha_n) \in \mathbb{N}_0^n$, then $D^\alpha u = \partial^{|\alpha|} u / \partial x_1^{\alpha_1} \ldots \partial x_n^{\alpha_n}, |\alpha| = \alpha_1 + \ldots + \alpha_n$.
If $\Omega \subset \mathbb{R}^n$, then $\partial \Omega =$ boundary of $\Omega, \overline{\Omega} =$ closure of $\Omega, \Omega^c = \mathbb{R}^n \backslash \Omega$.
dist $(x, \partial \Omega) =$ distance from x to $\partial \Omega$.
$T\!\upharpoonright_G$: Restriction of T to G.
$f^+ = \max(f, 0), f^- = -\min(f, 0)$.
$A \subset B$ for sets A, B allows for $A = B$.
$\omega_n = \pi^{n/2} / \Gamma(1 + \tfrac{1}{2} n) =$ volume of unit ball in $\mathbb{R}^n$.
$$\text{sgn}(x) := \begin{cases} 1, & \text{if } x > 0, \\ 0, & \text{if } x = 0, \\ -1, & \text{if } x < 0. \end{cases}$$

Chapter 1
Preliminaries

1.1 Integration

We assume that the reader is familiar with the theory of the Lebesgue integral on measurable subsets of $\mathbb{R}^n$. With convenience of reference in mind, we give here statements of results that will prove useful subsequently: proofs may be found in [52], [183] and [195], for example. By Ω we shall usually mean a measurable subset of $\mathbb{R}^n$; its Lebesgue $n-$measure will be denoted by $|\Omega|_n$, or even by $|\Omega|$ if the meaning is clear. All functions are assumed to be extended real-valued.

Given any extended real-valued function f on Ω, we write

$$f^+ = \max\{f, 0\}, \; f^- = -\min\{f, 0\}.$$

Theorem 1.1.1 *(The monotone convergence theorem) Let $(f_k)_{k \in \mathbb{N}}$ be a non-decreasing sequence of extended real-valued, measurable functions on Ω such that $\int_\Omega f_k^-(x)dx < \infty$ for some k. Then*

$$\lim_{k \to \infty} \int_\Omega f_k(x)dx = \int_\Omega \lim_{k \to \infty} f_k(x)dx.$$

Theorem 1.1.2 *(Fatou's lemma) Let $(f_k)_{k \in \mathbb{N}}$ be a sequence of non-negative, extended real-valued, measurable functions on Ω. Then*

$$\int_\Omega \liminf_{k \to \infty} f_k(x)dx \le \liminf_{k \to \infty} \int_\Omega f_k(x)dx.$$

Theorem 1.1.3 *(Lebesgue's dominated convergence theorem) Let $(f_k)_{k \in \mathbb{N}}$ be a sequence of measurable functions on Ω such that for almost all $x \in \Omega$, $\lim_{k \to \infty} f_k(x) = f(x)$. Suppose there exists a function g with finite Lebesgue integral over Ω such that*

$$|f_k(x)| \le g(x) \text{ for all } k \in \mathbb{N} \text{ and almost all } x \in \Omega.$$

Then f and each f_k have finite integrals over Ω and

$$\int_\Omega f(x)dx = \lim_{k\to\infty} \int_\Omega f_k(x)dx.$$

Theorem 1.1.4 *(Fubini's theorem) Let Ω_i be a measurable subset of $\mathbb{R}^{n_i}$ ($i = 1, 2$), put $\Omega = \Omega_1 \times \Omega_2$ and suppose $f : \Omega \to \mathbb{R}$ is such that $\int_\Omega f(x,y)dxdy < \infty$. Then $\int_{\Omega_1} f(x,y)dx$ exists for almost all $y \in \Omega_2$, $\int_{\Omega_2} f(x,y)dy$ exists for almost all $x \in \Omega_1$ and*

$$\int_\Omega f(x,y)dxdy = \int_{\Omega_1}\left(\int_{\Omega_2} f(x,y)dy\right)dx = \int_{\Omega_2}\left(\int_{\Omega_1} f(x,y)dx\right)dy.$$

We shall also need connections between various types of convergence of functions.

Theorem 1.1.5 *Let $p \in [1, \infty)$, let Ω be a measurable subset of $\mathbb{R}^n$ and suppose that f, f_k ($k \in \mathbb{N}$) are functions on Ω such that*

$$\int_\Omega |f(x)|^p < \infty, \int_\Omega |f_k(x)|^p\, dx < \infty\ (k \in \mathbb{N})$$

and

$$\int_\Omega |f(x) - f_k(x)|^p\, dx \to 0\ as\ k \to \infty.$$

Then:
(i) There is a subsequence of (f_k) that converges pointwise a.e. to f.
*(ii) The sequence (f_k) **converges in measure** to f: that is, given any $\varepsilon > 0$,*

$$\lim_{k\to\infty} |\{x \in \Omega : |f_k(x) - f(x)| > \varepsilon\}| = 0.$$

1.2 Functional Analysis

We take for granted the fundamental concepts and results concerning metric spaces.

Let X be a linear space over the real or complex field. A *norm* on X is a map $\|\cdot\|_X = \|\cdot\| : X \to [0, \infty)$ such that
(i) $\|x\| = 0$ if and only if $x = 0$,
(ii) $\|\lambda x\| = |\lambda|\,\|x\|$ for all scalars λ and all $x \in X$,
(iii) $\|x + y\| \le \|x\| + \|y\|$ for all $x, y \in X$.
When it is necessary to exhibit the dependence on X in its norm to avoid confusion, we write $\|u\|_X$ or $\|u|X\|$.

A linear space equipped with a norm is called a *normed linear space*. Every norm $\|\cdot\|$ on X defines a metric d on X by the rule $d(x, y) = \|x - y\|$ $(x, y \in X)$. A *Banach space* is a normed linear space X that is complete with respect to the metric

1.2 Functional Analysis

d. This means that given any Cauchy sequence (x_n) in X (so that $\|x_n - x_m\| \to 0$ as $m, n \to \infty$), there exists $x \in X$ such that $\|x_n - x\| \to 0$ as $n \to \infty$ (written $x_n \to x$).

Standard examples of Banach spaces are, with the natural definitions of addition and multiplication by scalars:

(i) $\mathbb{R}^n$ and $\mathbb{C}^n$ ($n \in \mathbb{N}$) with norm given by $\|x\| = \left(\sum_{j=1}^{n} |x_j|^2\right)^{1/2}$, $x = (x_1, \ldots, x_n)$.

(ii) l_p, the space of all sequences $x = (x_j)_{j \in \mathbb{N}}$ of scalars such that

$$\|x\|_p := \left(\sum_{j=1}^{\infty} |x_j|^p\right)^{1/p} < \infty \quad (1 \leq p < \infty),$$

and

$$\|x\|_\infty := \sup_j |x_j| < \infty \quad (p = \infty).$$

(iii) $L_p(\Omega)$, the linear space of all (Lebesgue) measurable functions f on a measurable subset Ω of $\mathbb{R}^n$, functions equal almost everywhere being identified, such that

$$\|f\|_{p,\Omega} = \left(\int_\Omega |f(x)|^p \, dx\right)^{1/p} < \infty \quad (1 \leq p < \infty),$$

and

$$\|f\|_{\infty,\Omega} = \operatorname*{ess\,sup}_{\Omega} |f(x)| < \infty \quad (p = \infty).$$

In chapter 12 we need the space $L_2(\Omega, \mathbb{C}^4)$ of $\mathbb{C}^4$-valued functions $f = (f_j)_{j=1}^{4}$ whose components f_j lie in $L_2(\Omega)$, and set

$$\|f\|_{L^2(\Omega,\mathbb{C}^4)} = \left(\int_\Omega |f(x)|^2_{\mathbb{C}^4} \, dx\right)^{1/2},$$

where $|f(x)|_{\mathbb{C}^4} = \left(\sum_{j=1}^{4} |f_j(x)|^2\right)^{1/2}$ is the $\mathbb{C}^4$ norm. To establish the norm property of $\|\cdot\|_{p,\Omega}$ when $p < \infty$, the *Minkowski inequality* is needed:

$$\|f + g\|_{p,\Omega} \leq \|f\|_{p,\Omega} + \|g\|_{p,\Omega} \quad \text{for all } f, g \in L_p(\Omega).$$

If $1 < p < \infty$, equality holds if and only if there are non-negative real numbers A, B (not both zero) such that $Af(x) = Bg(x)$ a.e. in Ω. Another important ingredient in the proof is *Hölder's inequality*: if $1 < p < \infty$ and $f \in L_p(\Omega)$, $g \in L_{p'}(\Omega)$, where $1/p' + 1/p = 1$, then

$$\int_\Omega |f(x)g(x)| \, dx \leq \|f\|_{p,\Omega} \|g\|_{p',\Omega},$$

with equality if and only if there are non-negative real numbers A, B (not both zero) such that $A\,|f(x)|^p = B\,|g(x)|^{p'}$ a.e. in Ω.

An *inner product* on a linear space X over the real or complex field Φ is a map $(\cdot,\cdot): X \times X \to \Phi$ such that
(i) $(\alpha x_1 + \beta x_2, y) = \alpha(x_1, y) + \beta(x_2, y)$ for all $\alpha, \beta \in \Phi$ and all $x_1, x_2, y \in X$;
(ii) $(x, y) = \overline{(y, x)}$ for all $x, y \in X$;
(iii) $(x, x) > 0$ if $x \in X \setminus \{0\}$.

A linear space equipped with an inner product is called an inner product space. *Schwarz's inequality* asserts that for all x, y in an inner product space X,

$$|(x, y)|^2 \leq (x, x)^{1/2} (y, y)^{1/2},$$

with equality if and only if x and y are linearly dependent. From this it follows easily that $x \longmapsto (x, x)^{1/2} := \|x\|$ is a norm on X. Elements x, y of an inner product space X such that $(x, y) = 0$ are said to be *orthogonal* and we then write $x \perp y$; two subsets M, N of X are called orthogonal, written $M \perp N$, if $(x, y) = 0$ for all $x \in M$ and $y \in N$; the *orthogonal complement* of a subset M of X is the set

$$M^\perp = \{x \in X : (x, y) = 0 \text{ for all } y \in M\};$$

a set $M \subset X$ is said to be *orthogonal* if $x \perp y$ for all $x, y \in M$ with $x \neq y$; an orthogonal set M such that $(x, x) = 1$ for all $x \in M$ is called *orthonormal*. An inner product space that is complete with respect to the norm $(\cdot, \cdot)^{1/2}$ is called a *Hilbert space*. The spaces l_2 and $L_2(\Omega)$ mentioned above are Hilbert spaces, with natural definitions of the inner products. The space $L_2(\Omega, \mathbb{C}^4)$ is a Hilbert space of $\mathbb{C}^4$-valued functions with inner product

$$(f, g)_{L_2(\Omega, \mathbb{C}^4)} = \int_\Omega \langle f(x), g(x) \rangle_{\mathbb{C}^4}\, dx,$$

where $\langle \cdot, \cdot \rangle_{\mathbb{C}^4}$ is the $\mathbb{C}^4$ inner product defined by $\langle a, b \rangle_{\mathbb{C}^4} = \sum_{j=1}^{4} a_j \overline{b}_j$ for $a = (a_1, a_2, a_3, a_4)$ and $b = (b_1, b_2, b_3, b_4)$ in $\mathbb{C}^4$. The closed unit ball $\{x \in X : \|x\| \leq 1\}$ of a Banach space X is denoted by B_X, and S_X will stand for the unit sphere $\{x \in X : \|x\| = 1\}$. Let X, Y be Banach spaces over the same field of scalars and let $T : X \to Y$ be linear. Then T is continuous if and only if

$$\|T\| = \sup\{\|Tx\|_Y : \|x\|_X \leq 1\} < \infty.$$

By $B(X, Y)$ is meant the linear space of all continuous linear maps from X to Y; the map $T \longmapsto \|T\|$ is a norm on this space endowed with which $B(X, Y)$ is a Banach space. We write $B(X)$ instead of $B(X, X)$. The *dual* X^* of X is the space $B(X, \Phi)$, where Φ is the underlying scalar field; if $x^* \in X^*$ and $x \in X$, we shall often denote $x^*(x)$ by $\langle x, x^* \rangle_X$, or even $\langle x, x^* \rangle$ if the context is clear. A sequence $(x_j)_{j \in \mathbb{N}}$ in X converges *strongly* to $x \in X$ (written $x_j \to x$) if and only if $\lim_{j \to \infty} \|x - x_j\| = 0$;

1.2 Functional Analysis

it converges *weakly* to x (written $x_j \rightharpoonup x$) if and only if

$$\lim_{j \to \infty} \langle x_j - x, x^* \rangle = 0 \text{ for all } x^* \in X^*.$$

If $T \in B(X, Y)$, its *adjoint* $T^* : Y^* \to X^*$ is defined by

$$\langle x, T^*y^* \rangle_X = \langle Tx, y^* \rangle_Y \text{ for all } x \in X \text{ and all } y^* \in Y^*.$$

It turns out that $T^* \in B(Y^*, X^*)$ and $\|T^*\| = \|T\|$.

The notation $X \hookrightarrow Y$ means that X can be identified with a subset of Y and that the natural embedding map from X to Y is continuous.

Let X be a Banach space. Its dual, X^*, is also a Banach space, the dual of which is denoted by X^{**} and called the bidual of X. There is a natural mapping J of X into X^{**} defined by

$$\langle x^*, Jx \rangle_{X^*} = \langle x, x^* \rangle_X \text{ for all } x \in X \text{ and all } x^* \in X^*;$$

J is an isometric isomorphism of X onto $J(X)$. If $J(X) = X^{**}$, then X is said to be *reflexive*. Every Hilbert space is reflexive; so are l_p and $L_p(\Omega)$ if $1 < p < \infty$: in fact, $(L_p(\Omega))^*$ is isometrically isomorphic to $L_{p'}(\Omega)$, where $1/p' + 1/p = 1$. An important property of any reflexive space X is that every bounded sequence in X contains a subsequence that is weakly convergent (to some point of X).

To give another useful class of Banach spaces we introduce the *modulus of convexity* of a Banach space X (with $\dim X \geq 2$). This is the map $\delta_X : (0, 2] \to [0, 1]$ defined by

$$\delta_X(\varepsilon) = \inf \left\{ 1 - \frac{1}{2} \|x + y\| : x, y \in B_X, \|x - y\| \geq \varepsilon \right\};$$

the space X is called *uniformly convex* if $\delta_X(\varepsilon) > 0$ for all $\varepsilon \in (0, 2]$. The spaces l_p and $L_p(\Omega)$ are uniformly convex if $1 < p < \infty$. Every uniformly convex Banach space is reflexive. A most useful fact (the Radon-Riesz or Kadets-Klee property) is that if (x_n) is a sequence in a uniformly convex space X such that $x_n \rightharpoonup x \in X$ and $\|x_n\| \to \|x\|$, then $\|x_n - x\| \to 0$.

Next we state some of the most fundamental theorems in functional analysis.

Theorem 1.2.1 (*The open mapping theorem*) *Let X, Y be Banach spaces and let $T \in B(X, Y)$ be surjective. Then $T(U)$ is open in Y whenever U is open in X.*

Corollary 1.2.2 (*The Banach inverse mapping theorem*) *Let X, Y be Banach spaces and let $T \in B(X, Y)$ be bijective. Then $T^{-1} \in B(Y, X)$.*

Theorem 1.2.3 (*The closed graph theorem*) *Let X, Y be Banach spaces; let $T : X \to Y$ be linear and such that its graph $G(T) := \{(x, Tx) : x \in X\}$ is closed in the product topology on $X \times Y$. Then T is continuous.*

Theorem 1.2.4 *(The Banach-Steinhaus theorem) Let X, Y be Banach spaces and let $(T_n)_{n \in \mathbb{N}}$ be a family of elements of $B(X, Y)$ such that*

$$\sup_{n \in \mathbb{N}} \|T_n x\| < \infty \text{ for every } x \in X.$$

Then $\sup_{n \in \mathbb{N}} \|T_n\| < \infty$. In particular, if $Tx := \lim_{n \to \infty} T_n x$ exists for each $x \in X$, then $T \in B(X, Y)$ and $\|T\| \leq \liminf_{n \to \infty} \|T_n\|$.

Theorem 1.2.5 *(A form of the Hahn-Banach theorem) Let Y be a linear subspace of a normed linear space X and suppose that $y^* \in Y^*$. Then there exists $x^* \in X^*$ such that $\|x^*\| = \|y^*\|$ and $x^* \!\upharpoonright_Y = y^*$. In particular,*

$$\|x\| = \sup\left\{|\langle x, x^* \rangle| : x^* \in X^*, \|x^*\| \leq 1\right\} \text{ for all } x \in X.$$

If X_0 is a proper, closed linear subspace of X and $x_0 \in X \setminus X_0$, there exists $x^* \in X^*$ such that $\langle x_0, x^* \rangle_X = 1$ and $\langle x, x^* \rangle_X = 0$ for all $x \in X_0$.

Theorem 1.2.6 *(The Riesz representation theorem) Let H be a Hilbert space with inner product $(\cdot, \cdot)$. Given any $y^* \in H^*$, there is a unique $y \in H$ such that $\langle x, y^* \rangle_H = (x, y)$ for all $x \in H$. The map $\sigma : y^* \longmapsto y$ thus defined is an isometry of H^* onto H such that for all scalars α, β and all $y_1^*, y_2^* \in H^*$,*

$$\sigma\left(\alpha y_1^* + \beta y_2^*\right) = \overline{\alpha}\sigma\left(y_1^*\right) + \overline{\beta}\sigma\left(y_2^*\right);$$

if the underlying field of scalars is the reals, then σ is linear and H^* may be identified with H.

We now give modifications of the last result that turn out to be very useful in connection with boundary-value problems.

Lemma 1.2.7 *(The Lax-Milgram lemma) Let H be a complex Hilbert space with inner product $(\cdot, \cdot)$, and let $B : H \times H \to \mathbb{C}$ be such that $g \mapsto B(f, g)$ is conjugate linear for each $f \in H$, and $g \mapsto B(g, f)$ is linear for each $f \in H$. Suppose there are positive numbers c and C such that for all $x, y \in H$,*

$$|B(x, y)| \leq C \|x\| \|y\|$$

and

$$|B(x, x)| \geq c \|x\|^2.$$

Then given any $F \in H^*$, there are unique elements $v, w \in H$ such that for all $x \in H$,

$$F(x) = B(x, v) = \overline{B(w, x)}.$$

Proof. For each fixed $v \in H$, the map $x \mapsto B(x, v)$ belongs to H^*, so that by the Riesz representation theorem, there is a unique $y \in H$ such that $B(x, v) = (x, y)$

for all $x \in H$. Evidently y depends linearly on v: we set $y = Av$, where A is linear. Since
$$|(x, Av)| = |B(x, v)| \leq C \|x\| \|v\|,$$
it follows that A is a bounded linear map of H to itself. Moreover,
$$c \|v\|^2 \leq |B(v, v)| = |(v, Av)| \leq \|v\| \|Av\|,$$
and so $c \|v\| \leq \|Av\|$, $v \in H$. Hence A is injective and has a bounded inverse. It easily follows that $A(H)$ is closed: in fact, $A(H) = H$. For if not, there exists $z \in H$, $z \neq 0$, with $z \perp A(H)$, that is, $(z, Av) = 0$ for all $v \in H$. Hence $0 = (z, Az) = B(z, z) \geq c \|z\|^2$, from which it follows that $z = 0$, and we have a contradiction.

Given $F \in H^*$, there is a unique $b \in H$ such that $F(x) = (x, b)$ for all $x \in H$, and there is a unique $v \in H$ such that $Av = b$. Thus
$$F(x) = (x, Av) = B(x, v) \text{ for all } x \in H.$$
As for $\overline{B(w, x)} = F(x)$, this follows from application of what has just been proved to the functional $\overline{B(w, x)}$. ∎

Lemma 1.2.7 was established in 1954 by Lax and Milgram [142]; see also Višik [211]. Independently, Fichera [79] gave a Banach space version in 1955; see also the much later paper [105]. This more general result concerns a bilinear functional on the product $X \times Y$ of two Banach spaces; that is, a map $B : X \times Y \to \mathbb{C}$ such that for all $x \in X$, $y \longmapsto B(x, y)$ is linear, and for all $y \in Y$, $x \longmapsto B(x, y)$ is linear. Such a functional is called bounded if there is a positive constant C such that $|B(x, y)| \leq C \|x\| \|y\|$ for all $x \in X$ and all $y \in Y$; it is said to be non-degenerate if only $x = 0$ has the property that $B(x, y) = 0$ for all $y \in Y$.

Theorem 1.2.8 *Let X and Y be Banach spaces, with Y reflexive, and let B be a bounded, non-degenerate, bilinear functional on $X \times Y$. Then every $F \in Y^*$ has a unique representation of the form $\langle y, F \rangle_Y = B(x, y)$ $(y \in Y)$ for some fixed $x \in X$ if and only if there exists $m > 0$ such that for each $x \in X$,*
$$\sup_{\|y\|=1} |B(x, y)| \geq m \|x\|.$$

Proof. Suppose that every $F \in Y^*$ may be uniquely represented in the way described. For each $y \in Y$, the bilinear functional B induces a bounded linear map $A : X \to Y^*$ by $B(x, y) = \langle y, Ax \rangle_Y$; since B is non-degenerate, A is injective. Moreover,
$$\|A\| = \|B\| = \sup_{\|x\|=\|y\|=1} |B(x, y)|.$$
As every element of Y^* has this representation, A is surjective and so has a bounded inverse, by Corollary 1.2.2. Thus there exists $m > 0$ such that $\|Ax\| \geq m \|x\|$ for all

$x \in X$. Since
$$\|Ax\| = \sup_{\|y\|=1} |\langle y, Ax\rangle_Y| = \sup_{\|y\|=1} |B(x, y)|,$$
the necessity of the condition follows.

Conversely, if the condition is satisfied, the map A defined above has a bounded inverse, so that the range of A is a closed linear subspace of Y^*. If $A(X) \neq Y^*$, there exists $y^* \in Y^* \backslash A(X)$, $y^* \neq 0$. By the Hahn-Banach theorem, there exists $y^{**} \in Y^{**}$ such that $\langle y^*, y^{**}\rangle_{Y^*} = 1$, $\langle A(X), y^{**}\rangle_{Y^*} = \{0\}$. Since Y is reflexive, a unique $y \in Y$ corresponds to $y^{**} \in Y^{**}$; thus $\langle y, y^*\rangle_Y = 1$. But $B(x, y) = \langle y, Ax\rangle_Y = 0$ for all $x \in X$, which implies that $y = 0$, as B is non-degenerate. This contradiction completes the proof. ∎

Let X, Y be Banach spaces. A linear map $T : X \to Y$ is said to be *compact* if for all bounded sets $B \subset X$, $\overline{T(B)}$ is compact. Plainly T is compact if and only if given any bounded sequence (x_n) in X, the sequence (Tx_n) contains a convergent subsequence. If T is compact, it is continuous: $T \in B(X, Y)$. The family of all compact linear maps from X to Y is denoted by $K(X, Y)$, or by $K(X)$ if $Y = X$. A useful result is that $T \in K(X, Y)$ if and only if $T^* \in K(Y^*, X^*)$.

We now present the basic results of the Fredholm-Riesz-Schauder theory of elements of $K(X)$, to prepare for which we recall that the *resolvent set* $\rho(T)$ of an operator $T \in B(X)$, with X complex, is defined to be
$$\rho(T) = \{\lambda \in \mathbb{C} : (\lambda I - T)^{-1} \text{ exists and belongs to } B(X)\};$$
here I stands for the identity map of X to itself. The *spectrum* of T is $\sigma(T) := \mathbb{C}\backslash\rho(T)$; by the *point spectrum* $\sigma_p(T)$ of T is meant the set of all eigenvalues of T, so that
$$\sigma_p(T) = \{\lambda \in \mathbb{C} : Tx = \lambda x \text{ for some } x \in X\backslash\{0\}\}.$$

If $\lambda \in \sigma_p(T)$, its geometric multiplicity is $\dim \ker(\lambda I - T)$; its algebraic multiplicity is
$$\dim \cup_{n=1}^{\infty} \ker(\lambda I - T)^n.$$

The following theorem contains some of the most fundamental results concerning compact operators acting in a complex Banach space.

Theorem 1.2.9 *Let X be a complex Banach space and let $T \in K(X)$. Then:*
(i) $\sigma_p(T)$ is at most countable and has no accumulation point except possibly 0.
(ii) Each point of $\sigma(T)\backslash\{0\}$ is an eigenvalue of finite algebraic multiplicity.
(iii) If $\lambda \in \mathbb{C}\backslash\{0\}$, then $\dim \ker(T - \lambda I) = \dim \ker(T^* - \overline{\lambda}I^*)$.
(iv) Suppose that $\lambda \in \mathbb{C}\backslash\{0\}$. Then the non-homogeneous equations
$$(T - \lambda I)x = y, \tag{1.2.1}$$
$$(T^* - \overline{\lambda}I^*)y^* = x^* \tag{1.2.2}$$

1.2 Functional Analysis

have unique solutions for any $y \in X$ and any $x \in X^*$ if and only if the homogeneous equations

$$(T - \lambda I)x = 0 \tag{1.2.3}$$

$$(T^* - \overline{\lambda} I^*)y^* = 0 \tag{1.2.4}$$

have only the zero solutions. If one of these homogeneous equations has a non-zero solution then they both have the same finite number of linearly independent solutions; and in this case, (1.2.1) and (1.2.2) have solutions if and only if y and x^* are orthogonal to all the solutions of (1.2.4) and (1.2.3) respectively in the sense that $\langle y, y^* \rangle_X = 0$ and $\langle x, x^* \rangle_X = 0$ for all y^* satisfying (1.2.4) and all x satisfying (1.2.3).

Proofs of the assertions made in this section may be found in numerous books on functional analysis, including [76], [198] and [222].

1.3 Function Spaces

1.3.1 Spaces of Continuous Functions

First, some standard notation. Throughout, Ω will stand for a non-empty open subset of $\mathbb{R}^n$ with boundary $\partial \Omega$ and closure $\overline{\Omega}$; a domain is a connected open set. Points of $\mathbb{R}^n$ will be denoted by $x = (x_i) = (x_1, ..., x_n)$ and we write $|x| = \left(\sum_{i=1}^n x_i^2 \right)^{1/2}$ and $(x, y) = \sum_{i=1}^n x_i y_i$; given $r > 0$, we put $B(x, r) = \{y \in \mathbb{R}^n : |x - y| < r\}$, abbreviating this to B_r if $x = 0$. If $\alpha = (\alpha_1, ..., \alpha_n) \in \mathbb{N}_0^n$, where $\mathbb{N}_0 = \mathbb{N} \cup \{0\}$, we write

$$\alpha! = \prod_{j=1}^n \alpha_j!, \quad |\alpha| = \sum_{j=1}^n \alpha_j, \quad x^\alpha = \prod_{j=1}^n x_j^{\alpha_j} \quad (x \in \mathbb{R}^n)$$

and

$$D^\alpha := \frac{\partial^{|\alpha|}}{\partial x_1^{\alpha_1} ... \partial x_n^{\alpha_n}} := \prod_{j=1}^n D_j^{\alpha_j}, \quad \text{where } D_j = \partial/\partial x_j;$$

it is to be understood that if some α_j is zero, then the corresponding term is to be omitted; if all α_j are zero, so that $\alpha = 0$, then $D^\alpha u = u$ for any appropriate function u. We remind the reader that when S is a measurable subset of $\mathbb{R}^n$, its Lebesgue n-measure will be denoted by $|S|_n$ or $|S|$ if the context is clear.

Given any $k \in \mathbb{N}_0$, by $C^k(\Omega)$ we mean the linear space of all real- or complex-valued functions u on Ω such that for all $\alpha \in \mathbb{N}_0^n$ with $|\alpha| \leq k$, the function $D^\alpha u$ exists and is continuous on Ω. The subspace of $C^k(\Omega)$ consisting of all those functions with compact support contained in Ω is denoted by $C_0^k(\Omega)$, and $C_0^\infty(\Omega) := \cap_{k=1}^\infty C_0^k(\Omega)$; recall that the *support* of a function u, supp u, is the closure of $\{x \in \Omega : u(x) \neq 0\}$. The function ϕ defined on $\mathbb{R}^n$ by

$$\phi(x) = \begin{cases} \exp\left(\frac{-1}{1-|x|^2}\right), & |x| < 1, \\ 0, & |x| \geq 1, \end{cases}$$

can easily be shown to be in $C_0^\infty(\mathbb{R}^n)$, with supp $\phi = \overline{B(0,1)}$ and $\int_{\mathbb{R}^n} \phi(x)dx > 0$, so that $\psi := \phi / \int_{\mathbb{R}^n} \phi(x)dx$ has the useful properties that $\psi \in C_0^\infty(\mathbb{R}^n)$ and $\int_{\mathbb{R}^n} \psi(x)dx = 1$.

We define $C^k(\overline{\Omega})$ to be the linear space of all bounded functions u in $C^k(\Omega)$ such that u and all its derivatives $D^\alpha u$ with $|\alpha| \leq k$ have bounded, continuous extensions to $\overline{\Omega}$: a norm $||| \cdot |||_{k,\Omega}$ is defined on this space by

$$|||u|||_{k,\Omega} := \max_{|\alpha| \leq k} \sup_{x \in \Omega} |D^\alpha u(x)|,$$

and it is a routine matter to verify that $C^k(\overline{\Omega})$ becomes a Banach space when given this norm. This notation is open to the objection that if Ω_1 and Ω_2 are different open sets with the same closure, then it may happen that $C^k(\overline{\Omega_1}) \neq C^k(\overline{\Omega_2})$. For example, if $n = 1$ and C is the Cantor subset of $[0,1]$, then $\Omega_1 := (0,1)$ and $\Omega_2 := (0,1) \setminus C$ have the same closure $[0,1]$, but consideration of the Cantor staircase function ψ shows that $\psi \restriction_{\Omega_1} \notin C^1(\overline{\Omega_1})$ and $\psi \restriction_{\Omega_2} \in C^1(\overline{\Omega_2})$, so that $C^k(\overline{\Omega_1}) \neq C^k(\overline{\Omega_2})$; we are indebted to an anonymous referee for this example. This cannot happen if the class of domains Ω is restricted to those for which the correspondence between Ω and $\overline{\Omega}$ is one-to-one, so that Ω is the interior of $\overline{\Omega}$, and for this reason various authors prefer to reserve the notation $C^k(\overline{\Omega})$ for such cases, using $\overline{C}^k(\Omega)$ for the general situation; see, for example, [205], p.30. Given an open set Ω, the space denoted by $C^k(\overline{\Omega})$ in our notation is well defined; a problem might arise only if we choose to write $C^k(K)$ where $K = \overline{\Omega}$. As we shall not do this in the present book, we shall continue to employ the rather more familiar notation given above.

Let $k \in \mathbb{N}_0$, $\lambda \in (0,1]$. We shall need various spaces of Hölder-continuous functions. First, $C^{0,\lambda}(\Omega)$ (often written as $C^\lambda(\Omega)$) will stand for the linear space of all continuous functions on Ω which satisfy a local Hölder condition on Ω; that is, given any compact subset K of Ω, there is a constant $C > 0$ such that

$$|u(x) - u(y)| \leq C|x-y|^\lambda \text{ for all } x, y \in K.$$

We also put

$$C^{k,\lambda}(\Omega) := \{u \in C^k(\Omega) : D^\alpha u \in C^\lambda(\Omega) \text{ for all } \alpha \in \mathbb{N}_0^n \text{ with } |\alpha| = k\}.$$

These spaces are not equipped with norms. However,

$$C^{k,\lambda}(\overline{\Omega}) := \{u \in C^k(\overline{\Omega}) : \text{given any } \alpha \in \mathbb{N}_0^n \text{ with } |\alpha| = k, \text{ there}$$
$$\text{exists } C > 0 \text{ such that for all } x, y \in \Omega, |D^\alpha u(x) - D^\alpha u(y)| \leq C|x-y|^\lambda\}$$

becomes a Banach space when provided with the norm

1.3 Function Spaces

$$|||u|||_{k,\lambda,\Omega} := |||u|||_{k,\Omega} + [u]_{k,\lambda,\Omega},$$

where

$$[u]_{k,\lambda,\Omega} = \max_{|\alpha|=k} \sup_{x,y \in \Omega, x \neq y} |D^\alpha u(x) - D^\alpha u(y)| / |x-y|^\lambda.$$

For convenience, when $\lambda \in (0,1)$, we write $C^\lambda(\overline{\Omega}) = C^{0,\lambda}(\overline{\Omega})$ and $||| \cdot |||_{\lambda,\Omega}$, $[\cdot]_{\lambda,\Omega}$ instead of $||| \cdot |||_{0,\lambda,\Omega}$, $[\cdot]_{0,\lambda,\Omega}$ respectively. By $C_0^{k,\lambda}(\Omega)$ will be meant the linear subspace of $C^{k,\lambda}(\Omega)$ consisting of all those functions with compact support contained in Ω. Note that if $u, v \in C^\lambda(\overline{\Omega})$, then

$$[uv]_{\lambda,\Omega} \leq |||u|||_{0,\Omega}[v]_{\lambda,\Omega} + |||v|||_{0,\Omega}[u]_{\lambda,\Omega};$$

and that if $u \in C^{\lambda_1}(\overline{\Omega})$, $v \in C^{\lambda_2}(\overline{\Omega})$ and Ω is bounded, then $uv \in C^\gamma(\overline{\Omega})$, where $\gamma = \min(\lambda_1, \lambda_2)$, and

$$|||uv|||_{\gamma,\Omega} \leq \max\left\{1, |\text{diam } \Omega|^{\lambda_1+\lambda_2-2\gamma}\right\} |||u|||_{\lambda_1,\Omega}|||v|||_{\lambda_2,\Omega}.$$

Useful properties relating these spaces of functions are given in the following theorem.

Theorem 1.3.1 *Let $k \in \mathbb{N}_0$, $0 < \nu < \lambda \leq 1$ and suppose that Ω is an open subset of $\mathbb{R}^n$. Then*

$$(i) \ C^{k+1}\left(\overline{\Omega}\right) \hookrightarrow C^k\left(\overline{\Omega}\right)$$

and

$$(ii) \ C^{k,\lambda}\left(\overline{\Omega}\right) \hookrightarrow C^{k,\nu}\left(\overline{\Omega}\right) \hookrightarrow C^k\left(\overline{\Omega}\right).$$

If Ω is bounded, both the embeddings in (ii) are compact. If Ω is convex, then

$$(iii) \ C^{k+1}\left(\overline{\Omega}\right) \hookrightarrow C^{k,1}\left(\overline{\Omega}\right)$$

and

$$(iv) \ C^{k+1}\left(\overline{\Omega}\right) \hookrightarrow C^{k,\nu}\left(\overline{\Omega}\right).$$

If Ω is bounded and convex, then the embeddings in (i) and (iv) are compact.

It is often desirable to extend functions defined on a given open set to a larger set, or to extend functions on $\partial\Omega$ to Ω. To help these procedures some conditions on the boundary of Ω may be useful. Let Ω be an open subset of $\mathbb{R}^n$ ($n \geq 2$) with non-empty boundary $\partial\Omega$, let $k \in \mathbb{N}_0$ and suppose that $\gamma \in [0, 1]$. Given $x_0 \in \partial\Omega$, $r > 0$, $\beta > 0$, local Cartesian coordinates $y = (y_1, ..., y_n) = (y', y_n)$ (where $y' = (y_1, ..., y_{n-1})$), with $y = 0$ at $x = x_0$, and a real continuous function $h : y' \mapsto h(y')$ ($|y'| < r$), we define a neighbourhood $U_{r,\beta,h}(x_0)$ of x_0 (an open subset of $\mathbb{R}^n$ containing x_0) by

$$U = U_{r,\beta,h}(x_0) = \left\{y \in \mathbb{R}^n : h(y') - \beta < y_n < h(y') + \beta, |y'| < r\right\}.$$

Then Ω is said to have boundary $\partial\Omega$ of class $C^{k,\gamma}$ if for each $x_0 \in \partial\Omega$ there are a local coordinate system, positive constants r and β and a function $h \in C^{k,\gamma}(B'_r)$ (where $B'_r = \{y' \in \mathbb{R}^{n-1} : |y'| < r\}$) such that

$$U_{r,\beta,h}(x_0) \cap \partial\Omega = \left\{y \in \mathbb{R}^n : y_n = h(y'), |y'| < r\right\}$$

and

$$U_{r,\beta,h}(x_0) \cap \Omega = \left\{y \in \mathbb{R}^n : h(y') - \beta < y_n < h(y'), |y'| < r\right\}.$$

In general, the constants r, β and the function h depend on x_0. However, if in addition Ω is bounded, there are points $x_1, \ldots, x_m \in \partial\Omega$, positive numbers r and β (independent of the x_j) and functions $h_1, \ldots, h_m$ such that the neighbourhoods $U_j = U_{r,\beta,h_j}(x_j)$ ($j = 1, \ldots, m$) cover $\partial\Omega$. When $\gamma = 0$ we simply write $\partial\Omega \in C^k$ (or $\partial\Omega \in C$ if $k = 0$). If $\partial\Omega \in C^{0,1}$ we shall say that the boundary is of Lipschitz class: if Ω is convex its boundary is of this class.

We shall also need spaces of functions defined on $\partial\Omega$, in connection with which the condition $\partial\Omega \in C^{k,\gamma}$ (for some $k \in \mathbb{N}_0$ and some $\gamma \in [0, 1]$) is useful. If $\partial\Omega \in C^{k,\gamma}$, then given $a \in \partial\Omega$, there are a neighbourhood $V = V(a)$ of a and a $C^{k,\gamma}$ homeomorphism Φ_a of V onto $B(0, 1)$ such that

$$\Phi_a(V \cap \Omega) = \{y \in B(0, 1) : y_n > 0\}, \ \Phi_a(V \cap \partial\Omega) = \{y \in B(0, 1) : y_n = 0\}.$$

Given a scalar-valued function f defined on $\partial\Omega$, we say that $f \in C^{k,\lambda}(\partial\Omega)$ if

$$f \circ \Phi_a^{-1} \in C^{k,\lambda}(\{y \in B(0, 1) : y_n = 0\}) \text{ for all } a \in \partial\Omega.$$

For further details and background information we refer to [160] and to [54], V.4, which also establishes the following Proposition.

Proposition 1.3.2 *Let $k \in \mathbb{N}$ and $\gamma \in [0, 1]$, suppose Ω is a bounded open subset of $\mathbb{R}^n$ with boundary $\partial\Omega$ of class $C^{k,\gamma}$ and let Ω_0 be an open set that contains $\overline{\Omega}$. Then:*
(i) If $u \in C^{k,\gamma}(\overline{\Omega})$, there is a function $U \in C_0^{k,\gamma}(\overline{\Omega_0})$ such that $U = u$ in Ω and

$$\|U\|_{k,\gamma,\Omega_0} \leq C \|u\|_{k,\gamma,\Omega},$$

where C depends only on k, Ω and Ω_0.
(ii) If $\phi \in C^{k,\gamma}(\partial\Omega)$, there is a function $\phi' \in C_0^{k,\gamma}(\Omega_0)$ such that $\phi' = \phi$ on $\partial\Omega$.

In view of (ii), it does not really matter whether we consider boundary values that belong to $C^{k,\lambda}(\partial\Omega)$ or $C_0^{k,\lambda}(\overline{\Omega})$. A norm may be defined on $C^{k,\lambda}(\partial\Omega)$ by

$$\|\|f\|\|_{k,\lambda,\partial\Omega} := \inf\left\{\|\|F\|\|_{k,\lambda,\Omega} : F \in C^{k,\lambda}(\overline{\Omega}) \text{ is an extension of } f\right\}.$$

1.3 Function Spaces

Spaces of functions that are integrable over $\partial\Omega$ come next. Suppose that Ω is bounded and $\partial\Omega \in C^{0,1}$; let U_j, h_j ($j = 1, ..., m$) be as above. There are functions $\phi_j \in C_0^\infty(\mathbb{R}^n)$ ($j = 1, ..., m$) such that supp $\phi_j \subset U_j$, $0 \leq \phi_j \leq 1$ and $\sum_{j=1}^m \phi_j(x) = 1$ for all $x \in \partial\Omega$: the ϕ_j form a **partition of unity** subordinate to the covering of the boundary by the U_j. Let $u : \partial\Omega \to \mathbb{R}$. Then

$$u(x) = \sum_{j=1}^m \phi_j(x) u(x) \quad (x \in \partial\Omega).$$

We say that u is integrable on $\partial\Omega$ if, for each $j \in \{1, ..., m\}$, the function

$$y' \longmapsto u\left(\phi_j\left(y', h_j(y')\right)\right) \phi_j\left(\phi_j\left(y', h_j(y')\right)\right) \left(1 + \left|\nabla' h_j(y')\right|^2\right)^{1/2}$$

is integrable on B_r': here ∇' stands for the gradient operator with respect to the variables $y' = (y_1, ..., y_n)$. By $d\sigma(y)$ we shall mean the "surface element" $\left(1 + \left|\nabla' h_j(y')\right|^2\right)^{1/2} dy'$. For such a function u,

$$\sum_{j=1}^m \int_{\partial\Omega \cap U_j} u \phi_j d\sigma$$
$$:= \sum_{j=1}^m \int_{B_r'} u\left(\phi_j\left(y', h_j(y')\right)\right) \phi_j\left(\phi_j\left(y', h_j(y')\right)\right) \left(1 + \left|\nabla' h_j(y')\right|^2\right)^{1/2} dy'$$

is called the surface integral of u over $\partial\Omega$ and is written $\int_{\partial\Omega} u\, d\sigma$. It can be shown that it depends only on u and $\partial\Omega$, and not on the particular choice of local coordinates. The surface measure $|\partial\Omega|$ of the boundary $\partial\Omega$ is the value of this integral when u is identically equal to 1. When $p \in [1, \infty)$, the space $L_p(\partial\Omega)$ is the space of all functions u defined on $\partial\Omega$ such that $x \longmapsto |u(x)|^p$ is integrable on $\partial\Omega$; endowed with the norm given by

$$\|u\|_{p,\partial\Omega} := \left(\int_{\partial\Omega} |u(y)|^p\, d\sigma(y)\right)^{1/p}$$

it is a Banach space. The space $L_\infty(\partial\Omega)$ is defined in the natural and analogous manner.

Let Ω have Lipschitz boundary. With the notation introduced above, it can be shown that

$$\left(1 + \left|\nabla' h_j(y')\right|^2\right)^{-1/2} \left(-\nabla' h_j(y'), 1\right) \tag{1.3.1}$$

is the exterior normal unit vector at $x \in \partial\Omega \cap U_j$, written in the new coordinates (y', y_n). Note that since h_j is a Lipschitz function, the gradient $\nabla' h_j(y')$ is well defined for almost all $y' \in B_r'$ and

$$\operatorname*{ess\,sup}_{y' \in B'_r} |\nabla' h_j(y')| < \infty (j = 1, ..., m).$$

Reverting to the original cordinates, (1.3.1) gives the uniquely defined exterior unit normal in the form $\nu(x) = (\nu_1(x), ..., \nu_n(x))$ for almost all $x \in \partial\Omega$. Further details will be found in [160].

1.3.2 Sobolev Spaces

We recall some basic facts about Sobolev spaces: for a systematic account of this topic and justification of the unproved assertions made below we refer to [54], V. Given any $f \in L_{1,loc}(\Omega)$ (that is, $f \in L_1(K)$ for every compact subset K of Ω), let us suppose that there is a function $g \in L_{1,loc}(\Omega)$ such that for all $\phi \in C_0^\infty(\Omega)$,

$$\int_\Omega f D_j \phi dx = -\int_\Omega g \phi dx.$$

Then g is said to be a weak derivative of f in Ω with respect to x_j, and we write $g = D_j f$ (weakly). More generally, if $\alpha \in \mathbb{N}_0^n$ and $h \in L_{1,loc}(\Omega)$ is such that

$$\int_\Omega f D^\alpha \phi dx = (-1)^{|\alpha|} \int_\Omega h \phi dx$$

for all $\phi \in C_0^\infty(\Omega)$, then we say that h is an αth weak derivative of f in Ω and write $h = D^\alpha f$ (weakly). It can be shown that weak derivatives, if they exists at all, are unique up to sets of zero Lebesgue n-measure. If $f \in C^{|\alpha|}(\Omega)$, then the classical and weak derivatives, up to and including those of order $|\alpha|$, coincide modulo sets of measure zero: weak differentiation extends classical differentiation. Given $m \in \mathbb{N}$ and $p \in [1, \infty]$, the Sobolev space $W_p^m(\Omega)$ is defined to be the linear space of all elements u of $L_p(\Omega)$ such that for all $\alpha \in \mathbb{N}_0^n$ with $|\alpha| \leq m$, the weak derivative $D^\alpha u$ exists and belongs to $L_p(\Omega)$. This linear space is made into a normed linear space by giving it the norm defined by

$$\|u\|_{m,p,\Omega} = \left(\sum\nolimits_{|\alpha| \leq m} \|D^\alpha u \mid L_p(\Omega)\|^p \right)^{1/p};$$

often we shall denote this by $\|\cdot\|_{m,p}$ if there is little chance of ambiguity. It can be shown that $W_p^m(\Omega)$ is a Banach space with this norm, and that it is uniformly convex (and hence reflexive) if $1 < p < \infty$. In fact, $W_2^m(\Omega)$ is even a Hilbert space, with inner product

$$(u, v)_{m,2} = (u, v)_{m,2,\Omega} = \int_\Omega \sum\nolimits_{|\alpha| \leq m} (D^\alpha u)(\overline{D^\alpha v}) dx.$$

1.3 Function Spaces

When $m = 0$ we shall write $(u, v)_{2,\Omega}$ or $(u, v)_2$ instead of $(u, v)_{0,2,\Omega}$. The closure of $C_0^\infty(\Omega)$ in $W_p^m(\Omega)$ is denoted by $\overset{0}{W}{}_p^m(\Omega)$. It is common practice to denote $W_2^m(\Omega)$ and $\overset{0}{W}{}_2^m(\Omega)$ by $H^m(\Omega)$ and $\overset{0}{H}{}^m(\Omega)$ respectively. In Chapter 12 we shall need the spaces $H^m(\Omega, \mathbb{C}^4)$ and $\overset{0}{H}{}^m(\Omega, \mathbb{C}^4)$ of $\mathbb{C}^4$-valued functions; their inner-product is

$$(u, v)_{m,2} = (u, v)_{m,2,\Omega} = \int_\Omega \sum_{|\alpha| \leq m} \langle D^\alpha u, D^\alpha v \rangle_{\mathbb{C}^4} \, dx,$$

where $\langle \cdot, \cdot \rangle_{\mathbb{C}^4}$ is the $\mathbb{C}^4$ inner-product. If $n = 1$ the elements of the various Sobolev spaces have special properties.

Theorem 1.3.3 *Let $a, b \in \mathbb{R}$, $a < b$, put $I = (a, b)$ and let $m \in \mathbb{N}$. If $p \in [1, \infty]$ and $u \in W_p^m(I)$, then $u^{(m-1)}$ is absolutely continuous on I. If $p \in (1, \infty)$ and $u \in \overset{0}{W}{}_p^m(I)$, then $u^{(m-1)}$ is absolutely continuous on $[a, b]$ and $u^{(j)}(a) = u^{(j)}(b) = 0$ for all $j \in \{0, 1, \ldots, m-1\}$.*

Next we give various embedding results.

Theorem 1.3.4 *Let Ω be a bounded open subset of $\mathbb{R}^n$ with Lipschitz boundary and suppose that $p \in [1, \infty)$ and $k \in \mathbb{N}$.*
(i) If $kp < n$, then

$$W_p^k(\Omega) \hookrightarrow L_s(\Omega) \text{ if } s \in [p, np/(n-kp)],$$

and the embedding is compact if $s \in [p, np/(n-kp))$.
(ii) If for some $l \in \mathbb{N}_0$ and $\gamma \in (0, 1]$ the inequality $(k - l - \gamma) p \geq n$ holds, then

$$W_p^k(\Omega) \hookrightarrow C^{l,\gamma}(\overline{\Omega}),$$

and the embedding is compact if $(k - l - \gamma) p > n$.
These results hold without any condition on $\partial\Omega$ when $W_p^k(\Omega)$ is replaced by $\overset{0}{W}{}_p^k(\Omega)$.

Further compact embeddings, under very weak or even no conditions on $\partial\Omega$, are given next.

Theorem 1.3.5 *Let Ω be a bounded open subset of $\mathbb{R}^n$.*
(i) If $k \in \mathbb{N}$ and $p \in (1, \infty)$, then for any $q \in [1, p)$,

$$W_p^k(\Omega) \text{ is compactly embedded in } W_q^{k-1}(\Omega).$$

(ii) If $p \in [1, \infty)$ and $k, l \in \mathbb{N}_0$ with $l > k$, then

$$\overset{0}{W}{}_p^l(\Omega) \text{ is compactly embedded in } \overset{0}{W}{}_p^k(\Omega).$$

If, in addition, $\partial\Omega$ is of class C, then

$$W_p^l(\Omega) \text{ is compactly embedded in } W_p^k(\Omega).$$

We now concentrate on spaces of the form $W_p^1(\Omega)$ or $\overset{0}{W}{}_p^1(\Omega)$ as these are of the greatest interest when studying boundary-value problems for second-order elliptic equations. For convenience of reference we give the principal embedding theorems for $\overset{0}{W}{}_p^1(\Omega)$ below, even though this involves some duplication of the results stated above.

Theorem 1.3.6 *Let Ω be a bounded open subset of $\mathbb{R}^n$.*
(i) If $1 \leq p < n$ and $1 \leq q < p^* := np/(n-p)$, then

$$\overset{0}{W}{}_p^1(\Omega) \text{ is compactly embedded in } L_q(\Omega).$$

(ii) Whenever $q \in [1, \infty)$,

$$\overset{0}{W}{}_n^1(\Omega) \text{ is compactly embedded in } L_q(\Omega).$$

(iii) If $n < p < \infty$ and $\lambda \in (0, 1 - n/p)$, then

$$\overset{0}{W}{}_p^1(\Omega) \text{ is compactly embedded in } C^\lambda\left(\overline{\Omega}\right).$$

Various inequalities are also extremely useful. The first we give is *Friedrichs' inequality*: it does not require the underlying open set to be bounded, but merely bounded in one direction.

Theorem 1.3.7 *Let Ω be an open subset of $\mathbb{R}^n$ that lies between two parallel coordinate hyperplanes at a distance l apart, and suppose that $p \in [1, \infty)$. Then for all $u \in \overset{0}{W}{}_p^1(\Omega)$,*

$$\|u\|_{p,\Omega} \leq l \, \||\nabla u|\|_{p,\Omega}.$$

Here and elsewhere

$$|\nabla u| = \left(\sum_{j=1}^n |D_j u|^2\right)^{1/2}.$$

We remark that if the parallel hyperplanes are not parallel to a coordinate hyperplane, then the result still holds but with l replaced by Cl for some constant C independent of u: this multiplicative constant arises from the desirability of performing a rotation of coordinates. An immediate consequence of this inequality is that, when Ω is bounded, for example,

$$u \longmapsto \||\nabla u|\|_{p,\Omega}$$

1.3 Function Spaces

is a norm on $\overset{0}{W}{}_p^1(\Omega)$ equivalent to that induced by the usual norm on $W_p^1(\Omega)$. For the case when Ω is bounded, we have

Theorem 1.3.8 *Let Ω be a bounded open subset of $\mathbb{R}^n$ and suppose that $p \in [1, \infty]$. Then for all $u \in \overset{0}{W}{}_p^1(\Omega)$,*

$$\|u\|_{p,\Omega} \leq (|\Omega|/\omega_n)^{1/n} \|\, |\nabla u|\, \|_{p,\Omega}.$$

If $1 \leq p < n$ and $q \in [1, np/(n-p)]$, then there is a constant C such that for all $u \in \overset{0}{W}{}_p^1(\Omega)$,

$$\|u\|_{q,\Omega} \leq C |\Omega|^{1/n + 1/q - 1/p} \|\, |\nabla u|\, \|_{p,\Omega}.$$

Counterparts of such results for $W_p^1(\Omega)$ follow. The first is usually referred to as *Poincaré's inequality*.

Theorem 1.3.9 *Let Ω be a bounded, convex open subset of $\mathbb{R}^n$ with diameter d and let $p \in [1, \infty]$. Then for all $u \in W_p^1(\Omega)$,*

$$\|u - u_\Omega\|_{p,\Omega} \leq (\omega_n/|\Omega|)^{1-1/n} d^n \|\, |\nabla u|\, \|_{p,\Omega},$$

where $u_\Omega = |\Omega|^{-1} \int_\Omega u(x)\,dx$.

A variant of this is

Theorem 1.3.10 *Let Ω be a bounded, convex open subset of $\mathbb{R}^n$, let $p \in [1, n)$ and $q \in [p, p^*]$, where $p^* = np/(n-p)$ (the Sobolev conjugate of p). Then there is a constant C such that for all $u \in W_p^1(\Omega)$,*

$$\|u - u_\Omega\|_{q,\Omega} \leq C \|\, |\nabla u|\, \|_{p,\Omega}.$$

The hypothesis of convexity in these last two theorems is for convenience of proof rather than necessity: versions of these results hold under weaker conditions on Ω.

A particularly useful property of the spaces $W_p^1(\Omega)$ and $\overset{0}{W}{}_p^1(\Omega)$, not possessed by $W_p^k(\Omega)$ or $\overset{0}{W}{}_p^k(\Omega)$ when $k > 1$, is given in the following

Theorem 1.3.11 *Let Ω be an open subset of $\mathbb{R}^n$, let $p \in (1, \infty)$ and suppose that u, v are real-valued functions in $W_p^1(\Omega)$. Then $\sup(u, v)$ and $\inf(u, v)$ belong to $W_p^1(\Omega)$. In particular, both $u^+ := \max(u, 0)$ and $u^- := \min(u, 0)$ belong to $W_p^1(\Omega)$, as does $|u|$; and for each $j \in \{1, \ldots, n\}$,*

$$D_j |u(x)| = \begin{cases} D_j u(x) & \text{if } u(x) > 0, \\ 0 & \text{if } u(x) = 0, \\ -D_j u(x) & \text{if } u(x) < 0. \end{cases}$$

The same holds with $W_p^1(\Omega)$ replaced by $\overset{0}{W}{}_p^1(\Omega)$.

For the remainder of this section we suppose that Ω is an open subset of $\mathbb{R}^n$ and $p \in (1, \infty)$.

Lemma 1.3.12 *Let $u \in W_p^1(\Omega)$ be non-negative and suppose that (u_k) is a real-valued sequence in $W_p^1(\Omega)$ that converges to u. Then (u_k^+) converges to u in $W_p^1(\Omega)$.*

Proof. For each $k \in \mathbb{N}$ put $v_k = u_k^+$ and

$$\Omega_k = \{x \in \Omega : u_k(x) < 0\}.$$

Since $|v_k - u| \leq |u_k - u|$ it follows that $v_k \to u$ in $L_p(\Omega)$. Hence we simply have to prove that $\int_\Omega |\nabla(v_k - u)|^p \, dx \to 0$ or, equivalently, $\int_{\Omega_k} |\nabla(v_k - u)|^p \, dx \to 0$. By Proposition V.2.6 of [54], $\nabla u = 0$ on $\{x \in \Omega : u(x) = 0\}$: thus it is enough to show that $|\Omega_k \cap \Omega^+| \to 0$ as $k \to \infty$, where $\Omega^+ = \{x \in \Omega : u(x) > 0\}$. As $u_k \to u$ in $L_p(\Omega)$, it follows from Theorem 1.1.5 that the sequence (u_k) converges to u in measure. This implies that given $\varepsilon > 0$, there exists $N \in \mathbb{N}$ such that

$$|\Omega_k \cap \{x \in \Omega : u(x) \geq \varepsilon\}| < \varepsilon \text{ if } k > N,$$

and so

$$\limsup_{k \to \infty} |\Omega_k \cap \Omega^+| \leq \varepsilon + \limsup_{k \to \infty} |\Omega_k \cap \{x \in \Omega : 0 < u(x) < \varepsilon\}| = O(\varepsilon).$$

Hence $\limsup_{k \to \infty} |\Omega_k \cap \Omega^+| = 0$ and the lemma follows. ∎

Corollary 1.3.13 *Let $u \in \overset{0}{W}{}_p^1(\Omega)$ be non-negative. Then there is a sequence of non-negative functions in $C_0^\infty(\Omega)$ that converges to u in $W_p^1(\Omega)$.*

Proof. Let (u_k) be a sequence in $C_0^\infty(\Omega)$ that converges to u in $\overset{0}{W}{}_p^1(\Omega)$. By Lemma 1.3.12, (u_k^+) also converges to u. Since each u_k^+ is continuous on Ω and has compact support in Ω, it is enough to deal with the case in which u has the same properties. Let ρ be a *mollifier*; that is, a non-negative function in $C_0^\infty(\mathbb{R}^n)$ with $\rho(x) = 0$ when $|x| \geq 1$ and $\int_{\mathbb{R}^n} \rho(x)dx = 1$. Given any sufficiently small $\varepsilon > 0$, put $u_\varepsilon(x) = \varepsilon^{-n} \int_\Omega \rho((x-y)/\varepsilon) u(y) dy = (\rho_\varepsilon * u)(x)$, where $\rho_\varepsilon(x) = \varepsilon^{-n} \rho(x/\varepsilon)$: then u_ε is a non-negative function in $C_0^\infty(\Omega)$ and $u_\varepsilon \to u$ in $W_p^1(\Omega)$ as $\varepsilon \to 0$ (see [54], Theorems V.1.5, V.1.6 and Lemma V.2.2). ∎

Theorem 1.3.14 *Let $u \in W_p^1(\Omega)$. Then there is a sequence $(u_k) \subset C_0^\infty(\mathbb{R}^n)$ such that $u_k \to u$ in $L_p(\Omega)$ and $\nabla u_k \to \nabla u$ in $L_p(\omega)$ for every open subset ω of $\mathbb{R}^n$ with compact closure contained in Ω.*

1.3 Function Spaces

Proof. This follows the same lines as in the last proof. Let ρ be a mollifier, given any real-valued function w on Ω, let $\bar{w}$ be the extension by zero of w to the whole of $\mathbb{R}^n$ and write $v_k = \rho_{1/k} * \bar{u}$. Then $v_k \in C^\infty(\mathbb{R}^n)$ and $v_k \to \bar{u}$ in $L_p(\mathbb{R}^n)$. Now let ω be an open subset of $\mathbb{R}^n$ with compact closure contained in Ω. We claim that $\nabla v_k \to \nabla u$ in $L_p(\omega)$. To establish this, let $\phi \in C_0^1(\Omega)$ be such that $0 \le \phi \le 1$ and $\phi = 1$ on a neighbourhood of ω. For large enough k,

$$\operatorname{supp}\left(\rho_{1/k} * \overline{\phi u} - \rho_{1/k} * \bar{u}\right)$$

$$\subset \operatorname{supp} \rho_{1/k} + \operatorname{supp} \overline{\left(1 - \phi\right) u} \subset B(0, 1/k) + \operatorname{supp} \overline{\left(1 - \phi\right)}$$

$$\subset \mathbb{R}^n \setminus \omega.$$

It follows that
$$\rho_{1/k} * \overline{\phi u} = \rho_{1/k} * \bar{u} \text{ on } \omega.$$

Moreover,
$$D_i \left(\rho_{1/k} * \overline{\phi u}\right) = \rho_{1/k} * \overline{(\phi D_i u + u D_i \phi)},$$

and so $D_i\left(\rho_{1/k} * \overline{\phi u}\right) \to \overline{(\phi D_i u + u D_i \phi)}$ in $L_p(\mathbb{R}^n)$, which implies that $D_i\left(\rho_{1/k} * \overline{\phi u}\right) \to D_i u$ in $L_p(\omega)$. Hence

$$D_i\left(\rho_{1/k} * \bar{u}\right) \to D_i u \text{ in } L_p(\omega).$$

Finally, let $\zeta \in C_0^\infty(\mathbb{R}^n)$ be such that $0 \le \zeta \le 1$ and $\zeta(x) = 1$ $(|x| \le 1)$, $\zeta(x) = 0$ $(|x| \ge 2)$; put $\zeta_k(x) = \zeta(x/k)$ $(k \in \mathbb{N})$. The sequence $(u_k) = (\zeta_k u_k)$ has the desired properties. ∎

As regards $\overset{0}{W}_p^1(\Omega)$, in addition to the Friedrichs inequality (Theorem 1.3.7), the following results should be noted.

Lemma 1.3.15 *If $u \in W_p^1(\Omega)$ and $\operatorname{supp} u$ is a compact subset of Ω, then $u \in \overset{0}{W}_p^1(\Omega)$.*

Proof. Let ω be a bounded open set such that $\operatorname{supp} u \subset \omega \subset \bar\omega \subset \Omega$ and let $\phi \in C_0^1(\Omega)$, with $\phi = 1$ on $\operatorname{supp} u$, so that $\phi u = u$. By Theorem 1.3.14, there is a sequence $(u_k) \subset C_0^\infty(\mathbb{R}^n)$ such that $u_k \to u$ in $L_p(\Omega)$ and $\nabla u_k \to \nabla u$ in $L_p(\omega)$. Hence $\phi u_k \to \phi u$ in $W_p^1(\Omega)$; thus $\phi u \in \overset{0}{W}_p^1(\Omega)$ and the result follows. ∎

Theorem 1.3.16 *Suppose that $u \in W_p^1(\Omega) \cap C(\bar{\Omega})$. Then*
(i) *if $u = 0$ on $\partial\Omega$, then $u \in \overset{0}{W}_p^1(\Omega)$;*
(ii) *if $\partial\Omega \in C^1$ and $u \in \overset{0}{W}_p^1(\Omega)$, then $u = 0$ on $\partial\Omega$.*

Proof. (i) Suppose first that supp u is bounded. Let $G \in C^1(\mathbb{R})$ be such that

$$|G(t)| \leq |t| \ (t \in \mathbb{R}), \ G(t) = 0 \ (|t| \leq 1), \ G(t) = t \ (|t| \geq 2),$$

and for each $k \in \mathbb{N}$ let $u_k := k^{-1}G(ku)$. By Lemma V.2.5 of [54], $u_k \in W_p^1(\Omega)$; we claim that $u_k \to u$ in $W_p^1(\Omega)$. For since $u_k = u$ on

$$\Omega^k := \{x \in \Omega : |u(x)| \geq 2/k\},$$

we have, with $\Omega_k := \Omega \setminus \Omega^k$,

$$\left(\int_\Omega |u_k - u|^p dx\right)^{1/p} = \left(\int_{\Omega_k} |u_k - u|^p dx\right)^{1/p}$$

$$\leq \left(\int_{\Omega_k} |u_k|^p dx\right)^{1/p} + \left(\int_{\Omega_k} |u|^p dx\right)^{1/p}$$

$$\leq 2\left(\int_{\Omega_k} |u|^p dx\right)^{1/p} \leq 4 |\text{supp } u|^{1/p} / k \to 0$$

as $k \to \infty$: hence $u_k \to u$ in $L_p(\Omega)$. Moreover, use of the chain rule shows that for $j = 1, ..., n$,

$$D_j u_k = G'(ku) D_j u.$$

If $k |u(x)| > 2$, then $G'(ku(x)) = 1$: hence $D_j u_k \to D_j u$ pointwise. Now use of the dominated convergence theorem shows that $D_j u_k \to D_j u$ in $L_p(\Omega)$, and our claim is justified. Since supp $u_k \subset \{x \in \Omega : |u(x)| \geq 1/k\}$, it follows that supp u_k is a compact subset of Ω. By Lemma 1.3.15, $u_k \in \overset{0}{W}{}_p^1(\Omega)$, and so $u \in \overset{0}{W}{}_p^1(\Omega)$. If supp u is unbounded, consider the functions $\zeta_k u \ (k \in \mathbb{N})$, where the ζ_k are as in the proof of Theorem 1.3.14. From the first part of the proof, $\zeta_k u \in \overset{0}{W}{}_p^1(\Omega)$, and since $\zeta_k u \to u$ in $W_p^1(\Omega)$, we see that $u \in \overset{0}{W}{}_p^1(\Omega)$.

(ii) Use of local coordinates reduces the problem to that in which

$$\Omega = \{x = (x', x_n) : |x'| < 1, x_n > 0\},$$

and we have to prove that if $u \in \overset{0}{W}{}_p^1(\Omega) \cap C(\overline{\Omega})$, then $u = 0$ on

$$\Omega_0 := \{x = (x', x_n) : |x'| < 1, x_n = 0\}.$$

Let $(u_k) \subset C_0^1(\Omega)$ be such that $u_k \to u$ in $W_p^1(\Omega)$. If $x = (x', x_n) \in \Omega$, then

$$|u_k(x', x_n)| \leq \int_0^{x_n} |D_n u_k(x', t)| dt;$$

1.3 Function Spaces

hence for $\varepsilon \in (0, 1)$,

$$\varepsilon^{-1} \int_{|x'|<1} \int_0^\varepsilon |u_k(x', x_n)| \, dx' dx_n \leq \int_{|x'|<1} \int_0^\varepsilon |D_n u_k(x', t)| \, dx' dt.$$

Hold ε fixed and let $k \to \infty$:

$$\varepsilon^{-1} \int_{|x'|<1} \int_0^\varepsilon |u(x', x_n)| \, dx' dx_n \leq \int_{|x'|<1} \int_0^\varepsilon |D_n u(x', t)| \, dx' dt.$$

Now let $\varepsilon \to 0$: since $u \in C(\overline{\Omega})$ and $D_n u \in L_1(\Omega)$, we have

$$\int_{|x'|<1} |u(x', 0)| \, dx' = 0.$$

Thus $u = 0$ on Ω_0. ∎

Next we discuss *nodal domains* of functions. For any continuous function $u : \Omega \to \mathbb{R}$, its zero set $\{x \in \Omega : u(x) = 0\}$ is denoted by $Z(u)$ and by its nodal domains we mean the connected components of $\Omega \setminus Z(u)$.

Theorem 1.3.17 *Let Ω be bounded, let $u \in \overset{\circ}{W}{}_p^1(\Omega) \cap C(\Omega)$ and suppose that N is a nodal domain of u. Then the function v defined by $v(x) = u(x)$ $(x \in N)$, $v(x) = 0$ $(x \in \Omega \setminus N)$, belongs to $\overset{\circ}{W}{}_p^1(N)$.*

Proof. We may suppose that $u > 0$ on N; by replacing u by u^+ we can assume that $u \geq 0$ on Ω. By decomposition of u into its positive and negative parts, we see it is sufficient to deal with the case in which $u \geq 0$. There is a sequence of $C_0^\infty(\Omega)$ functions that converges to u in $W_p^1(\Omega)$: using Corollary 1.3.13 we see that the non-negative parts of these functions form a sequence $(u_k) \subset \overset{\circ}{W}{}_p^1(\Omega) \cap C(\Omega)$ with $u_k \geq 0$, supp u_k compact $(\subset \Omega)$ and $u_k \to u$ in $\overset{\circ}{W}{}_p^1(\Omega)$. Let $v_k = \min(u, u_k)$: the sequence (v_k) has the same properties as (u_k); since v_k has compact support, $v_k \in C(\overline{\Omega})$. We claim that $v_k(x) = 0$ for all $x \in \partial N$. To justify this, note that if $x \in \Omega \cap \partial N$, then $u(x) = 0$ as N is a nodal domain, and so $v_k(x) = 0$; if $x \in \partial \Omega \cap \partial N$, then $v_k(x) = 0$ because v_k has compact support in Ω. By Theorem 1.3.16, each v_k belongs to $\overset{\circ}{W}{}_p^1(N)$ and hence to $\overset{\circ}{W}{}_p^1(\Omega)$ (see, for example, [54], p. 253). The result now follows. ∎

Remark 1.3.18

Since the connectivity of the components is nowhere used, the theorem also holds for the sets $\{x \in \Omega : u(x) > 0\}$ and $\{x \in \Omega : u(x) < 0\}$.

A natural question arising in connection with the embedding theorems just discussed is that of *sharpness*. Consider, for example, the embedding $\overset{0}{W}{}_p^1(\Omega) \hookrightarrow L_{p^*}(\Omega)$, where Ω is a bounded open subset of $\mathbb{R}^n$, $1 < p < n$ and $p^* = np/(n-p)$. Can $\overset{0}{W}{}_p^1(\Omega)$ be embedded in a space smaller than $L_{p^*}(\Omega)$? Is there a space larger than $\overset{0}{W}{}_p^1(\Omega)$ that can be embedded in $L_{p^*}(\Omega)$? Such questions naturally depend on the class of competing spaces being considered. Thus if the domain space $\overset{0}{W}{}_p^1(\Omega)$ is replaced by the larger space $\overset{0}{W}{}_q^1(\Omega)$ based on the Lebesgue space $L_q(\Omega)$ with $1 < q < p$, then it is not true that $\overset{0}{W}{}_q^1(\Omega) \hookrightarrow L_{p^*}(\Omega)$, as simple examples show. Similarly, replacement of the target space $L_{p^*}(\Omega)$ by the smaller Lebesgue space $L_r(\Omega)$, where $r > p^*$, is not possible as $\overset{0}{W}{}_p^1(\Omega)$ is not embedded in $L_r(\Omega)$. Hence, if we restrict ourselves to target spaces that are Lebesgue spaces, and domain spaces that are Sobolev spaces based on Lebesgue spaces, the embedding $\overset{0}{W}{}_p^1(\Omega) \hookrightarrow L_{p^*}(\Omega)$ is optimal in the sense that neither the domain space nor the target space can be improved. This leaves open the question of optimality in a class of spaces wider than those involving the Lebesgue scale.

In this connection the family of *rearrangement-invariant* spaces arises naturally. To explain what these are, for simplicity suppose that Ω is a bounded open subset of $\mathbb{R}^n$; χ_E will stand for the characteristic function of $E \subset \Omega$. Given a function $f \in \mathcal{M}(\Omega)$, the family of all extended scalar-valued measurable functions on Ω, its non-increasing rearrangement f^* is defined by

$$f^*(t) = \inf\{\lambda > 0 : |\{x \in \Omega : |f(x)| > \lambda\}| \leq\}, t \in [0, |\Omega|).$$

A Banach space $X(\Omega)$ of functions belonging to $\mathcal{M}(\Omega)$, with norm $\|\cdot\|$, is said to be a Banach function space if the following conditions are satisfied:

$$\|f\| \leq \|g\| \text{ if } f \leq g \text{ a.e.;} \tag{P1}$$

$$\|f_k\| \uparrow \|f\| \text{ if } 0 \leq f_k \uparrow f \text{ a.e.;} \tag{P2}$$

$$\|\chi_E\| < \infty \text{ for every } E \subset \Omega \text{ with } |E| < \infty; \tag{P3}$$

if $E \subset \Omega$ and $|E| < \infty$, there is a constant C_E such that (P4)
$$\int_E f(x)dx \leq C_E \|f\| \text{ for every } f \in X(\Omega).$$

If, in addition,

1.3 Function Spaces

$$\|f\| = \|g\| \text{ whenever } f^* = g^*, \tag{P5}$$

then $X(\Omega)$ is said to be rearrangement-invariant(r.-i.). If every f in the Banach function space $X(\Omega)$ has the property that $\|f\chi_{G_k}\| \to 0$ whenever $\{G_k\}_{k\in\mathbb{N}}$ is a decreasing sequence of subsets of Ω with $|G_k| \to 0$, then X is said to have absolutely continuous norm.

The Lebesgue space $L_p(\Omega)$ $(1 \leq p \leq \infty)$, with norm $\|\cdot\|_{p,\Omega}$, is an r.i. space; so is the Lorentz space $L_{p,q}(\Omega)$ defined, for all $p, q \in [1, \infty]$, to be the space of all $f \in \mathcal{M}(\Omega)$ such that $\|f\|_{p,q,\Omega} < \infty$, where

$$\|f\|_{p,q,\Omega} := \left\|t^{1/p-1/q} f^*(t)\right\|_{q,(0,|\Omega|)}.$$

For all $p \in [1, \infty]$, $L_{p,p}(\Omega) = L_p(\Omega)$. The dependence of the Lorentz spaces on the first index is given by

$$L_{r,s}(\Omega) \hookrightarrow L_{p,q}(\Omega) \text{ if } 1 \leq p < r \leq \infty \text{ and } q, s \in [1, \infty],$$

while as regards the second index we have

$$L_{p,q}(\Omega) \hookrightarrow L_{p,r}(\Omega) \text{ if } p \in [1, \infty] \text{ and } 1 \leq q < r \leq \infty.$$

For proofs of these assertions and further details see, for example, [55]. The well-known space BMO is not r.i.

It turns out that if $1 \leq p < n$, then the smallest r.i. space $X(\Omega)$ such that $\overset{0}{W}{}^1_p(\Omega) \hookrightarrow X(\Omega)$ is $X(\Omega) = L_{p^*,p}(\Omega)$, where $p^* = np/(n-p)$, so that the embedding

$$\overset{0}{W}{}^1_p(\Omega) \hookrightarrow L_{p^*,p}(\Omega)$$

is optimal within the class of r.i. spaces, so far as the target space is concerned. As for optimality of the domain space, we need to consider Sobolev spaces based on a general r.i. space $X(\Omega)$ rather than simply a Lebesgue space $L_p(\Omega)$: by $W^1(X(\Omega))$ we shall mean the space of all $u \in X(\Omega)$ such that $D_j u \in X(\Omega)$ for all $j \in \{1, ..., n\}$. This is a Banach space when endowed with the norm $u \mapsto \|u\|_{X(\Omega)} + \|\,|\nabla u|\,\|_{X(\Omega)}$; the closure of $C_0^\infty(\Omega)$ in it is denoted by $\overset{0}{W}{}^1(X(\Omega))$. What emerges is that if $n/(n-1) < p < \infty$, then the Lorentz space $L_{np/(n+p),p}(\Omega)$ is the largest r.i. space $X(\Omega)$ such that $\overset{0}{W}{}^1(X(\Omega)) \hookrightarrow L_p(\Omega)$, so that the imbedding

$$\overset{0}{W}{}^1\left(L_{np/(n+p),p}(\Omega)\right) \hookrightarrow L_p(\Omega)$$

is optimal within the class of r.i. space, so far as the domain space is concerned. For these results and many others see [39], [60], [125], [126] and [165].

Turning now to spaces on the boundary, suppose that Ω is a bounded open subset of $\mathbb{R}^n$ with Lipschitz boundary and let $\beta \in (0,1)$, $p \in (1,\infty)$. The Sobolev space $W_p^\beta(\partial\Omega)$ is defined to be the linear space of all $u \in L_p(\partial\Omega)$ for which

$$\|u\|_{\beta,p,\partial\Omega} := \left(\|u\|_{p,\partial\Omega}^p + \int_{\partial\Omega}\int_{\partial\Omega} \frac{|u(x)-u(y)|^p}{|x-y|^{n-1+\beta p}} d\sigma(x) d\sigma(y) \right)^{1/p} < \infty.$$

Endowed with the norm $\|\cdot\|_{\beta,p,\partial\Omega}$ it is a Banach space. These spaces are important in connection with the *trace operator*, by which is meant a bounded, surjective linear map

$$\Gamma : W_p^1(\Omega) \to W_p^{1-1/p}(\partial\Omega)$$

such that

$$\Gamma u = u|_{\partial\Omega} \text{ for all } u \in C^\infty(\overline{\Omega}).$$

That such a map really exists is established in [160], Chap. 2. The function Γu is called the *trace* of u with respect to $W_p^1(\Omega)$ and will usually be denoted by $u|_{\partial\Omega}$ to underline the connection with the case when u is smooth. With this understanding, it emerges that (see [160], Chap. 2)

$$\overset{0}{W}_p^1(\Omega) = \left\{ u \in W_p^1(\Omega) : u|_{\partial\Omega} = 0 \right\}.$$

We shall also need Green's formula, which asserts that if $u \in C^\infty(\overline{\Omega})$ and $v \in C^\infty(\overline{\Omega})^n$, then

$$\int_\Omega u \operatorname{div} v\, dx = \int_{\partial\Omega} u\nu.v d\sigma - \int_\Omega (\nabla u).v dx,$$

where $\nu(x) = (\nu_1(x), ..., \nu_n(x))$ denote the exterior normal to $\partial\Omega$ at x. This can be extended to all $u \in W_p^1(\Omega)$ and all $v \in W_{p'}^1(\Omega)^n$: note that $u|_{\partial\Omega} \in W_p^{1-1/p}(\partial\Omega)$ and $\nu.v|_{\partial\Omega} \in W_{p'}^{1-1/p'}(\partial\Omega)$ so that $u\nu.v|_{\partial\Omega} \in L_1(\partial\Omega)$. For details of arguments that underpin these assertions see [160], [71] and [224]. The special case of Green's formula when $u = 1$ is the **divergence theorem**.

1.4 The Hilbert and Riesz Transforms

These are prominent in the theory of singular integral operators and and their main properties are established in many textbooks, such as [94]. Our need for them in this book is dictated by their usefulness in obtaining estimates of the Lebesgue norms of second-order derivatives by that of the Laplacian, and so we are content here to summarise those of their properties needed for this very specific objective.

1.4 The Hilbert and Riesz Transforms

First some standard notation and results. By $\mathcal{S}(\mathbb{R}^n)$ we denote the Schwartz space of rapidly decreasing functions; that is, the space of all $f : \mathbb{R}^n \to \mathbb{C}$ such that

$$\rho_{\alpha,\beta}(f) := \sup_{x \in \mathbb{R}^n} \left| x^\alpha D^\beta f(x) \right| < \infty$$

for all $\alpha, \beta \in \mathbb{N}_0^n$. The $\rho_{\alpha,\beta}$ form a family of semi-norms on $\mathcal{S}(\mathbb{R}^n)$; endowed with these the Schwartz space becomes a Fréchet space (a complete, metrisable, locally convex topological vector space), the dual of which is denoted by $\mathcal{S}'(\mathbb{R}^n)$ and is called the space of tempered distributions. A linear functional u on $\mathcal{S}(\mathbb{R}^n)$ belongs to $\mathcal{S}'(\mathbb{R}^n)$ if and only if there exist $C > 0$ and $k, m \in \mathbb{N}_0$ such that for all $f \in \mathcal{S}(\mathbb{R}^n)$, the value $\langle f, u \rangle$ of u at f satisfies

$$|\langle f, u \rangle| \leq C \sum_{|\alpha| \leq m, |\beta| \leq k} \rho_{\alpha,\beta}(f).$$

Given $f \in \mathcal{S}(\mathbb{R}^n)$, the Fourier transform $\widehat{f}$ of f is defined by

$$\widehat{f}(\xi) = \int_{\mathbb{R}^n} f(x) e^{-2\pi i x \cdot \xi} dx \quad (\xi \in \mathbb{R}^n),$$

where $x \cdot \xi = \sum_{j=1}^n x_j \xi_j$. The Fourier transform maps $\mathcal{S}(\mathbb{R}^n)$ onto itself, and can be extended to a unitary isomorphism of $L_2(\mathbb{R}^n)$ onto itself. The map $f \mapsto \check{f}$ (where $f \in \mathcal{S}(\mathbb{R}^n)$ and $\check{f}(x) = \widehat{f}(-x)$) is called the inverse Fourier transform. Precise references from [54] will be given for well-known properties of the Fourier transform used in the remainder of the chapter.

Turning now to the Riesz transforms, for each $j \in \{1, \ldots, n\}$ and each $\phi \in \mathcal{S}(\mathbb{R}^n)$, define $W_j \in \mathcal{S}'(\mathbb{R}^n)$ by

$$\langle \phi, W_j \rangle = c(n) \lim_{\varepsilon \to 0} \int_{|y| \geq \varepsilon} \frac{y_j}{|y|^{n+1}} \phi(y) dy,$$

where

$$c(n) = \frac{\Gamma\left(\frac{n+1}{2}\right)}{\pi^{(n+1)/2}}.$$

To check that W_j is well-defined and belongs to $\mathcal{S}'(\mathbb{R}^n)$, note that

$$\left| \int_{\varepsilon \leq |y| \leq 1} \frac{y_j}{|y|^{n+1}} \phi(y) dy \right| = \left| \int_{\varepsilon \leq |y| \leq 1} \frac{y_j}{|y|^{n+1}} (\phi(y) - \phi(0)) dy \right|$$

$$\leq \|\nabla \phi\|_\infty \int_{|y| \leq 1} \frac{|y_j|}{|y|^n} dy \leq n \omega_n \|\nabla \phi\|_\infty,$$

and that

$$\left|\int_{|y|>1} \frac{y_j}{|y|^{n+1}} \phi(y) dy\right| \le \int_{|y|>1} \frac{|y|^n}{|y|^{2n}} |\phi(y)| dy \le C \int_{|y|>1} \frac{1}{|y|^{2n}} \sum_{|\beta|=n} |y^\beta| |\phi(y)| dy$$

$$\le C \left(\sum_{|\beta|=n} \sup_{z \in \mathbb{R}^n} |z^\beta \phi(z)| \right) \int_{|y|>1} \frac{dy}{|y|^{2n}}.$$

Limits of integrals such as that in the definition of W_j above are called *principal value integrals* and denoted by the symbol p.v.

Definition 1.4.1 For each $j \in \{1, \ldots, n\}$ the jth **Riesz transform** of $f \in \mathcal{S}(\mathbb{R}^n)$ is the function $R_j f$ given by

$$(R_j f)(x) = (f * W_j)(x) = \frac{\Gamma\left(\frac{n+1}{2}\right)}{\pi^{(n+1)/2}} \text{ p.v.} \int_{\mathbb{R}^n} \frac{x_j - y_j}{|x-y|^{n+1}} f(y) dy.$$

When $n = 1$ the Riesz transform is called the **Hilbert transform** and is denoted by the symbol H:

$$(Hf)(x) = \frac{1}{\pi} \text{ p.v.} \int_{-\infty}^{\infty} \frac{f(y)}{x-y} dy = \frac{1}{\pi} \text{ p.v.} \int_{-\infty}^{\infty} \frac{f(x-y)}{y} dy.$$

It is convenient to have the following characterisation of the Riesz transforms by means of the Fourier transform.

Proposition 1.4.2 For any $f \in \mathcal{S}(\mathbb{R}^n)$ and any $j \in \{1, \ldots, n\}$,

$$(R_j f)(x) = - \left(\frac{i\xi_j}{|\xi|} \widehat{f}(\xi) \right)^{\vee}(x) \quad (x \in \mathbb{R}^n).$$

Proof. Let $f \in \mathcal{S}(\mathbb{R}^n)$. Then, by [54], Lemma V.1.18,

$$\langle f, \widehat{W_j} \rangle = \langle \widehat{f}, W_j \rangle = c(n) \lim_{\varepsilon \to 0} \int_{|\xi| \ge \varepsilon} \frac{\xi_j}{|\xi|^{n+1}} \widehat{f}(\xi) d\xi$$

$$= c(n) \lim_{\varepsilon \to 0} \int_{\mathbb{R}^n} f(x) \int_{\varepsilon \le |\xi| \le 1/\varepsilon} e^{-2\pi i x \cdot \xi} \frac{\xi_j}{|\xi|^{n+1}} d\xi dx$$

$$= c(n) \lim_{\varepsilon \to 0} \int_{\mathbb{R}^n} f(x) \int_{S^{n-1}} \int_{\varepsilon}^{1/\varepsilon} e^{-2\pi i r x \cdot \theta} r^{-1} \theta_j dr d\theta dx.$$

Using the elementary facts that

$$\left| \int_a^b \frac{\sin y}{y} dy \right| \le 4 \ (0 < a < b < \infty), \quad \int_{-\infty}^{\infty} \frac{\sin(\lambda y)}{y} dy = \pi \text{sgn} \lambda,$$

it follows from the dominated convergence theorem that integration and the limiting process may be interchanged, giving

1.4 The Hilbert and Riesz Transforms

$$\langle f, \widehat{W_j} \rangle = -ic(n) \int_{\mathbb{R}^n} f(x) \int_{S^{n-1}} \int_0^\infty \sin(2\pi r x.\theta) r^{-1} \theta_j \, dr \, d\theta \, dx$$

$$= -i\frac{\pi}{2} c(n) \int_{\mathbb{R}^n} f(x) \int_{S^{n-1}} \mathrm{sgn}(x.\theta) \theta_j \, d\theta \, dx.$$

Now the identity

$$-i\frac{\pi}{2} c(n) \int_{S^{n-1}} \mathrm{sgn}(x.\theta) \theta_j \, d\theta = -i\frac{x_j}{|x|}$$

(see [94], Lemma 4.1.15) shows that

$$\langle f, \widehat{W_j} \rangle = \int_{\mathbb{R}^n} -if(x) \frac{x_j}{|x|} dx.$$

Thus $\widehat{W_j}$ can be identified with the function $x \longmapsto -ix_j/|x|$, and so $\left(\widehat{R_j f}\right)(\xi) = \widehat{f}(\xi) \widehat{W_j}(\xi)$: the Proposition follows from [54], Theorem V.1.14. ∎

Application of the Proposition to the Hilbert transform H shows that for all $f \in S(\mathbb{R})$,

$$Hf(x) = \left(\widehat{f}(\xi)(-i \, \mathrm{sgn}\, \xi)\right)^{\vee}(x)$$

and hence

$$\|Hf\|_2 = \|f\|_2.$$

Thus H can be extended to $L_2(\mathbb{R})$: this extension is also denoted by H and is an isometry. It is a routine matter to show that the Hilbert space adjoint H^* of H is simply $-H$, and that

$$H^2 = -I,$$

where I is the identity map of $L_2(\mathbb{R})$ to itself.

Note that for all $f \in S(\mathbb{R})$, the function F_f defined by

$$F_f(z) = \frac{i}{\pi} \int_{-\infty}^\infty \frac{f(t)}{z - t} dt$$

is analytic on the upper half-plane $\mathbb{C}_+ := \{z \in \mathbb{C} : \mathrm{im}\, z > 0\}$. Moreover, considerations involving the Poisson kernel show that when $\mathrm{im}\, z \to 0$, $F_f(z) \to f(x) + iHf(x)$, where $x = \mathrm{re}\, z$: for details see [94], pp. 253-255. Thus we may think of

$$z \longmapsto \frac{i}{\pi} \int_{-\infty}^\infty \frac{f(t)}{z - t} dt := u(z) + iv(z) \; (u, v \text{ real-valued})$$

as an analytic extension of $f + iHf$ to $\mathbb{C}_+$. Using this, it is next shown that the Hilbert transform H can be extended to $L_p(\mathbb{R})$ and that this extension (still denoted by H) is a bounded linear map of $L_p(\mathbb{R})$ to itself.

Theorem 1.4.3 *Let $p \in (1, \infty)$. Then there is a positive constant C_p such that for all $f \in \mathcal{S}(\mathbb{R})$,*
$$\|Hf\|_p \leq C_p \|f\|_p.$$

Proof. It is enough to deal with the case in which f is real-valued. Since $f + iHf$ has an analytic extension to $\mathbb{C}_+$, so does
$$(f + iHf)^2 = f^2 - H(f)^2 + 2ifHf.$$

Hence $f^2 - H(f)^2$ has a harmonic extension u to the upper half-plane whose conjugate harmonic function v must have boundary values $H\left(f^2 - H(f)^2\right)$ (see [94], pp. 254-255). It follows that
$$H\left(f^2 - H(f)^2\right) = 2fHf,$$
so that, using the fact that $H^2(f) = -f$,
$$H(f)^2 = f^2 + 2H(fHf).$$

Now suppose that $p = 2^k$ for some $k \in \mathbb{N}$. We use induction on k. When $k = 1$ the result is already known and the norm of H on $L_2(\mathbb{R})$ is 1. Suppose that for some $k \in \mathbb{N}$, the norm of H on $L_{2^k}(\mathbb{R})$ is bounded above by c_k. Then
$$\|Hf\|_{2^{k+1}} = \|H(f)^2\|_{2^k}^{1/2} \leq \left(\|f^2\|_{2^k} + \|2H(fHf)\|_{2^k}\right)^{1/2}$$
$$\leq \left(\|f\|_{2^{k+1}}^2 + 2c_k \|fHf\|_{2^k}\right)^{1/2}$$
$$\leq \left(\|f\|_{2^{k+1}}^2 + 2c_k \|f\|_{2^{k+1}} \|Hf\|_{2^{k+1}}\right)^{1/2}.$$

Thus
$$\left(\frac{\|Hf\|_{2^{k+1}}}{\|f\|_{2^{k+1}}}\right)^2 - 2c_k \frac{\|Hf\|_{2^{k+1}}}{\|f\|_{2^{k+1}}} - 1 \leq 0,$$
and so
$$\frac{\|Hf\|_{2^{k+1}}}{\|f\|_{2^{k+1}}} \leq c_k + \left(c_k^2 + 1\right)^{1/2}.$$

Hence the inductive step is complete and so $H \in B(L_{2^k}(\mathbb{R}))$ for all $k \in \mathbb{N}$. By interpolation, $H \in B(L_p(\mathbb{R}))$ for all $p \geq 2$; and since $H \in B(L_{2^k}(\mathbb{R}))$ has adjoint $-H$, the claimed result also follows for all $p \in (1, 2)$. ∎

In fact, the exact norm of $H \in B(L_p(\mathbb{R}))$ is known. This is a result of Pichorides [164]: it hinges on two lemmas, which we state in the form used by Grafakos [93] in his version of the proof.

1.4 The Hilbert and Riesz Transforms

Lemma 1.4.4 *Let $p \in (1, 2]$. Then for all $a, b \in \mathbb{R}$,*

$$|b|^p \le |a|^p \tan^p (\pi/2p) - B_p \,\mathrm{re}\, \{(|a| + ib)^p\},$$

where

$$B_p = \sin^{p-1}(\pi/2p)/\sin((p-1)\pi/2p) > 0.$$

Lemma 1.4.5 *Let $p \in (1, 2]$. Then the function $g : \mathbb{R}^2 \to \mathbb{R}$ defined by $g(x, y) = \mathrm{re}\,\{(|x| + iy)^p\}$ is subharmonic.*

For proofs of these technical results we refer to [164] and [93]; the definition and basic properties of subharmonic functions are given in Chapter 2 below. We can now give the result concerning the norm of $H \in B\left(L_p(\mathbb{R})\right)$.

Theorem 1.4.6 *Let $p \in (1, \infty)$. Then*

$$\|H : L_p(\mathbb{R}) \to L_p(\mathbb{R})\| = \cot\left(\frac{\pi}{2p^*}\right), \text{ where } p^* = \max\{p, p'\}.$$

Proof. Since $H^* f = -Hf$ ($f \in \mathcal{S}(\mathbb{R})$) it is enough to deal with the case in which $p \in (1, 2]$. Let $f \in C_0^\infty(\mathbb{R})$, use Lemma 1.4.4 with $a = f(x), b = (Hf)(x)$ and integrate over $\mathbb{R}$:

$$\int_{-\infty}^{\infty} |(Hf)(x)|^p \, dx \le \tan^p\left(\frac{\pi}{2p}\right) \int_{-\infty}^{\infty} |f(x)|^p \, dx$$
$$- B_p \int_{-\infty}^{\infty} \mathrm{re}\, \{(|f(x)| + i(Hf)(x))^p\} \, dx. \quad (1.4.1)$$

We claim that

$$\int_{-\infty}^{\infty} \mathrm{re}\, \{(|f(x)| + i(Hf)(x))^p\} \, dx \ge 0. \quad (1.4.2)$$

To do this, consider the analytic extension of $f + iHf$ to the upper half-plane given by

$$u(z) + iv(z) = \frac{i}{\pi} \int_{-\infty}^{\infty} \frac{f(t)}{z - t} dt,$$

and note that since the function g of Lemma 1.4.5 is subharmonic and $z \mapsto u(z) + iv(z)$ is analytic, the function $z \mapsto g(u(z), v(z))$ is subharmonic on the upper half-plane; on the x-axis it has the values

$$\mathrm{re}\, \{(|f(x)| + i(Hf)(x))^p\}.$$

For large R let C_R be the circle with centre $(0, R)$ and radius $R - 1/R$; let C_R^U and C_R^L be the upper and lower semicircles that make up C_R. The subharmonicity of $g(u(z), v(z))$ implies that (see Theorem 2.1.2 below)

$$\int_{C_R^U} g\,(u(z),v(z))\,ds + \int_{C_R^L} g\,(u(z),v(z))\,ds \geq 2\pi(R - 1/R)g\,(u(iR),v(iR)).$$
(1.4.3)

Moreover,

$$|u(x+iy)|, |v(x+iy)| \leq \frac{\|f\|_\infty \,|\operatorname{supp} f|}{\pi R} \text{ if } |y| \geq R;$$
(1.4.4)

hence

$$|(R - 1/R)g\,(u(iR),v(iR))| \leq (R - 1/R)\left(\frac{2\|f\|_\infty \,|\operatorname{supp} f|}{\pi R}\right)^p \to 0 \text{ as } R \to \infty,$$
(1.4.5)

and

$$\left|\int_{C_R^U} g\,(u(z),v(z))\,ds\right| \leq \pi(R - 1/R)\left(\frac{2\|f\|_\infty \,|\operatorname{supp} f|}{\pi R}\right)^p \to 0 \text{ as } R \to \infty.$$
(1.4.6)

Now let $R \to \infty$ in (1.4.3) and use (1.4.5) and (1.4.6): since $g\,(u(z),v(z))$ has boundary values $\operatorname{re}\{(|f(x)| + i(Hf)(x))^p\}$ on the x–axis this gives (1.4.2). From (1.4.1) we thus have

$$\int_{-\infty}^{\infty} |(Hf)(x)|^p\,dx \leq \tan^p\left(\frac{\pi}{2p}\right)\int_{-\infty}^{\infty} |f(x)|^p\,dx,$$

so that

$$\|H : L_p(\mathbb{R}) \to L_p(\mathbb{R})\| \leq \tan\left(\frac{\pi}{2p}\right) = \cot\left(\frac{\pi}{2p^*}\right).$$

It remains to show that this estimate is sharp. To do this it is enough to consider the function f defined by

$$f(x) = (x+1)^{-1}\,|x+1|^{2\gamma/\pi}\,|x-1|^{-2\gamma/\pi}\cos\gamma,$$

where $\gamma \in \left(\frac{\pi}{2p'},\frac{\pi}{2p}\right)$, and to show that

$$(Hf)(x) = \begin{cases} \frac{1}{x+1}\left|\frac{x+1}{x-1}\right|^{2\gamma/\pi}\sin\gamma, & |x| > 1, \\ -\frac{1}{x+1}\left|\frac{x+1}{x-1}\right|^{2\gamma/\pi}\sin\gamma, & |x| < 1. \end{cases}$$

For details of these calculations we refer to [164]. ∎

We now return to the Riesz transforms with the object of showing that they can be extended to be bounded linear maps of $L_p(\mathbb{R}^n)$ to itself for every $p \in (1,\infty)$. First we introduce the directional Hilbert transform given, for each $\theta \in S^{n-1}$, by

1.4 The Hilbert and Riesz Transforms

$$H_\theta f(x) = \frac{1}{\pi} \text{p.v.} \int_{-\infty}^{\infty} f(x - t\theta) \frac{dt}{t} \quad (f \in \mathcal{S}(\mathbb{R}^n), x \in \mathbb{R}^n).$$

Let $e_1, \ldots, e_n$ be the standard basis vectors in $\mathbb{C}$. Then H_{e_1} is obtained by applying the Hibert transform in the first variable followed by the identity operator in the remaining variables: thus H_{e_1} is bounded on $L_p(\mathbb{R}^n)$ with norm equal to that of the Hilbert transform. Since

$$H_{A(e_1)} f(x) = H_{e_1}(f \circ A)(A^{-1} x)$$

for every $A \in O(n)$ (the set of all real, symmetric, invertible $n \times n$ matrices), the L_p boundedness of H_θ can be reduced to that of H_{e_1}: hence $H_\theta \in B(L_p(\mathbb{R}^n))$ and has norm bounded by that of the Hilbert transform, for every $\theta \in S^{n-1}$. Given ε, N with $0 < \varepsilon < N$, put

$$H_\theta^{(\varepsilon, N)}(f)(x) = \frac{1}{\pi} \int_{\varepsilon \le |t| \le N} f(x - t\theta) \frac{dt}{t}.$$

By Minkowski's integral inequality,

$$\left\| H_\theta^{(\varepsilon, N)}(f) \right\|_p \le \frac{2}{\pi} \|f\|_p \log(N/\varepsilon).$$

Put

$$\Omega(\theta) = \frac{\Gamma\left(\frac{n+1}{2}\right) \theta_j}{\pi^{(n+1)/2}} \quad (\theta \in S^{n-1}).$$

Then

$$\int_{\varepsilon \le |y| \le N} \frac{\Omega(y/|y|)}{|y|^n} f(x - y) dy = \int_{S^{n-1}} \Omega(\theta) \int_\varepsilon^N f(x - r\theta) \frac{dr}{r} d\theta$$

$$= -\int_{S^{n-1}} \Omega(\theta) \int_\varepsilon^N f(x + r\theta) \frac{dr}{r} d\theta$$

$$= \frac{1}{2} \int_{S^{n-1}} \Omega(\theta) \int_\varepsilon^N \frac{f(x - r\theta) - f(x + r\theta)}{r} dr d\theta$$

$$= \frac{\pi}{2} \int_{S^{n-1}} \Omega(\theta) H_\theta^{(\varepsilon, N)}(f)(x) d\theta.$$

Using the Lebesgue dominated convergence theorem the limits as $\varepsilon \to 0$ and $N \to \infty$ can be taken inside the integral when $f \in \mathcal{S}(\mathbb{R}^n)$, giving

$$R_j(f)(x) = \frac{\pi}{2} \int_{S^{n-1}} \Omega(\theta) H_\theta(f)(x) d\theta.$$

Now use Minkowski's integral inequality to conclude that R_j may be extended to be an element of $B(L_p(\mathbb{R}^n))$.

Note that by use of the Fourier transform it may be shown that

$$\sum\nolimits_{j=1}^{n} R_j^2 = -I,$$

where I is the identity operator.

As for the norms of the Riesz transforms R_j in $L_p(\mathbb{R}^n)$, it turns out that they coincide with that of the Hilbert transform.

Theorem 1.4.7 *Let $p \in (1, \infty)$. Then for every $j \in \{1, ..., n\}$,*

$$\|R_j : L_p(\mathbb{R}^n) \to L_p(\mathbb{R}^n)\| = \cot\left(\frac{\pi}{2p^*}\right), \text{ where } p^* = \max\{p, p'\}.$$

This important result was proved by Iwaniec and Martin [110] by the so-called method of rotations, and by Bañuelos and Wang [18] using martingale theory: considerations of space prevent us from giving details of these very different proofs. Certain vectorial versions of this theorem are also known: for example, it is shown in [110] that if $p \in [2, \infty)$, then for all $f \in L_p(\mathbb{R}^n)$,

$$\left\|\left(\sum\nolimits_{j=1}^{n} |R_j f|^2\right)^{1/2}\right\|_p \leq \sqrt{2} \cot\left(\frac{\pi}{2p}\right) \|f\|_p.$$

As a simple consequence of Theorem 1.4.7 we have the following estimate of L_p norms of second derivatives by that of the Laplacian Δ (see 2.1 below).

Corollary 1.4.8 *Let $p \in (1, \infty)$ and suppose that $u \in C_0^2(\mathbb{R}^n)$. Then for all $j, k \in \{1, ..., n\}$,*

$$\|D_j D_k u\|_p \leq \cot^2\left(\frac{\pi}{2p^*}\right) \|\Delta u\|_p.$$

Proof. On using [54], Lemma V.1.15, we have

$$(D_j D_k u)^\wedge(\xi) = -4\pi^2 \xi_j \xi_k \widehat{u}(\xi)$$
$$= -\left(\frac{i\xi_j}{|\xi|}\right)\left(\frac{i\xi_k}{|\xi|}\right)(-4\pi^2 |\xi|^2)\widehat{u}(\xi) = -(R_j R_k \Delta u)^\wedge(\xi).$$

It follows that

$$D_j D_k u = -R_j R_k \Delta u$$

and Theorem 1.4.7 completes the proof. ∎

Note that in view of this Corollary,

1.4 The Hilbert and Riesz Transforms

$$\left(\sum_{j,k=1}^{n} \|D_j D_k u\|_p^p\right)^{1/p} \leq n^{2/p} \cot^2\left(\frac{\pi}{2p^*}\right) \|\Delta u\|_p. \tag{1.4.7}$$

When $p = 2$ this gives

$$\left(\sum_{j,k=1}^{n} \|D_j D_k u\|_2^2\right)^{1/2} \leq n \|\Delta u\|_2,$$

which contrasts rather sadly with the equality

$$\left(\sum_{j,k=1}^{n} \|D_j D_k u\|_2^2\right)^{1/2} = \|\Delta u\|_2 \tag{1.4.8}$$

that follows immediately from use of the Fourier transform. The passage from the sharp 'scalar' inequality given in Corollary 1.4.8 to a sharp 'vector-valued' inequality of the type of (1.4.7) seems difficult: for some progress in this direction see [18] and [110], but as their results are not sharp even for $p = 2$ it seems that a genuine problem remains. The papers [17] and [162] contain much interesting material on related topics.

Chapter 2
The Laplace Operator

2.1 Mean Value Inequalities

We remind the reader that for any $p \in [1, \infty]$, the norm on the usual Lebesgue space $L_p(\Omega)$ will be denoted by $\|\cdot\|_{p,\Omega}$, or even by $\|\cdot\|_p$ if it is clear from the context what is intended. By $L_{p,loc}(\Omega)$ we shall mean the space of all functions f that belong to $L_p(B)$ for all compact subsets B of Ω. All functions appearing in this chapter are assumed to be real-valued and Ω is a non-empty, open subset of $\mathbb{R}^n$.

If $u \in C^2(\Omega)$, the **Laplacian** of u, written as Δu, is defined by

$$\Delta u = \sum_{j=1}^{n} D_j^2 u.$$

Definition 2.1.1 *A function* $u : \Omega \to \mathbb{R}$ *in* $C^2(\Omega)$ *is called* **superharmonic in** Ω *or* **subharmonic in** Ω *according as* $\Delta u \leq 0$ *or* $\Delta u \geq 0$ *in* Ω. *If* $\Delta u = 0$ *in* Ω *we say that* u *is* **harmonic in** Ω.

Note that if u is harmonic in Ω and $p \in [2, \infty)$, then since

$$\Delta \left(|u|^p\right) = p|u|^{p-2} u \Delta u + p(p-1) |u|^{p-2} \sum_{j=1}^{n} |D_j u|^2,$$

$|u|^p$ is subharmonic in Ω.

We now give some of the most interesting simple properties of such functions, beginning with the mean-value inequalities. These involve the volume ω_n of the unit ball in $\mathbb{R}^n$, the explicit value of which may be quickly determined as follows. For each $y > 0$,

$$\int_{|x|^2 < y} dx = \text{volume of ball in } \mathbb{R}^n \text{ of radius } y^{1/2} = \omega_n y^{n/2};$$

put
$$F = \omega_n \int_0^\infty y^{n/2} e^{-y} dy = \omega_n \Gamma\left(1 + \frac{n}{2}\right).$$

Since
$$F = \int_0^\infty e^{-y} \left(\int_{|x|^2 < y} dx\right) dy = \int_{\mathbb{R}^n} \left(\int_{|x|^2}^\infty e^{-y} dy\right) dx$$
$$= \int_{\mathbb{R}^n} e^{-|x|^2} dx = \pi^{n/2},$$

it follows that
$$\omega_n = \pi^{n/2} / \Gamma\left(1 + \frac{n}{2}\right).$$

As the surface area of the ball of radius r is of the form Ar^{n-1}, we see that $\omega_n = \int_0^1 Ar^{n-1} dr = A/n$, and hence $A = n\omega_n$.

We denote by $B(x, \delta)$ the ball with centre x and radius δ in $\mathbb{R}^n$.

Theorem 2.1.2 *Let u be subharmonic in Ω, let $x \in \Omega$ and suppose that $\delta > 0$ is such that $\overline{B(x, \delta)} \subset \Omega$. Then*

$$u(x) \leq \frac{1}{n\omega_n \delta^{n-1}} \int_{\partial B(x,\delta)} u(y) d\sigma(y) \tag{2.1.1}$$

and

$$u(x) \leq \frac{1}{\omega_n \delta^n} \int_{B(x,\delta)} u(y) dy, \tag{2.1.2}$$

where $d\sigma$ denotes the surface measure.

Proof. By the divergence theorem $\left(\int_B \text{div } \mathbf{w} \, dx = \int_{\partial B} \mathbf{w}.\nu d\sigma;\ \text{see 1.3}\right)$ we find, on the introduction of radial and angular coordinates $r = |y - x|$ and $\theta = (y - x)/r$,

$$0 \leq \int_{B(x,r)} \Delta u(y) dy = r^{n-1} \int_{\partial B(0,1)} \frac{\partial}{\partial r} u(x + r\theta) d\sigma(\theta)$$

provided that $0 < r \leq \delta$. Thus

$$0 \leq \int_0^\delta \int_{\partial B(0,1)} \frac{\partial}{\partial r} u(x + r\theta) d\sigma(\theta) dr.$$

Hence
$$0 \leq \int_{\partial B(0,1)} \{u(x + \delta\theta) - u(x)\} d\sigma(\theta),$$

2.1 Mean Value Inequalities

from which (2.1.1) follows. To derive (2.1.2) we use the inequality just proved in the form

$$u(x) \le \frac{1}{n\omega_n} \int_{\partial B(0,1)} u(x+\rho\delta y) d\sigma(y),$$

for $0 < \rho \le 1$. Then

$$nu(x)\omega_n \int_0^1 \rho^{n-1} d\rho \le \int_0^1 \int_{\partial B(0,1)} \rho^{n-1} u(x+\rho\delta y) d\sigma(y) d\rho,$$

and (2.1.2) follows immediately. ∎

It is clear that Theorem 2.1.2 holds with the inequality signs reversed if u is superharmonic in Ω, and from this we derive the familiar result that the value of a harmonic function at a point is equal to the mean of its values on any sufficiently small ball or sphere centred at that point. More formally:

Corollary 2.1.3 (*The mean-value theorem*) *Let u be harmonic in Ω and let $x \in \Omega$. Then*

$$u(x) = \frac{1}{n\omega_n \delta^{n-1}} \int_{\partial B(x,\delta)} u(y) d\sigma(y) = \frac{1}{\omega_n \delta^n} \int_{B(x,\delta)} u(y) dy$$

for any $\delta > 0$ such that $\overline{B(x,\delta)} \subset \Omega$.

A simple consequence of the mean-value theorem is the result that if u is harmonic in $\mathbb{R}^n$ and belongs to $L_p(\mathbb{R}^n)$ for some $p \in (1, \infty)$, then it is the zero function. For by Corollary 2.1.3, for all $r > 0$ and all $x_0 \in \mathbb{R}^n$,

$$|u(x_0)| \le \frac{1}{\omega_n r^n} \int_{B(x_0,r)} |u(x)| dx \le \left(\frac{1}{\omega_n r^n}\right)^{1/p} \|u\|_{p,\mathbb{R}^n} \to 0 \text{ as } r \to \infty.$$

The mean-value property leads quickly to a number of interesting results concerning harmonic functions. Indeed, it actually characterises them, as we now show.

Theorem 2.1.4 *Let $u \in C(\Omega)$ and suppose that given any $x \in \Omega$ and any $\delta > 0$ such that $\overline{B(x,\delta)} \subset \Omega$,*

$$u(x) = \frac{1}{n\omega_n \delta^{n-1}} \int_{\partial B(x,\delta)} u(y) d\sigma(y).$$

Then $u \in C^\infty(\Omega)$ and u is harmonic in Ω. In particular, functions which are harmonic in Ω are in $C^\infty(\Omega)$.

Proof. Let ϕ be the standard test function given by

$$\phi(x) = \begin{cases} C \exp\left(\frac{1}{|x|^2-1}\right) & \text{if } |x| < 1, \\ 0 & \text{if } |x| \ge 1, \end{cases}$$

where C is a constant chosen so that $\int_{\mathbb{R}^n} \phi(x)dx = 1$. Given any $\varepsilon > 0$, put $\phi_\varepsilon(x) = \varepsilon^{-n}\phi(x/\varepsilon)$ and $\Omega_\varepsilon = \{x : \overline{B(x,\varepsilon)} \subset \Omega\}$. Then if $x \in \Omega_\varepsilon$, the function $y \mapsto \phi_\varepsilon(x-y)$ has support contained in Ω, and

$$\int_\Omega u(y)\phi_\varepsilon(x-y)dy = \int_{B(0,\varepsilon)} u(x-y)\varepsilon^{-n}\phi_\varepsilon(y/\varepsilon)dy = \int_{B(0,1)} u(x-\varepsilon y)\phi(y)dy$$

$$= \int_0^1 \int_{\partial B(0,1)} u(x-\varepsilon rz)\phi(rz)r^{n-1}d\sigma(z)dr$$

$$= n\omega_n u(x) \int_0^1 r^{n-1}\phi(rz)dr,$$

the final step following from the mean-value property and the independence of $\phi(rz)$ of the particular z on the unit sphere. The last expression equals

$$u(x)\int_0^1 \int_{\partial B(0,1)} \phi(rz)r^{n-1}d\sigma(z)dr = u(x)\int_{B(0,1)} \phi(y)dy = u(x),$$

and we conclude that if $x \in \Omega_\varepsilon$,

$$u(x) = \int_\Omega u(y)\phi_\varepsilon(x-y)dy.$$

Hence $u \in C^\infty(\Omega_\varepsilon)$ (see, for example, [54], p. 208), which implies that $u \in C^\infty(\Omega)$ as ε may be chosen arbitrarily small.

To finish the proof, take any $x \in \Omega$ and any $\delta > 0$ such that $\overline{B(x,\delta)} \subset \Omega$, and use the divergence theorem:

$$\int_{B(x,\delta)} \Delta u(y)dy = \delta^{n-1}\int_{\partial B(0,1)} \frac{\partial}{\partial \rho}u(x+\rho\theta)\big|_{\rho=\delta}\, d\sigma(\theta)$$

$$= \delta^{n-1}\frac{d}{d\rho}\int_{\partial B(0,1)} u(x+\rho\theta)d\sigma(\theta)\big|_{\rho=\delta}$$

$$= \delta^{n-1}\frac{d}{d\rho}(n\omega_n u(x)) = 0.$$

Since this is true for all small enough $\delta > 0$, $\Delta u(x) = 0$, and as x is an arbitrary point of Ω, u is harmonic in Ω. ∎

2.1 Mean Value Inequalities

Note that the mean-value property

$$u(x) = \frac{1}{n\omega_n \delta^{n-1}} \int_{\partial B(x,\delta)} u(y) d\sigma(y)$$

in Theorem 2.1.4 is equivalent to

$$u(x) = \frac{1}{\omega_n \delta^n} \int_{B(x,\delta)} u(y) dy.$$

For the second follows from the first by integration, while the first can be obtained from the second by differentiation.

Corollary 2.1.5 *Let $u \in L_{1,loc}(\Omega)$ be such that for all $x \in \Omega$,*

$$u(x) = \frac{1}{\omega_n \delta^n} \int_{B(x,\delta)} u(y) dy$$

whenever $\overline{B(x,\delta)} \subset \Omega$. Then u coincides almost everywhere with a function that is harmonic on Ω.

Proof. Elementary considerations show that $u \in C(\Omega)$, and so Theorem 2.1.4 applies. ∎

Corollary 2.1.6 *Let (u_k) be a sequence of functions, each of which is harmonic in Ω, and suppose that the sequence converges, uniformly on every compact subset of Ω, to a function u. Then u is harmonic in Ω.*

Proof. Each u_k has the mean value property:

$$u_k(x) = \frac{1}{n\omega_n \delta^{n-1}} \int_{\partial B(x,\delta)} u_k(y) d\sigma(y)$$

for all balls $B(x, \delta)$ with closure contained in Ω. By the uniformity of convergence we may pass to the limit under the integral sign; thus u must have the mean value property. Since $u \in C(\Omega)$, the result follows from Theorem 2.1.4. ∎

The strong maximum principle is an easy consequence of Theorem 2.1.2.

Theorem 2.1.7 *(The strong maximum principle) Let u be subharmonic (superharmonic) in Ω and suppose that there is a point $x_0 \in \Omega$ such that $u(x_0) = \sup_{x \in \Omega} u(x)$ ($u(x_0) = \inf_{x \in \Omega} u(x)$). Then if Ω is connected, u is constant in Ω.*

Proof. It is enough to deal with the case in which u is subharmonic. Evidently $\Omega' := \{x \in \Omega : u(x) = u(x_0)\}$ is a relatively closed subset of Ω. By Theorem 2.1.2, if $u(x) = u(x_0)$ for some $x \in \Omega$, then $u(y) = u(x_0)$ for all y in a small enough ball centred at x, so that Ω' is also relatively open. Since Ω is connected, $\Omega' = \Omega$. ∎

Corollary 2.1.8 *If Ω is connected, a non-constant function which is harmonic in Ω cannot have a maximum or minimum in Ω.*

Corollary 2.1.9 *(The weak maximum principle) Let Ω be bounded and let $u \in C(\overline{\Omega})$ be harmonic in Ω. Then the maximum and minimum values of u in $\overline{\Omega}$ are attained on $\partial\Omega$. If u is constant on $\partial\Omega$ it is constant throughout Ω.*

Proof. Since $\overline{\Omega}$ is compact the maximum M of u in $\overline{\Omega}$ is attained at some point $x_0 \in \overline{\Omega}$. If $x_0 \in \Omega$, Theorem 2.1.7 shows that u equals M on the connected component $C(x_0)$ of x_0 in $\overline{\Omega}$. We claim that $C(x_0) \cap \partial\Omega \neq \emptyset$: granted this, it follows that the value M is attained on $\partial\Omega$. To justify this claim, note that $C(x_0)$ is closed, and that if $C(x_0) \subset \Omega$, then given $y \in \partial C(x_0)$, there exists $\delta > 0$ with $B(y, \delta) \subset \Omega$; since $C(x_0) \cup B(y, \delta)$ is connected, this contradicts the maximality of the component $C(x_0)$. The rest of the Corollary is now clear. ∎

Note that the same argument shows that if the hypotheses are weakened by assuming that u is superharmonic (or subharmonic) in Ω, rather than harmonic in Ω, then the minimum (maximum) values of u on $\overline{\Omega}$ are attained on $\partial\Omega$.

Corollary 2.1.10 *Let Ω be bounded and connected, and suppose that $u, v \in C(\overline{\Omega})$ are such that u is harmonic in Ω, v is subharmonic in Ω and $u = v$ on $\partial\Omega$. Then $v \leq u$ in Ω.*

Proof. The function $v - u$ is subharmonic in Ω and vanishes on $\partial\Omega$; it has a maximum in $\overline{\Omega}$ since $\overline{\Omega}$ is compact. By Theorem 2.1.7, $v - u \leq 0$ in Ω. ∎

This last corollary justifies the use of the word *subharmonic*: a function which is subharmonic in a bounded domain Ω is dominated by any function which is harmonic in Ω and has the same boundary values.

A further simple consequence of the maximum principle is the uniqueness of solutions of the *Dirichlet problem* for Poisson's equation $\Delta u = f$ in Ω : this problem asks whether there is a function $u \in C^2(\Omega) \cap C(\overline{\Omega})$ such that

$$\Delta u = f \text{ in } \Omega, \ u = g \text{ on } \partial\Omega,$$

where $f \in C(\Omega)$ and $g \in C(\partial\Omega)$ are given. The following corollary shows that if there is such a function it is unique, provided that Ω is bounded.

Corollary 2.1.11 *Let Ω be a bounded domain and suppose that $u, v \in C^2(\Omega) \cap C(\overline{\Omega})$ are such that $\Delta u = \Delta v$ in Ω, $u = v$ on $\partial\Omega$. Then $u = v$ in Ω.*

Proof. Since $\Delta(u - v) = 0$ in Ω and $u - v = 0$ on $\partial\Omega$, the result follows from Corollary 2.1.9. ∎

Boundedness of Ω is essential here, for if we take Ω to be the infinite strip in $\mathbb{R}^2$, $\Omega = \{(x_1, x_2) : x_1 \in \mathbb{R}, \ 0 < x_2 < \pi\}$, the Dirichlet problem $\Delta u = 0$ in Ω, $u = 0$

2.1 Mean Value Inequalities

on $\partial\Omega$ has the two distinct solutions $u_1 = 0$ and $u_2(x_1, x_2) = e^{x_1} \sin x_2$. Uniqueness of solutions of the Dirichlet problem in unbounded domains can be achieved by the imposition of suitable limitations on the growth of solutions; that is, uniqueness holds within a class of solutions of appropriately limited growth. This follows from theorems of Phragmén-Lindelöf type, for which we refer to [173] and [81].

The maximum principle can be used to give an *a priori* estimate for solutions of the Dirichlet problem.

Theorem 2.1.12 *Let Ω be a bounded domain and let $u \in C^2(\Omega) \cap C(\overline{\Omega})$ satisfy $\Delta u = f$ in Ω, $u = \phi$ on $\partial\Omega$, where $f \in L_\infty(\Omega) \cap C(\Omega)$ and $\phi \in C(\partial\Omega)$ are given. Then there is a constant M, depending only on Ω, such that*

$$\|u\|_{\infty,\Omega} \leq \|\phi\|_{\infty,\partial\Omega} + M \|f\|_{\infty,\Omega}.$$

Proof. Without loss of generality we may, and shall, assume that Ω is contained in the strip $\{x \in \mathbb{R}^n : 0 \leq x_1 \leq d\}$. Let $g(x) = e^d - e^{x_1}$ ($x \in \mathbb{R}^n$) : in Ω,

$$e^d - 1 \geq g(x) \geq 0 \text{ and } \Delta g(x) = -e^{x_1} \leq -1.$$

Now put

$$h(x) = \|\phi\|_{\infty,\partial\Omega} + g(x) \|f\|_{\infty,\Omega}$$

and observe that in Ω,

$$\Delta h = \|f\|_{\infty,\Omega} \Delta g \leq -\|f\|_{\infty,\Omega}.$$

Thus $\Delta(u - h) = f - \Delta h \geq 0$ in Ω, while $u - h \leq 0$ on $\partial\Omega$, so that by the strong maximum principle, $u \leq h$ in Ω. If instead of h we use $-h$ we obtain $u \geq -h$, from which we see that $|u| \leq h$ and the result follows. ∎

While it is entirely natural that the maximum principle should have implications for the uniqueness of solutions of the Dirichlet problem, it may be less clear that it has anything to say about questions of existence. To illustrate its potential in this direction, first let $\Omega \subset \mathbb{R}^n$ ($n \geq 2$) be bounded and open, let $x_0 \in \partial\Omega$ and suppose that u is harmonic and bounded on Ω, continuous on $\overline{\Omega}\setminus\{x_0\}$ and zero on $\partial\Omega\setminus\{x_0\}$. We claim that u is zero on Ω. To establish this, let $a > 0$ be so small that $\Omega_a := \Omega\setminus B(x_0, a) \neq \emptyset$ and let b be so large that $\overline{\Omega} \subset B(x_0, b)$. Let $M := \sup_{x\in\Omega} |u(x)|$ and set

$$v(x) = \begin{cases} M \log(b/|x - x_0|)/\log(b/a) & \text{if } n = 2, \\ M a^{n-2}/|x - x_0|^{n-2} & \text{if } n > 2. \end{cases}$$

Then $w := v - u \in C^2(\Omega_a) \cap C(\overline{\Omega_a})$ and $\Delta w = 0$ in Ω_a. Since $v = M$ on $\partial B(x_0, a)$ and $v > 0$ elsewhere on $\partial\Omega_a$, it follows from Corollary 2.1.9 that $w \geq 0$ on $\overline{\Omega_a}$: thus $u \leq v$ on $\overline{\Omega_a}$. Given $x \in \Omega$ and $\varepsilon > 0$, choose $a > 0$ so small that $x \in \Omega_a$ and $v(x) < \varepsilon$. Then $u(x) \leq v(x) < \varepsilon$; as this holds for all $\varepsilon > 0$, $u(x) \leq 0$. Since

$-u$ satisfies the same conditions as u, it follows that $u(x) = 0$, and the claim is established.

Now suppose that $\partial\Omega$ has an isolated point, x_0. Define $g : \partial\Omega \to \mathbb{R}$ by

$$g(x) = \begin{cases} 1, & x = x_0, \\ 0, & \text{otherwise.} \end{cases}$$

Then $g \in C(\partial\Omega)$. If the Dirichlet problem

$$\Delta u = 0 \text{ in } \Omega, \ u = g \text{ on } \partial\Omega, \ u \in C^2(\Omega) \cap C(\overline{\Omega})$$

had a solution u, then in view of the above claim, u would be zero in Ω and hence zero in $\overline{\Omega}$, contradicting the fact that $u(x_0) = g(x_0) = 1$. In particular, this shows that if $\Omega = B(0,r)\setminus\{0\}$, then there are continuous boundary data such that the corresponding Dirichlet problem for the Laplace operator does not have a solution.

2.2 Representation of Solutions

To obtain more information about harmonic functions it is convenient to study certain special solutions of Laplace's equation. It is easy to verify that the function u defined by

$$u(x) = \begin{cases} |x|^{2-n} & \text{if } n > 2, \\ \log|x| & \text{if } n = 2 \end{cases}$$

is harmonic in $\mathbb{R}^n \setminus \{0\}$. For technical reasons which will become clear presently we introduce a multiple K of this function, namely

$$K(x) = \begin{cases} -\frac{1}{n\omega_n(n-2)} |x|^{2-n} & \text{if } n > 2, \\ \frac{1}{2\pi} \log|x| & \text{if } n = 2, \end{cases}$$

and call it the *fundamental solution* of Laplace's equation. This function K is most useful in the representation of harmonic functions; to be more precise, it is the function $x \mapsto K(x - y)$, where y is some fixed point of Ω, which is useful.

The idea is to use Green's second identity

$$\int_\Omega (v\Delta u - u\Delta v)dx = \int_{\partial\Omega} \left(v\frac{\partial u}{\partial \nu} - u\frac{\partial v}{\partial \nu} \right) d\sigma, \qquad (2.2.1)$$

where it is assumed that Ω is a bounded domain with C^1 boundary (see 1.3) and $u, v \in C^2(\Omega) \cap C^1(\overline{\Omega})$; ν denotes the exterior normal to $\partial\Omega$ and $d\sigma$ is the surface measure. This identity follows from the divergence theorem

$$\int_\Omega \operatorname{div} \mathbf{w} \, dx = \int_{\partial\Omega} \mathbf{w} \cdot \nu \, d\sigma$$

2.2 Representation of Solutions

on taking **w** first equal to v grad u, then equal to u grad v, and subtracting the results. We cannot take $v(x) = K(x - y)$ in (2.2.1) as this function is singular at $x = y$; indeed we have, for $i, j = 1, 2, \ldots, n$ and $x \neq y$,

$$\left. \begin{array}{l} n\omega_n D_i K(x - y) = (x_i - y_i)|x - y|^{-n}, \\ n\omega_n D_i D_j K(x - y) = |x - y|^{-n-2}\{|x - y|^2 \delta_{ij} - n(x_i - y_i)(x_j - y_j)\}, \end{array} \right\} \quad (2.2.2)$$

where δ_{ij} is the Kronecker delta, so that

$$\left. \begin{array}{l} n\omega_n |D_i K(x - y)| \leq |x - y|^{1-n}, \\ \omega_n |D_i D_j K(x - y)| \leq |x - y|^{-n}. \end{array} \right\} \quad (2.2.3)$$

In the same way it can be shown that for some constant C (independent of x and y),

$$|D_i D_j D_k K(x - y)| \leq C |x - y|^{-n-1} \quad (2.2.4)$$

for all $i, j, k \in \{1, 2, \ldots, n\}$. To avoid the problem caused by the singularity we apply (2.2.1) with Ω replaced by $\Omega \backslash B(y, \delta)$ for small enough $\delta > 0$. It follows that

$$\int_{\Omega \backslash B(y, \delta)} K \Delta u \, dx = \int_{\partial \Omega} \left(K \frac{\partial u}{\partial \nu} - u \frac{\partial K}{\partial \nu} \right) d\sigma(x)$$

$$+ \int_{\partial B(y, \delta)} \left(K \frac{\partial u}{\partial \nu} - u \frac{\partial K}{\partial \nu} \right) d\sigma(x). \quad (2.2.5)$$

To deal with the integral over $\partial B(y, \delta)$ we note that, with $\widetilde{K}(\delta) := K(x - y)$ for $|x - y| = \delta$,

$$\left| \int_{\partial B(y, \delta)} K \frac{\partial u}{\partial \nu} d\sigma(x) \right| = \left| \widetilde{K}(\delta) \int_{\partial B(y, \delta)} \frac{\partial u}{\partial \nu} d\sigma(x) \right|$$

$$\leq \widetilde{K}(\delta) n\omega_n \delta^{n-1} \sup_{\partial B(y, \delta)} |\text{grad } u(x)|$$

$$\to 0 \text{ as } \delta \to 0;$$

while

$$\int_{\partial B(y, \delta)} u \frac{\partial K}{\partial \nu} d\sigma(x) = -\widetilde{K}'(\delta) \int_{\partial B(y, \delta)} u \, d\sigma(x) = -\frac{1}{n\omega_n \delta^{n-1}} \int_{\partial B(y, \delta)} u \, d\sigma(x)$$

$$\to -u(y) \text{ as } \delta \to 0.$$

Thus if we let $\delta \to 0$ in (2.2.5) we obtain

$$u(y) = \int_{\partial\Omega} \left(u(x) \frac{\partial}{\partial \nu} K(x-y) - K(x-y) \frac{\partial u(x)}{\partial \nu} \right) d\sigma(x)$$

$$+ \int_\Omega K(x-y) \Delta u(x) dx, \qquad (2.2.6)$$

which is *Green's representation formula*. Notice that if u is also harmonic in Ω, (2.2.6) becomes

$$u(y) = \int_{\partial\Omega} \left(u(x) \frac{\partial}{\partial \nu} K(x-y) - K(x-y) \frac{\partial u(x)}{\partial \nu} \right) d\sigma(x),$$

which in view of the analyticity of the integrand with respect to y shows that u is analytic in Ω. It follows that every harmonic function is analytic on its domain of definition.

A more general version of the representation formula (2.2.6) can be obtained by the introduction of an arbitrary function $w \in C^2(\Omega) \cap C^1(\overline{\Omega})$ which is harmonic in Ω. From (2.2.1) we have

$$\int_\Omega w \Delta u \, dx = \int_{\partial\Omega} \left(w \frac{\partial u}{\partial \nu} - u \frac{\partial w}{\partial \nu} \right) d\sigma$$

which, when combined with (2.2.6), gives

$$u(y) = \int_{\partial\Omega} \left(u(x) \frac{\partial}{\partial \nu} G(x, y) - G(x, y) \frac{\partial u(x)}{\partial \nu} \right) d\sigma(x)$$

$$+ \int_\Omega G(x, y) \Delta u(x) dx, \qquad (2.2.7)$$

where $G(x, y) = K(x - y) + w(x)$. We summarise our results as follows:

Theorem 2.2.1 *Let Ω be a bounded domain with C^1 boundary and let $u \in C^2(\Omega) \cap C^1(\overline{\Omega})$ satisfy $\Delta u = f$ in Ω. Then for all $y \in \Omega$,*

$$u(y) = \int_{\partial\Omega} \left(u(x) \frac{\partial}{\partial \nu} G(x, y) - G(x, y) \frac{\partial u(x)}{\partial \nu} \right) d\sigma(x) + \int_\Omega G(x, y) f(x) dx. \qquad (2.2.8)$$

If $G(x, y) = 0$ for all $x \in \partial\Omega$ and all $y \in \Omega$, this theorem shows that

$$u(y) = \int_{\partial\Omega} u(x) \frac{\partial}{\partial \nu} G(x, y) d\sigma(x) + \int_\Omega G(x, y) f(x) dx, \qquad (2.2.9)$$

2.2 Representation of Solutions

from which we see that u may be determined from its values on $\partial\Omega$ and from f and G. Such a function G is called *Green's function of the first kind*; Corollary 2.1.10 shows that Green's function is unique, if it exists at all. If it does exist, then for all $x, y \in \Omega$ with $x \neq y$ we have

$$G(x, y) = G(y, x) < 0.$$

To establish this, let $\varepsilon, \varepsilon_1 > 0$ be so small that $\overline{B(x, \varepsilon)}$ and $\overline{B(y, \varepsilon_1)}$ are disjoint and both contained in Ω. Apply (2.2.1) to $\left(\Omega \setminus \overline{B(x, \varepsilon)}\right) \setminus \overline{B(y, \varepsilon_1)}$: since $z \longmapsto G(z, x)$ and $z \longmapsto G(z, y)$ are harmonic in this set,

$$\int_{\partial\Omega \cup \partial B(x,\varepsilon) \cup \partial B(y,\varepsilon_1)} \left\{ G(z, x) \frac{\partial}{\partial \nu} G(z, y) - G(z, y) \frac{\partial}{\partial \nu} G(z, x) \right\} d\sigma(z) = 0.$$

Since $G(z, x) = G(z, y) = 0$ when $z \in \partial\Omega$, we simply have to consider this integral over $\partial B(x, \varepsilon) \cup \partial B(y, \varepsilon_1)$. Consider the integral over $\partial B(x, \varepsilon)$: near x, $G(z, y)$ is well-behaved while $G(z, x) - K(x - z)$ is harmonic. As in the proof of Green's representation formula, the integral over $\partial B(x, \varepsilon) \to G(x, y)$ as $\varepsilon \to 0$; as $\varepsilon_1 \to 0$, the integral over $\partial B(x, \varepsilon_1) \to -G(y, x)$. Hence $G(x, y) = G(y, x)$. To show that $G(x, y) < 0$, fix $x \in \Omega$: $G(x, y) = 0$ if $y \in \partial\Omega$, and $G(x, y) \to -\infty$ as $y \to x$. Application of the maximum principle to $\Omega \setminus B(x, \varepsilon)$ now gives the result.

For certain domains, direct computation can be used to show that Green's function exists. The simplest and most important case is when Ω is a ball, and it is a routine matter to verify that if $\Omega = B(0, R)$ then Green's function G is given by

$$G(x, y) = \begin{cases} K(x - y) - K\left(\frac{|y|}{R}\left(x - \frac{R^2}{|y|^2} y\right)\right), & y \neq 0, \\ K(x) - \tilde{K}(R), & y = 0. \end{cases} \quad (2.2.10)$$

Note that for all $x, y \in \overline{B(0, R)}$, it can be directly checked that $G(x, y) = G(y, x)$ and $G(x, y) \leq 0$, while if $x \in \partial B(0, R)$, the normal derivative of G is

$$\frac{\partial}{\partial \nu} G(x, y) = \frac{R^2 - |y|^2}{n \omega_n R} |x - y|^{-n}.$$

Substitution of these expressions in (2.2.9) shows that if $u \in C^2(B(0, R)) \cap C^1(\overline{B(0, R)})$ is harmonic in $B(0, R)$, then

$$u(y) = \frac{R^2 - |y|^2}{n \omega_n R} \int_{\partial B(0,R)} \frac{u(x)}{|x - y|^n} d\sigma(x), \quad (2.2.11)$$

which is *Poisson's integral formula*. This formula enables us to solve the Dirichlet problem for Laplace's equation in a ball.

Theorem 2.2.2 *Let $B = B(0, R)$ and let $\phi : \partial B \to \mathbf{R}$ be continuous. Then the function u defined by*

$$n\omega_n R u(x) = \begin{cases} (R^2 - |x|^2) \int_{\partial B(0,R)} \phi(y) |x - y|^{-n} \, d\sigma(y), & x \in B, \\ n\omega_n R \phi(x), & x \in \partial B, \end{cases}$$

is harmonic in B and continuous on $\overline{B}$.

Proof. If $x \in B$ it is clear that the derivatives of u at x may be obtained by differentiating under the integral sign, and as G is harmonic in x, so is $\partial G/\partial \nu$; thus u is harmonic in B. To prove that $u \in C(\overline{B})$, note that (2.2.11) with $u(x) = 1$ in $\overline{B}$ gives

$$\int_{\partial B} P(x, y) d\sigma(y) = 1$$

for all $x \in B$, where P is the *Poisson kernel* given by

$$P(x, y) = \frac{R^2 - |x|^2}{n\omega_n R |x - y|^n}.$$

Given any $x_0 \in \partial B$ and any $\varepsilon > 0$, there exists $\delta > 0$ such that $|\phi(x) - \phi(x_0)| < \varepsilon$ if $|x - x_0| < \delta$, $x \in \partial B$; moreover, $|\phi|$ is bounded on ∂B, by M, say. It follows that if $|x - x_0| < \delta/2$, $x \in B$, we have

$$|u(x) - u(x_0)| = \left| \int_{\partial B} P(x, y) \{\phi(y) - \phi(x_0)\} d\sigma(y) \right|$$

$$\leq \int_{|y-x_0|<\delta} P(x, y) |\phi(y) - \phi(x_0)| d\sigma(y)$$

$$+ \int_{|y-x_0|\geq\delta} P(x, y) |\phi(y) - \phi(x_0)| d\sigma(y)$$

$$= I_1 + I_2, \text{ say.}$$

It is easy to see that $I_1 < \varepsilon$. As for I_2, observe that

$$I_2 \leq 2M \int_{|y-x_0|\geq\delta} P(x, y) |\phi(y) - \phi(x_0)| d\sigma(y)$$

$$= \frac{2M(R^2 - |x|^2)}{n\omega_n R} \int_{|y-x_0|\geq\delta} \frac{d\sigma(y)}{|x - y|^n}.$$

2.2 Representation of Solutions

Since $|y - x| \geq |y - x_0| - |x_0 - x| \geq \delta - \delta/2 = \delta/2$, we see that

$$I_2 \leq 2M(R^2 - |x|^2)R^{n-2}(2/\delta)^n,$$

so that

$$|u(x) - u(x_0)| < \varepsilon + 2M(R^2 - |x|^2)R^{n-2}(2/\delta)^n.$$

For small enough $|x_0 - x|$ it follows that $|u(x) - u(x_0)| < 2\varepsilon$; thus u is continuous at x_0 and hence on $\overline{B}$. ∎

Another important consequence of Poisson's integral formula is the following inequality, due to Harnack.

Theorem 2.2.3 *Let u be harmonic in $B = B(0, R)$ and continuous and non-negative in $\overline{B}$. Then for all $x \in B$,*

$$\frac{1 - |x|/R}{(1 + |x|/R)^{n-1}} u(0) \leq u(x) \leq \frac{1 + |x|/R}{(1 - |x|/R)^{n-1}} u(0). \tag{2.2.12}$$

Proof. Let $x \in B$ and choose $\varepsilon > 0$ so that $|x| < R - \varepsilon$. Put $B_1 = B(0, R - \varepsilon)$, $R_1 = R - \varepsilon$. By Theorem 2.2.2,

$$u(x) = \frac{R_1^2 - |x|^2}{n\omega_n R_1} \int_{\partial B_1} \frac{u(y)}{|x - y|^n} d\sigma(y).$$

For all $y \in \partial B_1$,

$$\frac{1}{(R_1 + |x|)^n} \leq \frac{1}{|x - y|^n} \leq \frac{1}{(R_1 - |x|)^n},$$

and hence

$$\frac{R_1^2 - |x|^2}{(R_1 + |x|)^n} \cdot \frac{1}{n\omega_n R_1} \int_{\partial B_1} u(y) d\sigma(y) \leq u(x)$$

$$\leq \frac{R_1^2 - |x|^2}{(R_1 - |x|)^n} \cdot \frac{1}{n\omega_n R_1} \int_{\partial B_1} u(y) d\sigma(y).$$

From the mean-value property of u,

$$u(0) = \frac{1}{n\omega_n R_1^{n-1}} \int_{\partial B_1} u(y) d\sigma(y);$$

we thus obtain (2.2.12) with $R - \varepsilon$ in place of R. Now let $\varepsilon \to 0$. ∎

Corollary 2.2.4 (Liouville's theorem) *Let u be harmonic in $\mathbb{R}^n$ and suppose that u is either bounded above or bounded below. Then u is constant.*

Proof. Suppose that u is bounded above, so that there is a constant C with $u(x) \leq C$ for all $x \in \mathbb{R}^n$. Then the function v defined by $v(x) = C - u(x)$ is harmonic and non-negative in $\mathbb{R}^n$, so that for all $R > 0$ and all $x \in B(0, R)$,

$$\frac{1 - |x|/R}{(1 + |x|/R)^{n-1}} v(0) \leq v(x) \leq \frac{1 + |x|/R}{(1 - |x|/R)^{n-1}} v(0).$$

Fix x and let $R \to \infty$: it follows that $v(x) = v(0)$ and so $u(x) = u(0)$ for all $x \in \mathbb{R}^n$. When u is bounded below, consider $-u$. ∎

Theorem 2.2.3 may be extended to give another Harnack inequality.

Theorem 2.2.5 *Let Ω be connected, let u be harmonic and non-negative in Ω, and let K be a compact subset of Ω. Then there is a positive constant C, which depends only on K and Ω, such that for all $x, y \in K$,*

$$C^{-1} u(x) \leq u(y) \leq C u(x).$$

Proof. If u is zero at any point of Ω, by the maximum principle it must be zero everywhere in Ω. We may therefore assume that u is positive in Ω, in which case it is enough to prove that $u(x)/u(y) \leq C$ for all $x, y \in K$. Suppose that no such constant C exists. Then given any $m \in \mathbf{N}$, there is a function u_m which is harmonic and positive in Ω, and there are points $x^{(m)}, y^{(m)} \in K$ such that $u_m(x^{(m)})/u_m(y^{(m)}) \geq m$. As K is compact, we may assume without loss of generality that the sequences $(x^{(m)})$ and $(y^{(m)})$ converge to points x and y respectively in K. Since Ω is path-connected, there is a path in Ω joining x to y, and evidently there is a family of open balls $B(z_1, \delta), ..., B(z_k, \delta)$ which covers the path and has the property that $x \in B(z_1, \delta)$, $y \in B(z_k, \delta)$, $B(z_i, \delta) \cap B(z_{i+1}, \delta) \neq \emptyset$ for $i = 1, 2, ..., k - 1$, and $B(z_i, 2\delta) \subset \Omega$ for $i = 1, 2, ..., k$. From Theorem 2.2.3 applied to $B(z_i, 2\delta)$ we see that if $x', y' \in B(z_i, \delta)$, then

$$\frac{1 - \frac{|x'-z_i|}{2\delta}}{\left(1 + \frac{|x'-z_i|}{2\delta}\right)^{n-1}} u(z_i) \leq u(x') \leq \frac{1 + \frac{|x'-z_i|}{2\delta}}{\left(1 - \frac{|x'-z_i|}{2\delta}\right)^{n-1}} u(z_i),$$

similar inequalities holding for $u(y')$. These show that

$$u(x') \leq \frac{3/2}{(1/2)^{n-1}} u(z_i) = 3 \cdot 2^{n-2} u(z_i),$$

and

$$u(z_i) \leq 2(3/2)^{n-1} u(y').$$

Combination of these inequalities gives

$$u(x') \leq 3 \cdot 2^{n-2} u(z_i) \leq 3^n u(y'),$$

2.2 Representation of Solutions

that is, $u(x')/u(y') \leq 3^n$. For each i let $w_i \in B(z_i, \delta) \cap B(z_{i+1}, \delta)$. Then for any $x' \in B(z_1, \delta)$ and $y' \in B(z_k, \delta)$,

$$\frac{v(x')}{v(y')} = \frac{v(x')}{v(w_1)} \cdot \frac{v(w_1)}{v(w_2)} \cdot \ldots \cdot \frac{v(w_{k-1})}{v(y')} \leq 3^{kn}$$

for any function v which is harmonic and positive in Ω. But if we choose $m \in \mathbb{N}$ with $m > 3^{kn}$ and $x^{(m)} \in B(z_1, \delta)$, $y^{(m)} \in B(z_k, \delta)$, we know that $u_m(x^{(m)})/u_m(y^{(m)}) \geq m > 3^{kn}$. This contradiction proves the theorem. ∎

Corollary 2.2.6 *Let (u_m) be a non-increasing or non-decreasing sequence of functions which are harmonic in a domain Ω, and suppose there is a point $x_0 \in \Omega$ such that $(u_m(x_0))$ converges. Then (u_m) converges uniformly on every compact subset of Ω, and the limit function is harmonic in Ω.*

Proof. It is enough to deal with the case in which (u_m) is non-decreasing. Let K be a compact subset of Ω. Since for all $k, l \in \mathbb{N}$, $u_{k+l} - u_k$ is harmonic and non-negative in Ω, we may apply Theorem 2.2.5 to the compact set $K \cup \{x_0\}$ to conclude that for all $x \in K$,

$$0 \leq u_{k+l}(x) - u_k(x) \leq C\{u_{k+l}(x_0) - u_k(x_0)\},$$

where C is a constant independent of x, k and l. Taking account of Corollary 2.1.6, the result follows immediately. ∎

The mean value property has another interesting consequence: it enables the derivatives of harmonic functions to be estimated.

Theorem 2.2.7 *Let Ω be bounded, let u be harmonic in Ω, let K be any compact subset of Ω and put $d =\mathrm{dist}\,(K, \partial\Omega)$. Then*

$$\sup_K |\mathrm{grad}\,u(x)| \leq nd^{-1} \sup_\Omega |u(x)|.$$

Proof. Let $y \in K$ and let $0 < d_1 < d$; put $B = B(y, d_1)$. Then by the divergence theorem we have $\int_B \mathrm{grad}\,u\,dx = \int_{\partial B} uv d\sigma$ (apply the fact that $\int_B \mathrm{div}\,\mathbf{w}\,dx = \int_{\partial B} \mathbf{w}.v d\sigma$ to vectors of the form $\mathbf{w} = (u, 0, ..., 0)$, etc). Together with the mean-value theorem this shows that

$$|\mathrm{grad}\,u(y)| = \frac{1}{\omega_n d_1^n}\left|\int_B \mathrm{grad}\,u\,dx\right| \leq \frac{n}{d_1} \sup_{\partial B} |u(x)|.$$

This gives the result. ∎

A routine induction argument shows that this inequality can be extended to higher-order derivatives: there is a constant $C = C(n)$ such that for all $\alpha \in \mathbb{N}_0^n$ with $|\alpha| = m$,

$$\sup_K |D^\alpha u(x))| \leq m! e^{m-1} C^k d^{-m} \sup_\Omega |u(x)|.$$

This extension enables the following result of Liouville type to be obtained: if u is harmonic in $\mathbb{R}^n$ and there are positive constants C and c such that $|u(x)| \le C(1+|x|^c)$ for all $x \in \mathbb{R}^n$, then u is a polynomial of degree $\le c$. For taking $\Omega = B_R$ and allowing R to become arbitrarily large we see that $D^\alpha u$ must be identically zero whenever $|\alpha| > c$.

Theorem 2.2.7 enables us to give the following result of Montel type.

Corollary 2.2.8 *Let (u_m) be a sequence of functions which are harmonic in Ω, and suppose that the sequence is uniformly bounded; that is, there is a constant M such that $|u_m(x)| \le M$ for all $m \in \mathbb{N}$ and all $x \in \Omega$. Then there is a subsequence of (u_m) which converges uniformly on every compact subset of Ω to a function harmonic in Ω.*

Proof. The last theorem implies that (u_m) is equicontinuous and so by the Arzelà-Ascoli theorem, there is a subsequence of (u_m) which converges uniformly on every compact subset of Ω. Now use Corollary 2.1.6 ∎

We next show that the maximum principle and the solubility of the Dirichlet problem in a ball may be used to discuss the singularities of a harmonic function.

Theorem 2.2.9 *Let $n \ge 3$, let $x_0 \in \Omega$, let u be harmonic in $\Omega\setminus\{x_0\}$ and suppose that $u(x) = o(|x - x_0|^{2-n})$ as $x \to x_0$. Then u may be extended so as to be harmonic in Ω.*

Proof. Without loss of generality we may assume that $x_0 = 0$. There is an open ball $B = B(0, \delta)$ with closure contained in Ω; let v be the function harmonic in B and equal to u on ∂B. For any $\varepsilon \in \mathbb{R}$, $\Delta(u - v + \varepsilon |x|^{2-n}) = 0$ in $B\setminus\{0\}$; if $\varepsilon > 0$, $u(x) - v(x) + \varepsilon |x|^{2-n} > 0$ on ∂B and $u(x) - v(x) + \varepsilon |x|^{2-n} > 0$ if $|x| \le \eta$, $x \ne 0$, for small enough η, say $0 < \eta < \delta$. Apply the maximum principle to $B\setminus B(0, \eta)$: it follows that $u(x) - v(x) + \varepsilon |x|^{2-n} > 0$ in $B\setminus\{0\}$. The same argument, but with $-\varepsilon$ in place of ε, shows that $u(x) - v(x) < \varepsilon |x|^{2-n}$, and we conclude that $|u(x) - v(x)| < \varepsilon |x|^{2-n}$ in $B\setminus\{0\}$. As this is true for all $\varepsilon > 0$, $u(x) = v(x)$ in $B\setminus\{0\}$. If we define $u(0) = v(0)$ we obtain the required extension. ∎

This result holds when $n = 2$ under the condition $u(x) = o(\log(x - x_0))$ as $x \to x_0$; the proof is virtually identical. Thus singularities of the form $o(\log(x - x_0))$ in two dimensions and $o(|x - x_0|^{2-n})$ in higher dimensions are removable.

To conclude this section we give a result due to Weyl. If u is harmonic in Ω, then integration by parts shows that for all $\phi \in C_0^2(\Omega)$,

$$\int_\Omega u \Delta \phi \, dx = 0.$$

Weyl's result goes in the opposite direction.

Theorem 2.2.10 *(Weyl's lemma) Suppose that $u \in L_{1,loc}(\Omega)$ and that $\int_\Omega u \Delta \phi \, dx = 0$ for all $\phi \in C_0^\infty(\Omega)$. Then u is equal a.e. to a function harmonic in Ω.*

2.2 Representation of Solutions

Proof. Let ϕ be the standard test function defined in the proof of Theorem 2.1.4, and given any $\varepsilon > 0$ let $\phi_\varepsilon(x) = \varepsilon^{-n}\phi(x/\varepsilon)$, $\Omega_\varepsilon = \left\{ x \in \Omega : \overline{B(x,\varepsilon)} \subset \Omega \right\}$ and $u_\varepsilon = u * \phi_\varepsilon$. Then (see, for example, [54], pp. 208–11) $u_\varepsilon \in C^\infty(\Omega_\varepsilon)$ and $\|u_\varepsilon\|_{1,\Omega_\varepsilon} \leq \|u\|_{1,\Omega_\varepsilon}$. Moreover, the assumption on u implies that

$$\Delta u_\varepsilon(x) = \int_\Omega u(y)(\Delta \phi_\varepsilon)(x-y)dy = 0,$$

from which it follows that u_ε is harmonic on Ω_ε.

Fix a sufficiently small $\varepsilon_0 > 0$ and let $U = \overline{\Omega_{\varepsilon_0}} \neq \emptyset$. Consider the family $\mathcal{F}$ of functions u_ε, where $0 < \varepsilon < \varepsilon_0/2$. This family is uniformly bounded on U: for given $x \in U$,

$$|u_\varepsilon(x)| \leq \frac{1}{|B(x,\varepsilon_0/2)|} \int_{B(x,\varepsilon_0/2)} |u_\varepsilon(y)|\,dy \leq \frac{1}{|B(x,\varepsilon_0/2)|} \|u\|_{1,U}.$$

It is also equicontinuous on U. To see this, note that given $\varepsilon \in (0, \varepsilon_0/2)$ and $x, y \in U$, we have for small enough $s > 0$,

$$|u_\varepsilon(x) - u_\varepsilon(y)| = \left| \frac{1}{|B(x,s)|} \int_{B(x,s)} u_\varepsilon(z)dz - \frac{1}{|B(y,s)|} \int_{B(y,s)} u_\varepsilon(z)dz \right|$$

$$\leq \frac{1}{|B(0,s)|} \left\{ \int_{B(x,s)\setminus B(y,s)} |u_\varepsilon(z)|\,dz + \int_{B(y,s)\setminus B(x,s)} |u_\varepsilon(z)|\,dz \right\},$$

from which the equicontinuity is a consequence of the uniform boundedness just established and the fact that, for fixed s, $|B(x,s)\setminus B(y,s)|$ depends only on $|x-y|$ and goes to zero as $|x-y| \to 0$. Thus by the Arzelà-Ascoli theorem, there is a sequence $(u_{\varepsilon(k)})$ (with $\varepsilon(k) \to 0$) that converges uniformly on U to a function $v \in C(U)$. Since each $u_{\varepsilon(k)}$ is harmonic on U it follows from Corollary 2.1.6 that v is harmonic on Ω_{ε_0}. However, $\|u - u_\varepsilon\|_{1,U} \to 0$ as $\varepsilon \to 0$, and so u coincides with the harmonic function v a.e. on U. The result is now clear. ∎

2.3 Dirichlet Problems: The Method of Perron

Throughout this section Ω will be an open subset of $\mathbb{R}^n$.

Definition 2.3.1 *A function $u \in C(\Omega)$ is called (C-) **subharmonic (superharmonic)** in Ω if given any open ball B with $\overline{B} \subset \Omega$ and any function $v \in C(\overline{B})$ which is harmonic in B and satifies $u \leq v$ ($u \geq v$) on ∂B, it follows that $u \leq v$ ($u \geq v$) in B.*

Use of the classical maximum principle (Theorem 2.1.7) shows that any function in $C^2(\Omega)$ which is subharmonic (superharmonic) in the classical sense is also subharmonic (superharmonic) in the sense of this new definition. We now establish some basic properties of these functions, starting with the strong maximum principle.

Lemma 2.3.2 *Let $u \in C(\Omega)$ be subharmonic (superharmonic) in Ω and suppose there is a point $x_0 \in \Omega$ such that $u(x_0) = \sup_\Omega u(x)$ ($\inf_\Omega u(x)$). Then if Ω is connected, u is constant in Ω.*

Proof. We deal with the case in which u is subharmonic; the other case is similar. Let $\delta > 0$ be such that $\overline{B(x_0, \delta)} \subset \Omega$ and let v be the unique function in $C(\overline{B(x_0, \delta)})$ which is harmonic on $B(x_0, \delta)$ and equal to u on $\partial B(x_0, \delta)$. Since u is subharmonic on Ω, $u \leq v$ in $B(x_0, \delta)$, which implies that v is harmonic in $B(x_0, \delta)$ and on $\overline{B(x_0, \delta)}$ has an interior maximum. It follows that v is constant on $B(x_0, \delta)$ and consequently that $u(x) = u(x_0)$ for all $x \in \partial B(x_0, \delta)$ (otherwise $v(x) = u(x) < u(x_0)$: contradiction). Repetition of this argument for all smaller balls centred at x_0 shows that $u(x) = u(x_0)$ for all $x \in B(x_0, \delta)$. Hence the set of all maximum points of u in Ω is open, and as the continuity of u shows that this set is also closed (relative to Ω), it must be the case that u is constant in Ω, since Ω is connected. ∎

Definition 2.3.3 *Let Ω be connected and let $u \in C(\overline{\Omega})$ be such that given any $x \in \Omega$, there is a ball $B(x_0, \delta)$, with $\overline{B(x_0, \delta)} \subset \Omega$, in which u is subharmonic. Then u is said to be **locally subharmonic** in Ω.*

Lemma 2.3.4 *If u is locally subharmonic in Ω, then it is subharmonic in Ω.*

Proof. It is clear from the proof of Lemma 2.3.2 that the strong maximum principle holds for functions which are locally subharmonic. Given any open ball B with $\overline{B} \subset \Omega$, let $v \in C(\overline{B})$ be any function which is harmonic in B and $\geq u$ on ∂B. Then $u - v$ is locally subharmonic in B and $u - v \leq 0$ on ∂B; by the maximum principle, $u - v \leq 0$ on B. Hence u is subharmonic in Ω. ∎

Lemma 2.3.5 *Let $u, v \in C(\overline{\Omega})$ be respectively subharmonic and superharmonic in Ω and suppose that $v \geq u$ on $\partial\Omega$. Then if Ω is bounded and connected, either $v > u$ in Ω or $v = u$ in Ω.*

Proof. Suppose the result is false. Then there exists $x_0 \in \Omega$ such that

$$u(x_0) - v(x_0) = \sup_\Omega (u - v)(x) := M \geq 0,$$

and there is a ball B, centred at x_0, such that $u - v$ is not identically equal to M on ∂B. Let $\tilde{u}, \tilde{v}$ be the functions harmonic on B and equal to u, v respectively on ∂B; then as in the proof of Lemma 2.3.2 we see that $\tilde{u} - \tilde{v} = M$ throughout B and hence $u - v = M$ on ∂B, giving a contradiction. ∎

Lemma 2.3.6 *Let $u_1, ..., u_m$ be subharmonic in Ω. Then so is $\max(u_1, ..., u_m)$.*

Proof. This follows immediately from Definition 2.3.1. ∎

Now let u be subharmonic in a domain Ω, let B be an open ball with $\overline{B} \subset \Omega$ and let $\tilde{u}$ be the function harmonic in B and equal to u on ∂B. The *harmonic lifting* of u in B is the function U defined by

2.3 Dirichlet Problems: The Method of Perron

$$U(x) = \begin{cases} \tilde{u}(x), & x \in B, \\ u(x), & x \in \Omega \setminus B. \end{cases}$$

This function U is subharmonic in Ω. To prove this, let B_1 be any open ball with $\overline{B_1} \subset \Omega$ and let $v \in C(\overline{B_1})$ be harmonic in B_1, with $v \geq U$ on ∂B_1. Since $u \leq U$ on B_1 ($u = U$ in $B_1 \setminus B$, $u \leq U$ in $B_1 \cap B$), it follows that $u \leq v$ on B_1 and hence $U \leq v$ in $B_1 \cap B$. Thus $U \leq v$ in B_1, which shows that U is subharmonic in Ω.

The existence of a solution of the Dirichlet problem will be established with the help of special subharmonic and superharmonic functions. Let Ω be bounded and connected, and let $\phi : \partial \Omega \to \mathbb{R}$ be bounded. A function $u \in C(\overline{\Omega})$ which is subharmonic (superharmonic) in Ω is called a *subfunction* (*superfunction*) *relative to ϕ* if $u \leq \phi$ ($u \geq \phi$) on $\partial \Omega$. By Lemma 2.3.5, every subfunction relative to ϕ is less than or equal to every superfunction relative to ϕ. Note that $\inf_{\partial \Omega} \phi$ ($\sup_{\partial \Omega} \phi$) is a subfunction (superfunction) relative to ϕ; thus the set S_ϕ of all subfunctions relative to ϕ is not empty. We can now give the fundamental existence result due to Perron.

Theorem 2.3.7 *Let Ω be bounded and connected and let $\phi : \partial \Omega \to \mathbb{R}$ be bounded. The function u on $\overline{\Omega}$ defined by*

$$u(x) = \sup\{v(x) : v \in S_\phi\}$$

is harmonic in Ω.

Proof. By Lemma 2.3.5, $v \leq \sup_{\partial \Omega} \phi$ for all $v \in S_\phi$; thus u is well defined. Let $x \in \Omega$ be fixed. There is a sequence $(v_m) \subset S_\phi$ such that $v_m(x) \to u(x)$; since v_m may be replaced by $\max(v_m, \inf_{\partial \Omega} \phi)$ it may be assumed that the sequence (v_m) is bounded. Let $\delta > 0$ be so small that $B = B(x, \delta)$ has closure contained in Ω, and let V_m be the harmonic lifting of v_m in B. It follows that $V_m \in S_\phi$ and $V_m(x) \to u(x)$; by Corollary 2.2.8 there is a subsequence $(V_{m(k)})$ of (V_m) which converges uniformly on any ball $B(x, \eta)$ with $\eta < \delta$ to a function v which is harmonic in B. It is clear that $v \leq u$ in B and $v(x) = u(x)$. In fact, $v = u$ throughout B. For if not, $v(z) < u(z)$ for some $z \in B$, and hence $v(z) < \tilde{u}(z)$ for some $\tilde{u} \in S_\phi$. Let $w_k = \max(\tilde{u}, v_{m(k)})$ and let W_k be the harmonic lifting of w_k in B; by the compactness result used above we see that there is a subsequence of (W_k) which converges to a function w which is harmonic in B and satisfies $v \leq w \leq u$ in B and $v(x) = w(x) = u(x)$. However, by the maximum principle, $v = w$ in B, and as this contradicts the definition of $\tilde{u}$ it follows that $u = v$ in B; that is, u is harmonic in B. Since the point x was arbitrarily chosen in Ω, u must be harmonic in Ω. ∎

Theorem 2.3.7 ensures the existence of a function which is harmonic in the arbitrary bounded domain Ω, but says nothing about its behaviour as the boundary of Ω is approached. Since the Dirichlet problem asks for a function which is harmonic in Ω and continuous on $\overline{\Omega}$, assuming that the prescribed boundary function ϕ is continuous on $\partial \Omega$, further investigation of this point is essential, and it turns out that some restriction on the geometric properties of $\partial \Omega$ helps to ensure that the boundary values can be assumed continuously. The vital concept in all this is that of a barrier function, which we now define.

Definition 2.3.8 *Let Ω be a domain and let $\xi \in \partial\Omega$. A function $w \in C(\overline{\Omega})$ is called a **barrier at** ξ (relative to Ω) if w is superharmonic in Ω, $w(\xi) = 0$ and $w > 0$ in $\overline{\Omega} \setminus \{\xi\}$. A function w is called a **local barrier at** ξ if there is a neighbourhood V of ξ such that w is a barrier at ξ relative to $\Omega \cap V$.*

Given a local barrier w at $\xi \in \partial\Omega$ it is easy to construct a barrier at ξ relative to Ω. All we need do is to take any open ball B with $\xi \in B$, $\overline{B} \subset V$, and to set

$$W(x) = \begin{cases} \min(m, w(x)), & x \in \overline{\Omega} \cap B, \\ m, & x \in \overline{\Omega} \setminus B, \end{cases}$$

where $m = \inf\{w(x) : x \in V \setminus B\} > 0$. Evidently $W \in C(\overline{\Omega})$ and it is superharmonic in two overlapping domains since it is equal to m on $\overline{\Omega} \setminus B$ and is the minimum of two superharmonic functions in $\overline{\Omega} \cap V$; because of this it is locally superharmonic in Ω and hence superharmonic in Ω, by Lemma 2.3.4. Since $W > 0$ in $\overline{\Omega} \setminus \{\xi\}$, is a barrier at ξ relative to Ω.

Definition 2.3.9 *Given a bounded domain Ω, a point $\xi \in \partial\Omega$ is said to be a **regular point** if there is a barrier at ξ relative to Ω.*

The importance of regular points is that the barrier function defined by the Perron method is continuous at such points if the boundary function ϕ is also continuous there.

Lemma 2.3.10 *Let Ω be a bounded domain, let $\xi \in \partial\Omega$ be regular, let $\phi : \partial\Omega \to \mathbb{R}$ be bounded and continuous at ξ, and let u be the function defined by Theorem 2.3.7. Then $u(x) \to \phi(\xi) = u(\xi)$ as $x \to \xi$.*

Proof. Let $M = \sup_{\partial\Omega} |\phi(x)|$. Since ξ is regular, there is a barrier w at ξ, and as ϕ is continuous at ξ, we see that given any $\varepsilon > 0$, there are constants δ and k such that

$$|\phi(x) - \phi(\xi)| < \varepsilon \text{ if } |x - \xi| < \delta, x \in \partial\Omega; \text{ and } kw(x) \geq 2M \text{ if } |x - \xi| \geq \delta, x \in \overline{\Omega}.$$

It is immediate that $\phi(\xi) + \varepsilon + kw$ and $\phi(\xi) - \varepsilon - kw$ are respectively a superfunction and a subfunction relative to ϕ, from which we have, using Lemma 2.3.5,

$$\phi(\xi) - \varepsilon - kw(x) \leq u(x) \leq \phi(\xi) + \varepsilon + kw(x), \quad x \in \overline{\Omega}.$$

Thus

$$|u(x) - \phi(\xi)| \leq \varepsilon + kw(x) \to \varepsilon \text{ as } x \to \xi;$$

hence $u(x) \to \phi(\xi)$ as $x \to \xi$. Also $|u(\xi) - \phi(\xi)| \leq \varepsilon$, which shows that $u(\xi) = \phi(\xi)$. ∎

2.3 Dirichlet Problems: The Method of Perron

The principal existence theorem now follows directly.

Theorem 2.3.11 *Let Ω be a bounded domain. All the Dirichlet problems*

$$\Delta u = 0 \, in \, \Omega, \, u = \phi \, on \, \partial\Omega, \, \phi \in C(\partial\Omega),$$

have solutions in $C^2(\Omega) \cap C(\overline{\Omega})$ if, and only if, all points of $\partial\Omega$ are regular.

Proof. If all points of the boundary are regular, Lemma 2.3.10 ensures the existence of the required kind of solution. Conversely, if the Dirichlet problem is soluble for all continuous boundary data, then given any $\xi \in \partial\Omega$, the function ϕ defined by $\phi(x) = |x - \xi|$ ($x \in \partial\Omega$) is continuous on $\partial\Omega$ and the harmonic function which is the solution of the Dirichlet problem with boundary values given by ϕ is a barrier at ξ. Hence ξ is regular, and so all points of the boundary must be regular. ∎

In view of this result it is clearly desirable to have a geometric criterion by means of which regular points can be detected. The following proposition gives one of the most useful tests of regularity.

Proposition 2.3.12 *Let Ω be bounded and let $\xi \in \partial\Omega$. Suppose there is an open ball B such that $\overline{B} \cap \overline{\Omega} = \{\xi\}$. Then ξ is a regular point.*

Proof. Let $B = B(x_0, R)$. The function w defined by

$$w(x) = \begin{cases} R^{2-n} - |x_0 - x|^{2-n}, & n \geq 3, \\ \log(|x_0 - x|/R), & n = 2, \end{cases}$$

is plainly a barrier at ξ. ∎

Domains Ω satisfying the hypothesis of this proposition at all points of $\partial\Omega$ are said to fulfil the *exterior ball condition*; examples of such domains are balls, annuli, cubes, regular polyhedra and all domains with boundary of class C^2. For all these domains the Dirichlet problem for the Laplace operator is soluble for arbitrarily given continuous boundary data. An example of a domain with a non-regular boundary point was provided in 1912 by Lebesgue [144], who took Ω to be the set

$$\{(x, y, z) \in \mathbb{R}^3 : r^2 + z^2 < 1, \, r > \exp(-1/(2z)), \, z > 0\},$$

where $r^2 = x^2 + y^2$, and showed that the inward cusp at $(0, 0, 0) \in \partial\Omega$ is not regular for Ω. For details of this we refer to [106], pp.175–6; [99], pp.124–5; and [81], p.85; see also [209].

To obtain further information about the regularity up to the boundary of solutions of the Dirichlet problem more detailed analysis is needed. We take up this question in Chapter 5, both for the Laplace operator and for the more general elliptic operators of order 2 introduced in the next chapter.

2.4 Notes

For additional information about the topics covered in this chapter see [90], [100] and [156].

Chapter 3
Second-Order Elliptic Equations

Throughout this chapter Ω will stand for a non-empty, open subset of $\mathbb{R}^n$ and all functions are assumed to be real-valued.

3.1 Basic Notions

Definition 3.1.1 *Let*

$$L = \sum_{j,k=1}^{n} a_{jk} D_j D_k + \sum_{j=1}^{n} b_j D_j + c,$$

*where the coefficients a_{jk}, b_j and c are given real-valued functions defined on Ω and $a_{jk} = a_{kj}$ for all $j, k \in \{1, 2, ..., n\}$. The second-order differential operator L is said to be **elliptic at a point** $x_0 \in \Omega$ if there is a constant $\mu(x_0) > 0$ such that for all $\xi = (\xi_j) \in \mathbb{R}^n$,*

$$\sum_{j,k=1}^{n} a_{jk}(x_0) \xi_j \xi_k \geq \mu(x_0) |\xi|^2.$$

*If L is elliptic at every point of Ω, it is called **elliptic in** Ω. If the coefficients a_{jk} are in $L_\infty(\Omega)$ and there is a constant $\mu > 0$ such that for all $x \in \Omega$ and all $\xi \in \mathbb{R}^n$,*

$$\sum_{j,k=1}^{n} a_{jk}(x) \xi_j \xi_k \geq \mu |\xi|^2,$$

*then L is said to be **uniformly elliptic in** Ω.*

Justification for this terminology may be obtained by noting that, corresponding to the case $n = 2$, the curve $\sum_{j,k=1}^{2} a_{jk}(x_0) \xi_j \xi_k = 1$ is an ellipse if L is elliptic at x_0. Note also that in this case there is a linear map, depending on x_0, mapping this ellipse onto the circle $\sum_{j=1}^{2} \eta_j^2 = 1$.

We see that the operators L which are elliptic in Ω are those for which the quadratic form $\sum_{j,k=1}^{n} a_{jk}(x)\xi_j\xi_k$ is positive definite in Ω, while if L is uniformly elliptic in Ω its quadratic form is bounded above and below in Ω by positive, constant multiples of $|\xi|^2$. The prototypical elliptic operator is, of course, the Laplace operator, which is uniformly elliptic on $\mathbb{R}^n$. Observe that the coefficients b_j and c have no influence on the classification of L as elliptic or not: all that matters is the *principal part* $\sum_{j,k=1}^{n} a_{jk} D_j D_k$ of L and the behaviour of the *principal symbol* $p(x,\xi) := \sum_{j,k=1}^{n} a_{jk}(x)\xi_j\xi_k$.

To some extent the theory of solutions of elliptic equations runs parallel to that of harmonic functions. We now give some results leading up to maximum principles. As universal assumptions throughout this development we shall suppose that

(i) Ω is open, bounded and connected;
(ii) $a_{jk}, b_j, c \in C(\overline{\Omega})$ for $j, k = 1, 2, ..., n$;
(iii) L is uniformly elliptic in Ω with constant μ as above.

3.2 Maximum Principles

Theorem 3.2.1 *Suppose that $u \in C^2(\Omega) \cap C(\overline{\Omega})$ satisfies $Lu > 0$ in Ω and that $c \leq 0$ in Ω. Then u cannot have a non-negative maximum in Ω.*

Proof. Suppose that u has a non-negative maximum in Ω, attained at $x_0 \in \Omega$. Then $D_j u(x_0) = 0$ ($j = 1, ..., n$) and $B := (D_j D_k u(x_0))$ is negative semi-definite, that is,

$$\sum_{j,k=1}^{n} D_j D_k u(x_0) \xi_j \xi_k \leq 0 \text{ for all } \xi \in \mathbb{R}^n.$$

[This is a standard fact from the theory of functions of several variables, but for a direct proof, suppose the the result is false. Then there exists $\xi \in \mathbb{R}^n$, with $|\xi| = 1$, such that $\sum_{j,k=1}^{n} D_j D_k u(x_0) \xi_j \xi_k = a > 0$. Taylor's formula

$$u(x_0 + h) = u(x_0) + \frac{1}{2} \sum_{j,k=1}^{n} D_j D_k u(x_0) h_j h_k + o(|h|^2)$$

with $h = \beta\xi$, $\beta > 0$, β sufficiently small, then gives

$$u(x_0 + h) = u(x_0) + \frac{1}{2}\beta^2 \sum_{j,k=1}^{n} D_j D_k u(x_0) \xi_j \xi_k + o(\beta^2)$$

$$> u(x_0) + \frac{1}{2}\beta^2 a + o(\beta^2) > u(x_0),$$

and we have a contradiction.]

3.2 Maximum Principles

By ellipticity, $A := (a_{jk}(x_0))$ is positive definite. Thus

$$(Lu)(x_0) = \sum_{j,k=1}^{n} a_{jk}(x_0) D_j D_k u(x_0) + c(x_0)u(x_0)$$

$$\leq \sum_{j,k=1}^{n} a_{jk}(x_0) D_j D_k u(x_0).$$

This last expression is the trace of AB, $\operatorname{tr}(AB)$, defined to be $\sum_{j=1}^{n}(AB)_{jj}$. We claim that $\operatorname{tr}(AB) \leq 0$, from which it follows that $(Lu)(x_0) \leq 0$, contradicting the assumption that $(Lu)(x_0) > 0$. It remains to justify our claim. Diagonalise the negative semi-definite symmetric matrix B, so that for some orthogonal matrix P and constants $c_{jj} \leq 0$ ($j = 1, ..., n$), $\Lambda := PBP^{-1}$ is a diagonal matrix with diagonal entries c_{jj}. Then

$$\operatorname{tr}(AB) = \operatorname{tr}(PABP^{-1}) := \operatorname{tr}(PAP^{-1}PBP^{-1}) = \operatorname{tr}(PAP^{-1}\Lambda)$$

$$= \sum_{j=1}^{n} c_{jj} \left(\sum_{k,m=1}^{n} a_{km}(x_0) p_{jk} p_{jm} \right) \leq 0$$

since $c_{jj} \leq 0$ and $\sum_{k,m=1}^{n} a_{km}(x_0) p_{jk} p_{jm} \geq 0$. ∎

Remark 3.2.2

If $c(x) = 0$ for all $x \in \Omega$, the condition $u(x_0) \geq 0$ is redundant.

Theorem 3.2.3 *(The weak maximum principle) Suppose that $u \in C^2(\Omega) \cap C(\overline{\Omega})$ satisfies $Lu \geq 0$ in Ω, with $c(x) \leq 0$ in Ω. Then if the maximum of u in $\overline{\Omega}$ is non-negative, it is attained on $\partial \Omega$.*

Proof. Let $\varepsilon > 0$ and put $w(x) = u(x) + \varepsilon e^{ax_1}$, where the constant a is to be fixed later. Then
$$Lw = Lu + \varepsilon e^{ax_1}(a_{11}a^2 + b_1 a + c).$$

By the ellipticity of L, $a_{11}(x) \geq \mu$ for all $x \in \Omega$; moreover, b_1 and c are bounded.. Hence if a is large enough,

$$a_{11}a^2 + b_1 a + c \geq \mu a^2 + b_1 a + c > 0$$

for all $x \in \Omega$, and so $Lw > 0$ in Ω. By Theorem 3.2.1, w attains its non-negative maximum only on $\partial \Omega$: thus
$$\sup_{\Omega} w \leq \sup_{\partial \Omega} w^+.$$

Hence
$$\sup_\Omega u \le \sup_\Omega w \le \sup_{\partial\Omega} w^+ \le \sup_{\partial\Omega} u^+ + \varepsilon \sup_{\partial\Omega} e^{ax_1}.$$

Now let $\varepsilon \to 0$. ∎

Corollary 3.2.4 *If $c(x) \le 0$ in Ω, there is at most one $u \in C^2(\Omega) \cap C(\overline{\Omega})$ such that*
$$Lu = f \text{ in } \Omega, \; u = \phi \text{ on } \partial\Omega,$$
where $f \in C(\Omega)$ and $\phi \in C(\partial\Omega)$ are given.

Proof. If there were two such functions, u_1 and u_2, then
$$L(u_1 - u_2) = 0 \text{ in } \Omega \text{ and } u_1 - u_2 = 0 \text{ on } \partial\Omega,$$
and so by Theorem 3.2.3, $u_1 - u_2 \le 0$ in Ω. Now repeat this argument for $u_2 - u_1$. ∎

Theorem 3.2.5 *Let B be an open ball in $\mathbb{R}^n$ with $x_0 \in \partial B$, and suppose that $u \in C^2(B) \cap C(B \cup \{x_0\})$ satisfies $Lu \ge 0$ in B. Assume also that $c \le 0$ in B, and that $u(x) < u(x_0)$ for all $x \in B$, with $u(x_0) \ge 0$. Then for each outward direction v at x_0 (that is, such that $v \cdot \mathbf{n}(x_0) > 0$, where $\mathbf{n}(x_0)$ is the outward normal to B at x_0),*
$$\liminf_{t \to 0} \{u(x_0) - u(x_0 - tv)\}/t > 0.$$

In particular, if $u \in C^1(B \cup \{x_0\})$, then $\frac{\partial u}{\partial v}(x_0) > 0$.

Proof. Without loss of generality we shall assume that $B = B(0, r)$. We may also assume that $u \in C(\overline{B})$ and $u(x) < u(x_0)$ for all $x \in \overline{B}\setminus\{x_0\}$, since we may construct a ball $B_1 \subset B$ which is tangent to B at x_0 and take this as our ball. Let $\varepsilon > 0$ and put $v(x) = u(x) + \varepsilon h(x)$, where $h(x) = e^{-a|x|^2} - e^{-ar^2}$ and a is to be chosen later. Put $\widetilde{B} = B \cap B(x_0, r/2)$. Then in $\widetilde{B}$,

$$(Lh)(x) = e^{-a|x|^2}\left\{4a^2 \sum\nolimits_{i,j=1}^n a_{ij}(x)x_i x_j - 2a \sum\nolimits_{i=1}^n [a_{ii}(x) + b_i(x)x_i] + c\right\}$$
$$- ce^{-ar^2}$$
$$\ge e^{-a|x|^2}\left\{4a^2 \sum\nolimits_{i,j=1}^n a_{ij}(x)x_i x_j - 2a \sum\nolimits_{i=1}^n [a_{ii}(x) + b_i(x)x_i] + c\right\}.$$

By ellipticity,
$$\sum\nolimits_{i,j=1}^n a_{ij}(x)x_i x_j \ge \mu |x|^2 > \mu(r/2)^2 \text{ in } \widetilde{B}.$$

Hence for large enough a, $Lh > 0$ in $\widetilde{B}$ and so for all $\varepsilon > 0$,
$$Lv = Lu + \varepsilon Lh > 0 \text{ in } \widetilde{B}.$$

3.2 Maximum Principles

By Theorem 3.2.1, v cannot have a non-negative maximum in $\widetilde{B}$.

Now we look at the behaviour of v on $\partial\widetilde{B}$. If $x \in \partial\widetilde{B} \cap B$, then since $u(x) < u(x_0)$, we must have $u(x) < u(x_0) - \delta$ for some $\delta > 0$. Take $\varepsilon > 0$ so small that $\varepsilon h < \delta$ on $\partial\widetilde{B} \cap B$. For such an ε, $v(x) < u(x_0)$ for $x \in \partial\widetilde{B} \cap B$. On the other hand, if $x \in \partial\widetilde{B} \cap \partial B$, then $h(x) = 0$ and $u(x) < u(x_0)$ for $x \neq x_0$. Hence $v(x) < u(x_0)$ on $\partial\widetilde{B} \cap \partial B \setminus \{x_0\}$ and $v(x_0) = u(x_0)$. Thus

$$\frac{v(x_0) - v(x_0 - t\nu)}{t} \geq 0 \text{ for all small enough } t > 0,$$

and so

$$\liminf_{t \to 0} \frac{1}{t}\{u(x_0) - u(x_0 - t\nu)\} \geq -\varepsilon \frac{\partial h}{\partial \nu}(x_0) > 0,$$

since from the definition of h we see that

$$\frac{\partial h}{\partial \nu}(x_0) = \operatorname{grad} h(x_0).\nu = -2a|x_0|e^{-a|x_0|^2}\mathbf{n}(x_0).\nu < 0.$$

The proof is complete. ∎

Theorem 3.2.6 *(The strong maximum principle) Let $u \in C^2(\Omega)$ satisfy $Lu \geq 0$ in Ω, with $c \leq 0$ in Ω. If there exists $x_0 \in \Omega$ such that*

$$u(x_0) = \max\{u(x) : x \in \Omega\} \geq 0,$$

then u is constant in Ω.

Proof. Let $S = \{x \in \Omega : u(x) = u(x_0)\}$: S is relatively closed in Ω. Suppose $S \neq \Omega$: then $\partial S \cap \Omega \neq \emptyset$, for otherwise $S \cap {}^cS = \emptyset$, so that S would be open as well as closed, and hence identical to the connected set Ω. There is an open ball $B \subset \Omega \setminus S$ with a point $q \in \partial B \cap S$: for example, one could choose $p \in \Omega \setminus S$ such that $\operatorname{dist}(p, S) < \operatorname{dist}(p, \partial\Omega)$ and take $B = B(p, \operatorname{dist}(p, S))$. Then

$$Lu \geq 0 \text{ in } B, \ u(x) < u(q) \text{ for all } x \in B, \ u(q) \geq 0.$$

By Theorem 3.2.5, $\frac{\partial u}{\partial \nu}(q) > 0$, where ν is the outward normal from B at q. But q is an interior maximum point of u and so $\operatorname{grad} u(q) = 0$: contradiction. Hence $S = \Omega$. ∎

The assumption that $c \leq 0$ in Ω may be omitted if the maximum value $u(x_0) = 0$: see [172], pp. 14-15 for details.

Corollary 3.2.7 *Suppose $u \in C^2(\Omega) \cap C(\overline{\Omega})$ satisfies $Lu \geq 0$ in Ω, with $c \leq 0$ in Ω. If $u \leq 0$ on $\partial\Omega$, then $u \leq 0$ on Ω; either $u < 0$ in Ω or $u = 0$ in Ω.*

Corollary 3.2.8 *Suppose $u \in C^2(\Omega) \cap C^1(\overline{\Omega})$ satisfies $Lu \geq 0$ in Ω, with $c \leq 0$ in Ω, and suppose that u has a non-negative maximum at $x_0 \in \Omega$. Suppose that Ω has* the interior ball property: that is, given any $z \in \partial\Omega$, there is a ball $B \subset \Omega$ with $z \in \partial B$. Then either u is constant in Ω, or $x_0 \in \partial\Omega$ and for any outward derivative ν to $\partial\Omega$ at x_0,

$$\frac{\partial u}{\partial \nu}(x_0) > 0.$$

The results so far have been proved on the assumption that $c \leq 0$ in Ω. There is a good reason for this: the function $u(x) = e^{-|x|^2}$ satisfies the equation

$$\Delta u + (2n - 4|x|^2)u = 0 \text{ in } \mathbb{R}^n,$$

but has a global maximum at 0. Nevertheless, it is possible to obtain useful results even when the sign condition on c is omitted, and the following theorem illustrates this.

Theorem 3.2.9 *Suppose $u \in C^2(\Omega) \cap C(\overline{\Omega})$ satisfies $Lu \geq 0$ in Ω and assume that $u \leq 0$ in Ω. Then either $u < 0$ in Ω or $u = 0$ in Ω.*

Proof. Suppose $u(x_0) = 0$ for some $x_0 \in \Omega$. Put $c = c^+ - c^-$ (that is, decompose c into its positive and negative parts). Then

$$\sum_{j,k=1}^{n} a_{jk} D_j D_k u + \sum_{j=1}^{n} b_j D_j u - c^- u \geq -c^+ u \geq 0 \text{ in } \Omega.$$

By Theorem 3.2.6, $u = 0$ in Ω. ∎

A standard use of the maximum principle is to obtain *a priori* estimates for solutions of the Dirichlet problem for L, as in the case of the Laplace operator.

Theorem 3.2.10 *Let $f \in C(\overline{\Omega})$, $g \in C(\partial\Omega)$, $c \leq 0$ and suppose that $u \in C^2(\Omega) \cap C(\overline{\Omega})$ satisfies $Lu = f$ in Ω, $u = g$ on $\partial\Omega$. Then there is a constant C, depending only on μ, diam Ω and $\max_{i,j} \max_{\overline{\Omega}} |a_{ij}| + \max_i \max_{\overline{\Omega}} |b_i|$, such that*

$$\sup_{\Omega} |u| \leq \max_{\partial\Omega} |g| + C \max_{\overline{\Omega}} |f|.$$

Proof. Suppose $\Omega \subset \{x \in \mathbb{R}^n : 0 < x_1 < d\}$; put $F = \max_{\overline{\Omega}} |f|$ and $G = \max_{\partial\Omega} |g|$; define $w(x) = G + (e^{\lambda d} - e^{\lambda x_1})F$, where $\lambda > 0$ is to be chosen later. Then

$$-Lw \geq (a_{11}\lambda^2 + b_1\lambda)F \geq (\mu\lambda^2 + b_1\lambda)F \geq F$$

if we choose λ so large that $\mu\lambda^2 + b_1(x)\lambda \geq 1$ for all $x \in \Omega$. Thus $L(w \pm u) = Lw \pm f \leq 0$ in Ω, and $w \pm u = w \pm g \geq 0$ on $\partial\Omega$. Hence by Corollary 3.2.7, $-w \leq u \leq w$ in Ω, so that $\sup_{\Omega} |u| \leq G + (e^{\lambda d} - 1)F$. ∎

3.2 Maximum Principles

Estimates of the gradient of u may be obtained by similar procedures.

For detailed treatment of maximum principles and allied topics we refer to [81] and [172].

To conclude our brief discussion of second-order elliptic differential operators we touch upon the question of the existence of solutions of the corresponding Dirichlet problems. We shall see soon that for uniformly elliptic operators such existence can be proved in domains with regular boundaries. Some indication of the dangers that are present if the ellipticity assumption is omitted is given by a striking example due to Fichera [80]. He takes $n = 2$ and defines

$$L = x_2^2 D_1^2 - 2x_1 x_2 D_1 D_2 + x_1^2 D_2^2 - 2x_1 D_1 - 2x_2 D_2 + c, \quad c < 0.$$

The principal symbol of this operator is

$$p(x, \xi) = x_2^2 \xi_1^2 - 2x_1 x_2 \xi_1 \xi_2 + x_1^2 \xi_2^2 = (x_2 \xi_1 - x_1 \xi_2)^2 \geq 0.$$

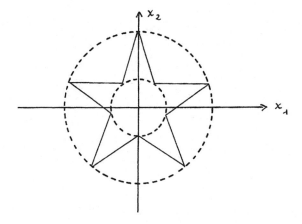

Hence $p(x, \xi) = 0$ if, and only if, $x_2 \xi_1 = x_1 \xi_2$; given any $x = (x_1, x_2) \in \mathbb{R}^2$, there exists $\xi = (\xi_1, \xi_2) \in \mathbb{R}^2 \setminus \{0\}$ with $p(x, \xi) = 0$. Hence L is not elliptic at any point of $\mathbb{R}^2$. Note that if $R > 0$ and ν is the outward normal to $B(0, R)$ at any point $(x_1, x_2) \in \partial B(0, R)$, then $x_1 \nu_2 = x_2 \nu_1$, so that $p(x, \nu) = 0$. Fichera shows that for the equation $Lu = 0$ in $B(0, R)$ specification of boundary data on $\partial B(0, R)$ is inadmissible: this gives an example of a problem without boundary conditions. However, if $B(0, R)$ is replaced by the star-shaped domain the value of u may be prescribed on the boundary.

Chapter 4
The Classical Dirichlet Problem for Second-Order Elliptic Operators

4.1 Preamble

A celebrated result is that existence of solutions of the classical Dirichlet problem for the uniformly elliptic operators studied in the last section can be obtained if a little additional smoothness is required of the coefficients of L and the boundary data, provided that the boundary itself is reasonably smooth. Our largely self-contained account of this is based on the approach given by M. König in [132], [133] and [134]; the Schauder boundary estimates are derived. Throughout this section we shall assume that Ω is a bounded domain in $\mathbf{R}^n$ with $n \geq 2$, and that all functions are real-valued.

4.2 The Poisson Equation

This is the equation $\Delta u = f$, where f is a prescribed function. When $f \in L_1(\Omega)$, the function w defined by

$$w(x) = \int_\Omega K(x-y) f(y) dy \ (x \in \Omega)$$

(where K is the fundamental solution of the Laplace equation given in 2.2) is called the *Newtonian potential* of f (on Ω). Its usefulness will soon become clear.

Proposition 4.2.1 *Let $R > 0$, suppose that $f \in C^\lambda\left(\overline{B_R}\right)$ for some $\lambda \in (0, 1)$ and let w be the Newtonian potential of f (on B_R). Then $w \in C^2(B_R) \cap C^1\left(\overline{B_R}\right)$ and $\Delta w = f$ in B_R.*

Proof. We shall write B instead of B_R. For each $i \in \{1, 2, ..., n\}$ define $v_i : \overline{B} \to \mathbb{R}$ by

$$v_i(x) = \int_B D_i K(x-y) f(y) dy;$$

in view of (2.2.3), v_i is well defined. In fact, $v_i = D_i w$. To verify this, let $\eta \in C^1(\mathbb{R})$ be such that $0 \leq \eta \leq 1$, $0 \leq \eta' \leq 2$, $\eta(t) = 0$ if $t \leq 1$ and $\eta(t) = 1$ when $t \geq 2$. For each $\varepsilon > 0$ let

$$w_\varepsilon(x) = \int_B K(x-y) \eta\left(|x-y|/\varepsilon\right) f(y) dy \quad (x \in \overline{B}).$$

Then $w_\varepsilon \in C^1(\overline{B})$, and since

$$v_i(x) - D_i w_\varepsilon(x) = \int_{|x-y| \leq 2\varepsilon, y \in B} D_i \{(1-\eta)(|x-y|/\varepsilon) K(x-y)\} f(y) dy,$$

we see that

$$|v_i(x) - D_i w_\varepsilon(x)| \leq \|f\|_{\infty, B} \int_{|x-y| \leq 2\varepsilon, y \in B} \left(|D_i K(x-y)| + \frac{2}{\varepsilon} |K(x-y)|\right) dy$$

$$\leq C(\varepsilon) \|f\|_{\infty, B},$$

where $C(\varepsilon) = 2n\varepsilon/(n-2)$ if $n > 2$, and $C(\varepsilon) = 4\varepsilon(1 + |\log 2\varepsilon|)$ if $n = 2$. Hence w_ε and $D_i w_\varepsilon$ converge uniformly in $\overline{B}$ to w and v_i respectively as $\varepsilon \to 0$, which implies that $w \in C^1(\overline{B})$ and $v_i = D_i w$.

Next, use of (2.2.3) and the fact that $f \in C^\lambda(\overline{B})$ shows that for all $i, j \in \{1, 2, ..., n\}$, the function u given by

$$u(x) = \int_B (f(x) - f(y)) D_i D_j K(x-y) dy - f(x) \int_{\partial B} D_i K(x-y) \nu_j(y) d\sigma(y)$$

(where $\nu = (\nu_k)$ is the unit outward normal to ∂B) is well defined on B. As before, let $v_i = D_i w$; for each $\varepsilon > 0$ put

$$V_\varepsilon(x) = \int_B \eta\left(|x-y|/\varepsilon\right) f(y) D_i K(x-y) dy.$$

Then $V_\varepsilon \in C^1(B)$ and given $x \in B$,

$$D_j V_\varepsilon(x) = \int_B D_j \{\eta(|x-y|/\varepsilon) D_i K(x-y)\} (f(x) - f(y)) dy$$

$$+ f(x) \int_B D_j \{\eta(|x-y|/\varepsilon) D_i K(x-y)\} dy$$

4.2 The Poisson Equation

$$= \int_B D_j \{\eta(|x-y|/\varepsilon) D_i K(x-y)\}(f(x) - f(y)) dy$$

$$- f(x) \int_{\partial B} \nu_j(y) D_i K(x-y) d\sigma(y)$$

provided that $0 < \varepsilon < (R - |x|)/2$. Thus $|u(x) - D_j V_\varepsilon(x)|$ is given by

$$\left| \int_{|x-y| \leq 2\varepsilon, y \in B} D_j \{(1 - \eta(|x-y|/\varepsilon)) D_i K(x-y)\}(f(x) - f(y)) dy \right|$$

which, using (2.2.3) and the properties of η, can be estimated from above by $(4 + n/\lambda)(2\varepsilon)^\lambda \|f\|_{\alpha,B}$ for small enough ε. It follows that $D_j V_\varepsilon$ converges to u uniformly on compact subsets of B as $\varepsilon \to 0$: since $V_\varepsilon \to v_i = D_i w$ uniformly in B, we see that $w \in C^2(B)$ and $u = D_i D_j w$. Finally, the above arguments show that $D_i D_j w(x)$ is given by

$$\int_B D_i D_j K(x-y)(f(x) - f(y)) dy - f(x) \int_{\partial B} D_i K(x-y) \nu_j(y) d\sigma(y)$$

for all $x \in B$. Clearly this also holds if B is replaced by a ball of small enough radius r centred at x: the harmonic property of K and (2.2.2) then shows that

$$\Delta w(x) = \frac{f(x)}{n \omega_n r^{n-1}} \int_{|x-y|=r} \sum_{i=1}^n \nu_i^2(y) d\sigma(y) = f(x),$$

which completes the proof. ∎

The sudden appearance of Hölder spaces in this Proposition may puzzle the reader, who might well wonder whether they merely reflect some deficiency in the proof: could not similar results be obtained under the assumption that $f \in C(\overline{B_R})$? Unfortunately this weaker assumption is not adequate: there are functions $f \in C(\overline{B_R})$ the Newtonian potentials of which are not in $C^2(B_R)$. For example, suppose that $n = 2$, $r \in (0, 1)$ and define $f : B_r \to \mathbb{R}$ by $f(x) = x_1 x_2 |x|^{-2} |\log |x||^{-1}$ ($x \neq 0$), $f(0) = 0$. It may be checked that while the Newtonian potential w of f belongs to $C^2(B_r) \setminus \{0\}$, nevertheless $D_1 D_2 w(x) \to \infty$ as $x \to 0$. Hence something more than mere continuity is needed: see [156], p. 54 for this example and a discussion of the relevance of Dini continuity.

Theorem 4.2.2 *Let $R > 0$ and suppose that $f \in C^\lambda(\overline{B_R})$ for some $\lambda \in (0, 1)$. Then there is exactly one function $u \in C^{2,\lambda}(\overline{B_R})$ that solves the Dirichlet problem*

$$\Delta u = f \text{ in } B_R \text{ and } u = 0 \text{ on } \partial B_R.$$

Proof. We suppose that $n > 2$: the argument when $n = 2$ follows similar lines. Let w be the Newtonian potential of f and put $v = u - w$. The stated Dirichlet problem is equivalent to the problem

$$\Delta v = 0 \text{ in } B_R \text{ and } v = -w \text{ on } \partial B_R,$$

which has a solution given by the Poisson integral formula of Theorem 2.2.2; Corollary 2.1.11 gives uniqueness. Together with Proposition 4.2.1 this shows that there is exactly one function $u \in C^2(B_R)$ that solves the given Dirichlet problem, and that this function is given by

$$u(x) = \int_B G(x, y) f(y) dy,$$

where G is Green's function for the ball $B = B_R$: see (2.2.9). Note that (see (2.2.10))

$$G(x, y) = -\frac{1}{(n-2)\omega_n} \left\{ |x-y|^{2-n} - \left(R^2 - 2(x, y) + \frac{|x|^2 |y|^2}{R^2} \right)^{(2-n)/2} \right\}$$

and that for a suitable constant C,

$$|G(x, y)| \le C |x-y|^{2-n}, |D_i G(x, y)| \le C |x-y|^{1-n},$$
$$|D_i D_j G(x, y)| \le C |x-y|^{-n}, |D_i D_j D_k G(x, y)| \le C |x-y|^{-(n+1)} \quad (4.2.1)$$

for all $i, j, k \in \{1, 2, ..., n\}$ and all $x, y \in B, x \ne y$. It remains to show that $u \in C^{2,\lambda}(\overline{B})$.

For all $x \in B$ and all $i, j \in \{1, 2, ..., n\}$,

$$D_i D_j u(x) = W(x) + f(x) D_i D_j v(x),$$

where

$$W(x) = \int_B \{f(y) - f(x)\} D_i D_j G(x, y) dy, \quad v(x) = \int_B G(x, y) dy.$$

Since G is Green's function for B, v is the solution of the boundary-value problem

$$\Delta v = 1 \text{ in } B, v(x) = 0 (x \in \partial B).$$

As $x \mapsto (|x|^2 - R^2)/(2n)$ is a solution of this problem, it must coincide with v in $\overline{B}$, and so certainly $f D_i D_j v \in C^\lambda(\overline{B})$. To deal with the other term in the expression for $D_i D_j u(x)$, let $x_1 \in B$, suppose that $0 < d < R - |x_1|$, let $x_2 \in B$ be such that $|x_1 - x_2| = d/2$ and put $Q = B(x_1, d)$. Then since for any $x \in Q$ and any $i, j \in \{1, ..., n\}$,

4.2 The Poisson Equation

$$\left| \int_{B \setminus Q} D_i D_j G(x, y) dy \right| = \left| D_i \int_{B \setminus Q} D_j G(x, y) dy \right| = \left| D_i \int_{B \setminus Q} \frac{\partial}{\partial y_j} G(x, y) dy \right|$$

$$= \left| D_i \int_{\partial (B \cap Q)} G(x, y) \nu_j(y) d\sigma(y) \right|$$

$$= \left| \int_{\partial (B \cap Q)} D_i G(x, y) \nu_j(y) d\sigma(y) \right|,$$

the estimates (4.2.1) of Green's function plus the Hölder continuity of f show that

$$|I_1| := \left| \int_{B \setminus Q} (f(x_1) - f(x_2)) D_i D_j G(x_1, y) dy \right| \leq C |x_1 - x_2|^\lambda.$$

Here, and below, C will stand for a constant, possibly different from line to line. Next, let

$$I_2 = \int_{B \setminus Q} \{D_i D_j G(x_1, y) - D_i D_j G(x_2, y)\} \{f(y) - f(x_2)\} dy.$$

By the mean-value theorem, for each $y \in B \setminus Q$ there exists a point $\widehat{x}$ on the line segment joining $x_1 = (x_1^k)$ and $x_2 = (x_2^k)$ such that

$$I_2 = \int_{B \setminus Q} \{f(y) - f(x_2)\} \sum_{k=1}^n \left(x_1^k - x_2^k \right) D_k D_i D_j G(\widehat{x}, y) dy.$$

When $y \in B \setminus Q$, since $|y - x_1| \geq 2 |x_1 - x_2|$ we have

$$|y - \widehat{x}| \geq |y - x_1| - |x_1 - \widehat{x}| \geq |y - x_1| - |x_1 - x_2| \geq |y - x_1|/2$$

and

$$|y - x_2| \leq |y - x_1| + |x_1 - x_2| \leq 3 |y - x_1|/2.$$

Thus

$$\left| \sum_{k=1}^n \left(x_1^k - x_2^k \right) D_k D_i D_j G(\widehat{x}, y) \right| \leq \frac{C |x_1 - x_2|}{|y - \widehat{x}|^{n+1}} \leq \frac{C |x_1 - x_2|}{|y - x_1|^{n+1}},$$

and hence

$$|I_2| \leq C |x_1 - x_2| \int_{B \setminus Q} \frac{|y - x_2|^\lambda}{|y - x_1|^{n+1}} dy \leq C |x_1 - x_2| \int_{B \setminus Q} \frac{dy}{|y - x_1|^{n+1-\lambda}}$$

$$\leq C |x_1 - x_2| \{|x_1 - x_2|^{\lambda-1} + c\} \leq C |x_1 - x_2|^\lambda.$$

Similar calculations show that for the quantities

$$I_3 := \int_Q (f(y) - f(x_1)) D_i D_j G(x_1, y) dy, \quad I_4 := \int_Q (f(y) - f(x_2)) D_i D_j G(x_2, y) dy$$

we have the estimates

$$|I_3| \leq C \int_{|x_1 - y| \leq d} |y - x_1|^{\lambda - n} dy \leq C d^\lambda \leq C |x_1 - x_2|^\lambda$$

and

$$|I_4| \leq C \int_{|x_1 - y| \leq d} |y - x_2|^{\lambda - n} dy \leq C \int_{|x_2 - y| \leq 2d} |y - x_2|^{\lambda - n} dy \leq C |x_1 - x_2|^\lambda.$$

Hence

$$|W(x_1) - W(x_2)| \leq \sum_{m=1}^{4} |I_m| \leq C |x_1 - x_2|^\lambda,$$

which completes the proof that $u \in C^{2,\lambda}\left(\overline{B}\right)$. ∎

Corollary 4.2.3 *With the notation of Theorem 4.2.2, there is a constant c such that for all $f \in C^\lambda\left(\overline{B_R}\right)$, the solution u satisfies*

$$|||u|||_{2,\lambda,B_R} \leq c |||f|||_{\lambda, B_R}.$$

Proof. Define $T : C^\lambda\left(\overline{B_R}\right) \to X := \{v \in C^{2,\lambda}\left(\overline{B_R}\right) : v = 0 \text{ on } \partial B_R\}$ by $Tf = u$, where u is the unique solution of the Dirichlet problem for the Poisson equation with zero boundary data corresponding to a given f; note that X is a closed linear subspace of $C^{2,\lambda}\left(\overline{B_R}\right)$. Since $T^{-1} : X \to C^\lambda\left(\overline{B_R}\right)$ is given by $T^{-1} v = \Delta v$, it is plainly continuous. Hence, by Banach's inverse mapping theorem (Corollary 1.2.2), T is continuous and the result follows. ∎

To conclude this section we consider the Poisson equation on more general bounded open sets. Throughout this discussion we shall suppose that Ω is a bounded open subset of $\mathbb{R}^n$ with $\partial \Omega \in C^{2,\lambda}$ for some $\lambda \in (0, 1]$; the outward unit normal to $\partial \Omega$ at y will be denoted by $\nu(y)$. We begin with two preparatory lemmas.

Lemma 4.2.4 *Let $f \in C^\lambda(\overline{\Omega})$ and $u \in C(\overline{\Omega}) \cap C^{2,\lambda}(\Omega)$ be such that $\Delta u = f$ in Ω and $u = 0$ on $\partial \Omega$. Then there exist positive constants a and c, depending only on n and $\partial \Omega$, such that for all $y \in \partial \Omega$ and all $t \in [0, a]$,*

$$|u(y - t\nu(y))| \leq ct |||f|||_{0,\Omega}.$$

Proof. Since $\partial \Omega \in C^{2,\lambda}$, given any $z \in \partial \Omega$, there is a ball contained in Ω, with radius $r(z) > 0$ and boundary that meets $\partial \Omega$ only at z; as $\partial \Omega$ is compact, there exists $r > 0$ such that $r(z) \geq r$ for all $z \in \partial \Omega$. Fix $y \in \partial \Omega$; by use of a linear mapping to change coordinates, if necessary, we may suppose that $y = 0$ and $\nu(y) = (0, ..., 0, -1)$. We shall find it convenient to write points $x = (x_1, ..., x_n)$

4.2 The Poisson Equation

of $\mathbb{R}^n$ in the form (x', x_n), where $x' = (x_1, \ldots, x_{n-1})$. As $\partial\Omega \in C^{2,\lambda}$, there are a $C^{2,\lambda}$ function q and a positive number δ such that

$$\{x \in \mathbb{R}^n : |x'| < \delta, x_n = q(x')\} \subset \partial\Omega;$$

for later convenience we take $\delta = (4n+1)^{1/2} r/(2n+1)$, reducing r if necessary. Let

$$S := \left\{ x \in \mathbb{R}^n : |x'| < \delta, q(x') < x_n < r/(2n+1) \right\}.$$

For the moment, suppose there is a function $w : \overline{S} \to [0, \infty)$ such that

$$\Delta(\pm u + w) \leq 0 \text{ on } S, \text{ while } \pm u + w \geq 0 \text{ on } \partial S.$$

Then by the discussion following Corollary 2.1.9, $\pm u + w \geq 0$ on $\overline{S}$, so that $|u| \leq w$ on $\overline{S}$. Elementary but tedious calculation shows that such a function is that defined by

$$w(x) = \{W(x) + 2Kv(x)\} |||u|||_{0,\Omega} + |||f|||_{0,\Omega} v(x),$$

where

$$W(x) = (2n+1)^2((4n+1)r^2)^{-1} |x'|^2 + (2n+1)r^{-1} \left(x_n + r - (r^2 - |x'|^2)^{1/2} \right),$$

$$K = (n-1)(2n+1)^2((4n+1)r^2)^{-1} + (2n+1)^2(2r^2)^{-1}$$

and

$$v(x) = -x_n^2 + rx_n(2n+1)^{-1} + (r - (r^2 - |x'|^2)^{1/2})(r(2n+1)^{-1} - x_n)$$
$$= \left\{ x_n - \left(r^2 - |x'|^2 \right)^{1/2} + r \right\} \left\{ r(2n+1)^{-1} - x_n \right\}.$$

Note that on $\overline{S}$, $v \geq 0$ and $\Delta v \leq -1$. Let $\underline{0}$ stand for the zero element of $\mathbb{R}^{n-1}$. Since

$$w(\underline{0}, x_n) = \left\{ (2n+1)r^{-1}x_n + 2Kx_n \left\{ r(2n+1)^{-1} - x_n \right\} \right\} |||u|||_{0,\Omega}$$
$$+ \left\{ r(2n+1)^{-1} - x_n \right\} x_n |||f|||_{0,\Omega}$$

and $|||u|||_{0,\Omega} \leq M |||f|||_{0,\Omega}$ by Theorem 2.1.12, the result now follows with $a := r/(2n+1)$. ∎

Lemma 4.2.5 *Let X be a Banach space and let $T : X \to C^{k,\lambda}(\overline{\Omega})$ be linear, where $k \in \mathbb{N}_0$; let $I : C^{k,\lambda}(\overline{\Omega}) \to C(\overline{\Omega})$ be the embedding map. Then if $I \circ T : X \to C(\overline{\Omega})$ is continuous, so is T.*

Proof. Let (u_m) be a sequence in X such that $u_m \to u$ and $Tu_m \to v$: since I is continuous, $Iv = \lim_{m \to \infty} I \circ Tu_m = I \circ Tu$. As I is injective, $v = Tu$, which implies that T is closed and hence continuous, by the closed graph theorem (Theorem 1.2.3). ∎

Theorem 4.2.6 *Let $f \in C^\lambda(\overline{\Omega})$ and $g \in C^{2,\lambda}(\partial\Omega)$; suppose that $u \in C(\overline{\Omega}) \cap C^{2,\lambda}(\Omega)$ satisfies $\Delta u = f$ in Ω, $u = g$ on $\partial\Omega$. Then $u \in C^{1,\gamma}(\overline{\Omega}) \cap C^{2,\lambda}(\Omega)$ for all $\gamma \in [0, 1)$.*

Proof. We suppose that $n > 2$, the argument when $n = 2$ being similar. Since g may be extended to a function $\tilde{g} \in C^{2,\lambda}(\overline{\Omega})$ with $C^{2,\lambda}(\overline{\Omega})$ norm bounded above by a constant times the $C^{2,\lambda}(\partial\Omega)$ norm of g, consideration of the function $u - \tilde{g}$ shows that it is enough to deal with the case $g = 0$. Let $z \in \partial\Omega$. As before, we may assume that $z = 0$ and that the outward unit normal to $\partial\Omega$ at 0 is $\nu = \nu(0) = (0, ..., 0, -1)$. Because $\partial\Omega \in C^{2,\lambda}$, there exists $r > 0$ such that the ball $B = B(-r\nu, r)$ meets $\partial\Omega$ only at 0. Let G be Green's function for B: specifically, $(n-2)\omega_n G(x,y)$ is given by

$$|x-y|^{2-n} - \left\{ r^2 - 2(x+r\nu, y+r\nu) + r^{-2}|x+r\nu|^2|y+r\nu|^2 \right\}^{(2-n)/2}$$

when $x \neq y$. As before, we have estimates for G and its derivatives of the form

$$\begin{cases} |G(x,y)| \leq C|x-y|^{2-n}, & |D_i G(x,y)| \leq C|x-y|^{1-n}, \\ |D_i D_j G(x,y)| \leq C|x-y|^{-n} \end{cases} \quad (4.2.2)$$

when $x \neq y$. In B the function u may be represented as

$$u(x) = \int_B G(x,y) f(y) dy$$
$$+ \left(r^2 - |x+r\nu|^2\right)(r\omega_n)^{-1} \int_{|y+r\nu|=r} u(y)|x-y|^{-n} d\sigma(y). \quad (4.2.3)$$

Put $(Tf)(x) = \int_B G(x,y) f(y) dy$ ($f \in C^\lambda(\overline{B})$, $x \in B$). Then $Tf \in C^{1,\gamma}(\overline{B})$ for any $\gamma \in [0, 1)$, and so T may be regarded as a linear map of $C^\lambda(\overline{B})$ into $C^{1,\gamma}(\overline{B})$. Since

$$|||Tf|||_{0,B} \leq C|||f|||_{\lambda,B},$$

it follows from Lemma 4.2.5 that $T : C^\lambda(\overline{B}) \to C^{1,\gamma}(\overline{B})$ is continuous: thus

$$|||Tf|||_{1,\gamma,B} \leq C|||f|||_{\lambda,\Omega}.$$

Now let a be the constant from Lemma 4.2.4 and put $b = \min\{a, r\}$, $S = \{x \in \partial B : \text{dist}(x, \partial\Omega) \leq 2|x_n| \leq b\}$. Decompose the second term on the right-hand side of (4.2.3) into terms $v_1(x)$, $v_2(x)$ obtained by taking the integral over the sets $\partial B \setminus S$ and S respectively. The function v_1 is of class C^∞ in some neighbourhood U of 0. Consideration of the linear map $u \restriction_{\partial B \setminus S} \to v_1 \restriction_U$ together with Lemma 4.2.5 shows that $|||v_1|||_{2,U} \leq C|||u|||_{0,\Omega}$. As for v_2, it is harmonic in B and

4.2 The Poisson Equation

$$r\omega_n D_j v_2(x) = -2(x_j - r\delta_{jn})\int_S u(y)|x-y|^{-n}\,d\sigma(y)$$
$$- n\left(r^2 - |x+r\nu|^2\right)\int_S u(y)(x_j - y_j)|x-y|^{-n-2}\,d\sigma(y).$$

Since $\text{dist}(y, \partial\Omega) \leq 2y_n \leq a$ for all $y \in S$, the estimate $|u(y)| \leq Cy_n|||f|||_{0,G}$ follows from Lemma 4.2.4. Because $\sum_{j=1}^n y_j^2 = (2r-y_n)y_n \geq (2r-r)y_n$ for all $y \in S$, we see that

$$|u(y)| \leq C|||f|||_{0,G} \sum_{j=1}^n y_j^2.$$

Thus for $x := -t\nu$ ($0 \leq t \leq r$),

$$\left|D_j v_2(-t\nu)\right| \leq C\delta_{jn}(t+r)|||f|||_{0,\Omega}\int_S |y+t\nu|^{2-n}\,d\sigma(y)$$
$$+ C(r^2 - (r-t)^2)\int_S |y+t\nu|^{1-n}\,d\sigma(y)$$
$$\leq C|||f|||_{0,G}.$$

A similar procedure shows that for all $k \in \{1, \ldots, n-1\}$ and $x := -t\nu$ ($0 \leq t \leq r$), $\left|D_j D_k v_2(-t\nu)\right| \leq C|||f|||_{0,G}$. Since v_2 is harmonic in B, this last estimate also holds when $j = k = n$. Thus in a neighbourhood of $\partial\Omega$ all the functions u, $D_j u$ and $D_j D_k u$ ($j, k \in \{1, \ldots, n\}$) are bounded, which implies that $u \in C^{1,\gamma}(\overline{\Omega})$ for all $\gamma \in [0, 1)$, as required. ∎

In fact, $u \in C^{2,\lambda}(\overline{\Omega})$: in the next section we establish this and so obtain an exact analogue of Theorem 4.2.2 (in which Ω was a ball).

4.3 More General Elliptic Operators

We begin with an operator that is a slight perturbation of the Laplacian.

Theorem 4.3.1 *Let $\lambda \in (0, 1)$, $R > 0$, put $B = B(0, R)$ and suppose that $L := \sum_{j,k=1}^n a_{jk} D_j D_k$ is uniformly elliptic in $\overline{B}$, with each $a_{jk} \in C^\lambda(\overline{B})$ and*

$$\sum_{j,k=1}^n |||a_{jk} - \delta_{jk}|||_{\lambda,B} \leq 1/(2c),$$

where c is the constant appearing in Corollary 4.2.3. Then given any $f \in C^\lambda(\overline{B})$, there is exactly one function $u \in C^{2,\lambda}(\overline{B})$ such that $Lu = f$ in B and $u = 0$ on ∂B.

Proof. If $v \in C^{2,\lambda}(\overline{B})$, then $\Delta v - Lv + f \in C^\lambda(\overline{B})$. By Theorem 4.2.2, there is exactly one function $w \in C^{2,\lambda}(\overline{B})$ such that $\Delta w = \Delta v - Lv + f$ in B and $w = 0$ on ∂B. A linear map $T : C^{2,\lambda}(\overline{B}) \to C^{2,\lambda}(\overline{B})$ is thus defined by $w = Tv$, and the desired result will follow if it can be shown that T has a unique fixed point. To do this, note that if $w_j = Tv_j$ ($j = 1, 2$), then

$$\Delta(w_1 - w_2) = \Delta(v_1 - v_2) - L(v_1 - v_2) \text{ on } B, \quad w_1 - w_2 = 0 \text{ on } \partial B.$$

Use of Corollary 4.2.3 shows that

$$|||w_1 - w_2|||_{2,\lambda,B} \leq c|||(\Delta - L)(v_1 - v_2)|||_{\lambda,B}$$
$$\leq c|||v_1 - v_2|||_{2,\lambda,B} \sum_{j,k=1}^n |||a_{jk} - \delta_{jk}|||_{\lambda,B}$$
$$\leq \frac{1}{2}|||v_1 - v_2|||_{2,\lambda,B}.$$

By the contraction mapping theorem, T has a unique fixed point u. ∎

Theorem 4.3.2 *Let Ω be a bounded open subset of $\mathbb{R}^n$ with $\partial\Omega \in C^{2,\lambda}$ for some $\lambda \in (0, 1)$ and let $f \in C^\lambda(\overline{\Omega})$; suppose that $\gamma \in [\lambda, 1)$. If $u \in C^{1,\gamma}(\overline{\Omega}) \cap C^{2,\lambda}(\Omega)$ is such that $\Delta u = f$ in Ω and $u = 0$ on $\partial\Omega$, then $u \in C^{2,\lambda}(\overline{\Omega})$.*

Proof. Let x_0 be an arbitrary point of $\partial\Omega$: as before we may suppose that $x_0 = 0$ and that the tangent plane $T(0, \Omega)$ to Ω at 0 (which exists because $\partial\Omega \in C^{2,\lambda}$) has unit normal $(0, ..., 0, 1)$ at 0 which points into Ω. Moreover, there is a neighbourhood $U \subset \mathbb{R}^n$ of 0 and a C^2 map $\phi : U \cap T(0, \Omega) \to \mathbb{R}$ with second derivatives that satisfy a uniform Hölder condition with exponent λ and graph $\{(x', \phi(x')) : x' \in U \cap T(0, \Omega)\} \subset \partial\Omega$; and there exists $R > 0$ such that the ball B of radius R centred at $(0, ..., 0, R)$ is contained in $\Omega \cap U$ and $\partial\Omega \cap \partial B = \{0\}$.

Define $\psi : \{x' \in \mathbb{R}^{n-1} : |x'| \leq R\} \to \mathbb{R}$ by $\psi(x') = \phi(x') - R + \left(R^2 - |x'|^2\right)^{1/2}$ and set $v(x', x_n) = u\left(x', x_n + \psi(x')\right)$: plainly $v \in C^{1,\gamma}(\overline{B}) \cap C^{2,\lambda}(B)$. If $x_n \leq R$ and $(x', x_n) \in \partial B$, then $v(x', x_n) = u(x', \phi(x')) = 0$. Routine calculations show that

$$\Delta u(x) = \sum_{j,k=1}^n \widetilde{a}_{jk}(x') D_j D_k v(x) + \widetilde{a}_n(x') D_n v(x),$$

where

$$\widetilde{a}_{jk}(x') = \delta_{jk}, \widetilde{a}_{jn}(x') = -2D_j\psi(x') \quad (j, k \in \{1, ..., n-1\}),$$

and

$$\widetilde{a}_n(x') = -\sum_{j=1}^{n-1} D_j^2 \psi(x'), \widetilde{a}_{nn}(x') = 1 + \sum_{j=1}^{n-1}(D_j\psi(x'))^2.$$

Since

$$D_j\psi(x') = D_j\phi(x') - x_j\left(R^2 - |x'|^2\right)^{-1/2},$$

we see that $D_j\psi(0) = 0$ and hence $\widetilde{a}_{jk}(0) = \delta_{jk}$ for all $j, k \in \{1, ..., n\}$. Moreover, as $\widetilde{a}_{jk} \in C^{1,\lambda}(\overline{B})$, there exists $r \in (0, R)$ such that

4.3 More General Elliptic Operators

$$\sum\nolimits_{j,k=1}^{n} |||\tilde{a}_{jk} - \delta_{jk}|||_{\lambda, \overline{B} \cap \overline{B(0,r)}} < 1/(2c),$$

where c is the constant appearing in Corollary 4.2.3.

Now define, for all $j, k \in \{1, ..., n\}$,

$$a_{jk}(x') = \begin{cases} \tilde{a}_{jk}(x') & \text{if } x \in \overline{B} \cap \overline{B(0,r)}, \\ \tilde{a}_{jk}(x'/r) & \text{if } x \in \overline{B} \setminus B(0,r). \end{cases}$$

These coefficients belong to $C^\lambda(\overline{B})$ and satisfy the hypotheses of Theorem 4.3.1. Let $\theta \in C_0^\infty(\mathbb{R}^n)$ be such that supp $\theta \subset B(0, 2r/3)$ and $\theta(x) = 1$ if $x \in B(0, r/2)$; define $F : \overline{B} \to \mathbb{R}$ by

$$F(x) = \theta(x) f(x', x_n + \psi(x')) + 2 \sum\nolimits_{j,k=1}^{n} a_{jk}(x') D_j v(x) D_k \theta(x)$$
$$+ v(x) \sum\nolimits_{j,k=1}^{n} a_{jk}(x') D_j D_k \theta(x) - \theta(x) \tilde{a}_n(x') D_n v(x).$$

Since $v \in C^{1,\gamma}(\overline{B} \cap \overline{B(0,r)})$ and supp $\theta \subset B(0, 2r/3)$, we see that $F \in C^\lambda(\overline{B})$. Thus by Theorem 4.3.1, the Dirichlet problem

$$\sum\nolimits_{j,k=1}^{n} a_{jk}(x') D_j D_k w(x) = F(x) (x \in B), \ w(x) = 0 \ (x \in \partial B),$$

has a unique solution $w \in C^{2,\lambda}(\overline{B})$. In $B \cap B(0,r)$ the function θv satisfies the differential equation

$$\sum\nolimits_{j,k=1}^{n} a_{jk}(x') D_j D_k (\theta v)(x) = F(x).$$

For $x \in B \cap B(0,r)$ we have $a_{jk}(x') = \tilde{a}_{jk}(x')$; on $\partial B \cap B(0,r)$, the function v is zero; and on $\overline{B} \setminus B(0,r)$, both θv and F are identically zero. Thus $w = \theta v$ in $\overline{B}$ and since $\theta(x) = 1$ if $x \in B(0, r/2)$, it follows that $v \in C^{2,\lambda}\left(\overline{B \cap B(0, r/2)}\right)$. Hence there is an $\mathbb{R}^n$-neighbourhood Q of 0 such that $u \in C^{2,\lambda}\left(Q \cap \overline{\Omega}\right)$: the proof is complete. ∎

As an immediate corollary of this theorem and Theorem 4.2.6 we have the following result, known in the literature as Kellogg's theorem (see [124]), the first complete proof of which was given by Schauder in [182].

Corollary 4.3.3 *Let Ω be a bounded, connected, open subset of $\mathbb{R}^n$ with $\partial \Omega \in C^{2,\lambda}$ for some $\lambda \in (0, 1)$, and let $f \in C^\lambda(\overline{\Omega})$, $g \in C^{2,\lambda}(\partial \Omega)$. Then the Dirichlet problem*

$$\Delta u = f \text{ in } \Omega, u = g \text{ on } \partial \Omega,$$

has a unique solution in $C^{2,\lambda}(\overline{\Omega})$.

This completes our study of the classical Dirichlet problem involving the Poisson equation. We now show that corresponding results for more general second-order elliptic operators can be obtained in a relatively straightforward manner, and begin with some useful lemmas. We always suppose that Ω is a bounded, connected, open subset of $\mathbb{R}^n$ with $\partial\Omega \in C^{2,\lambda}$ for some $\lambda \in (0, 1)$.

Lemma 4.3.4 *Let $L_0 := \sum_{i,j=1}^n b_{ij} D_i D_j$ be elliptic, with constant coefficients b_{ij}. Then given any $f \in C^\lambda(\overline{\Omega})$ and $g \in C^{2,\lambda}(\partial\Omega)$, the Dirichlet problem*

$$L_0 u = f \text{ in } \Omega, u = g \text{ on } \partial\Omega,$$

has exactly one solution $u \in C^{2,\lambda}(\overline{\Omega})$; and there exists a constant k, depending on Ω and the b_{ij} but independent of f and g, such that

$$|||u|||_{2,\lambda,\Omega} \le k \left\{ |||f|||_{0,\lambda,\Omega} + |||g|||_{2,\lambda,\partial\Omega} \right\}.$$

Proof. As L_0 is elliptic, there is a nonsingular linear map $S : \mathbb{R}^n \to \mathbb{R}^n$ under the action of which the given Dirichlet problem is transformed into the problem $\Delta v = \widetilde{f}$ in $S(\Omega)$, $v = \widetilde{g}$ on $\partial(S(\Omega))$, in which $\widetilde{f}, \widetilde{g}$ have the same properties as f, g respectively. Since $S(\Omega)$ is bounded, open and connected with $C^{2,\lambda}$ boundary, Corollary 4.3.3 shows that this transformed problem has a unique solution $v \in C^{2,\lambda}(\overline{S(\Omega)})$: the first part of the Lemma follows. The linear map $T : C^\lambda(\overline{\Omega}) \times C^{2,\lambda}(\partial\Omega) \to C^{2,\lambda}(\overline{\Omega})$ defined by $T(f, g) = u$ is injective and its inverse, given by $u \mapsto (L_0 u, u \lceil_{\partial\Omega})$, is continuous. Use of the inverse mapping theorem (Corollary 1.2.2) shows that T is continuous, which completes the proof. ∎

To deal with more general elliptic operators we need a little preparation.

Lemma 4.3.5 *(Ehrling's lemma). For $i = 1, 2, 3$, let X_i be a Banach space with norm $\|\cdot\|_i$; suppose that $K : X_1 \to X_2$ is compact and linear, and let $T : X_2 \to X_3$ be linear, injective and continuous. Then given any $\varepsilon > 0$, there exists a positive constant $c(\varepsilon)$ such that for all $x \in X_1$,*

$$\|Kx\|_2 \le \varepsilon \|x\|_1 + c(\varepsilon) \|(T \circ K)x\|_3.$$

Proof. Suppose the claim is false. Then there exist $\varepsilon_0 > 0$ and a sequence (x_n) of points of X_1 such that for all $n \in \mathbb{N}$,

$$\|Kx_n\|_2 > \varepsilon_0 \|x_n\|_1 + n \|(T \circ K)x_n\|_3.$$

We may assume that $\|x_n\|_1 = 1$ for all $n \in \mathbb{N}$. The continuity of K implies that $\|Kx_n\|_2 \le \|K\|$; hence $\|(T \circ K)x_n\|_3 < (\|K\| - \varepsilon_0)/n$, and so $(T \circ K)x_n \to 0$ as $n \to \infty$. As K is compact, there is a subsequence of (x_n), again denoted by (x_n) for simplicity, such that $Kx_n \to y$ for some $y \in X_2$. Thus $(T \circ K)x_n \to Ty$, which implies that $Ty = 0$: because T is injective, $y = 0$, and so $Kx_n \to 0$. But $\|Kx_n\|_2 > \varepsilon_0 > 0$: contradiction. ∎

4.3 More General Elliptic Operators

Corollary 4.3.6 *Let $l \in \mathbb{N}$, $\lambda \in (0, 1)$, take $X_1 = C^{l,\lambda}(\overline{\Omega})$, $X_2 = C^l(\overline{\Omega})$, $X_3 = C(\overline{\Omega})$, and let $K : X_1 \to X_2$ and $T : X_2 \to X_3$ be the natural embeddings. Then given any $\varepsilon > 0$, there is a positive constant $c(\varepsilon)$ such that for all $u \in C^{l,\lambda}(\overline{\Omega})$,*

$$|||u|||_{l,0,\Omega} \leq \varepsilon |||u|||_{l,\lambda,\Omega} + c(\varepsilon) |||u|||_{0,0,\Omega}.$$

Proof. Since K is compact and T is continuous and injective (see Theorem 1.3.1), the result is immediate from Lemma 4.3.5. ■

Theorem 4.3.7 *Let*

$$L := \sum_{i,j=1}^{n} a_{ij} D_i D_j + \sum_{i=1}^{n} a_i D_i + a$$

be uniformly elliptic in $\overline{\Omega}$ with coefficients $a_{ij}, a_i, a \in C^\lambda(\overline{\Omega})$ and $a(x) \leq 0$ in Ω; suppose that $f \in C^\lambda(\overline{\Omega})$, $g \in C^{2,\lambda}(\partial\Omega)$. Then there exists a unique $u \in C^{2,\lambda}(\overline{\Omega})$ such that

$$Lu = f \text{ in } \Omega, u = g \text{ on } \partial\Omega.$$

Proof. We first claim that there is a positive constant k such that for all $t \in [0, 1]$ and all $u \in C^{2,\lambda}(\overline{\Omega})$,

$$|||u|||_{2,\lambda,\Omega} \leq k \left\{ ||| (1-t)\Delta u + tLu |||_{\lambda,\Omega} + |||u|||_{2,\lambda,\partial\Omega} \right\}. \quad (4.3.1)$$

Let $t \in [0, 1]$ and for each $y \in \overline{\Omega}$ put

$$L_y = \sum_{i,j=1}^{n} \left\{ (1-t)\delta_{ij} + ta_{ij}(y) \right\} D_i D_j.$$

By Lemma 4.3.4, given $y \in \overline{\Omega}$, there is a constant $c(y, t)$ such that for all $v \in C^{2,\lambda}(\overline{\Omega})$,

$$|||v|||_{2,\lambda,\Omega} \leq c(y,t) \left\{ |||L_y v|||_{\lambda,\Omega} + |||v|||_{2,\lambda,\partial\Omega} \right\}. \quad (4.3.2)$$

Put

$$P(y) = c(y,t) \sum_{i,j=1}^{n} |||a_{ij}|||_{\lambda,\Omega}, \ O(y) = \left\{ x \in \mathbb{R}^n : |x-y|^\lambda < 1/(2P(y)) \right\}.$$

The compact set $\overline{\Omega}$ is covered by a finite family $\left\{O(x^j)\right\}_{j \in M}$; let $\phi_j \in C_0^\infty(\mathbb{R}^n)$ ($j \in M$) form a partition of unity subordinate to this covering, with $\mathrm{supp}\ \phi_j \subset O(x^j)$. Let $j \in \{1, ..., M\}$. Since

$$L_{x^j}(\phi_j u) = t \sum_{i,l=1}^{n} \left(a_{il}(x^j) - a_{il}(x) \right) D_i D_l(\phi_j u) + L_x(\phi_j u),$$

we see from (4.3.2) that there is a constant $k(x^j, t)$ such that $|| \phi_j u |||_{2,\lambda,\Omega} / k(x^j, t)$ is bounded above by

$$|||t\sum_{i,l=1}^{n}\left(a_{il}(x^j) - a_{il}(x)\right) D_i D_l(\phi_j u) + L_x(\phi_j u)|||_{\lambda,\Omega} + |||\phi_j u|||_{2,\lambda,\partial\Omega}$$

$$\leq (2P(x^j))^{-1}\sum_{i,l=1}^{n} |||a_{il}|||_{\lambda,\Omega}|||\phi_j u|||_{2,\lambda,\partial\Omega}$$

$$+ 2\sum_{i,l=1}^{n} |||a_{il}|||_{\lambda,\Omega}|||\phi_j u|||_{2,\lambda,\partial\Omega} + |||L_x(\phi_j u)|||_{\lambda,\Omega} + |||\phi_j u|||_{2,\lambda,\partial\Omega}.$$

From Corollary 4.3.6, the estimate

$$|||w|||_{k,0,\Omega} \leq \varepsilon |||w|||_{k,\lambda,\Omega} + C(\varepsilon)|||w|||_{0,\Omega} \quad (4.3.3)$$

holds for all $\varepsilon > 0$ and all $w \in C^{k,\alpha}(\overline{\Omega})$. Using Theorem 3.2.10 we have

$$|||\phi_j u|||_{0,\Omega} \leq C\left\{|||L_x(\phi_j u)|||_{0,\Omega} + |||\phi_j u|||_{0,\partial\Omega}\right\}, \quad (4.3.4)$$

and so

$$|||\phi_j u|||_{2,\lambda,\Omega} \leq \widetilde{c}(x^j)\left\{|||L_x(\phi_j u)|||_{\lambda,\Omega} + |||\phi_j u|||_{2,\lambda,\partial\Omega}\right\}.$$

Hence

$$|||u|||_{2,\lambda,\Omega} = |||\sum_{j=1}^{M}\phi_j u|||_{2,\lambda,\Omega} \leq \sum_{j=1}^{M}|||\phi_j u|||_{2,\lambda,\Omega}$$

$$\leq \sum_{j=1}^{M}\widetilde{c}(x^j)\left\{|||L_x(\phi_j u)|||_{\lambda,\Omega} + |||\phi_j u|||_{2,\lambda,\partial\Omega}\right\}. \quad (4.3.5)$$

Thus after writing out the derivatives of $\phi_j u$ and using the triangle inequality we obtain, with suitable constants k_1 and k_2,

$$|||u|||_{2,\lambda,\Omega} \leq k_1|||(1-t)\Delta u + tLu|||_{\lambda,\Omega} + k_2\sum_{i,j=1}^{n}|||a_{ij}|||_{2,\lambda,\Omega}|||u|||_{2,\Omega}$$

$$+ k_2\left\{\left(\sum_{i=1}^{n}|||a_i|||_{\lambda,\Omega} + |||a|||_{\lambda,\Omega}\right)|||u|||_{1,\lambda,\Omega} + |||u|||_{2,\lambda,\partial\Omega}\right\}.$$

Taking into account (4.3.4) and (4.3.5) this gives, for all $u \in C^{2,\lambda}(\overline{\Omega})$, the estimate

$$|||u|||_{2,\lambda,\Omega} \leq k(t)\left\{|||(1-t)\Delta u + tLu|||_{\lambda,\Omega} + |||u|||_{2,\lambda,\partial\Omega}\right\}. \quad (4.3.6)$$

We wish to show that this holds with a constant k on the right-hand side that is independent of t. Let $u \in C^{2,\lambda}(\overline{\Omega})$ and $t_0 \in [0,1]$. From (4.3.6) we see that for all $t \in [0,1]$ with $|t - t_0| < \delta(t_0)$, where

$$1/\delta(t_0) = 2k(t_0)\left\{n + \sum_{i,j=1}^{n}|||a_{ij}|||_{2,\lambda,\Omega} + \sum_{i=1}^{n}|||a_i|||_{\lambda,\Omega} + |||a|||_{\lambda,\Omega}\right\},$$

4.3 More General Elliptic Operators

we have

$$||| u |||_{2,\lambda,\Omega} \leq k(t_0) |t - t_0| ||| \Delta u - Lu |||_{\lambda,\Omega}$$
$$+ k(t_0) ||| (1 - t)\Delta u + tLu |||_{\lambda,\Omega} + ||| u |||_{2,\lambda,\partial\Omega},$$

$$||| u |||_{2,\lambda,\Omega} \leq \frac{1}{2} ||| u |||_{2,\lambda,\Omega} + k(t_0) \{||| (1 - t)\Delta u + tLu |||_{\lambda,\Omega} + ||| u |||_{2,\lambda,\partial\Omega}\},$$

and so

$$||| u |||_{2,\lambda,\Omega} \leq 2k(t_0) \{||| (1 - t)\Delta u + tLu |||_{\lambda,\Omega} + ||| u |||_{2,\lambda,\partial\Omega}\}. \quad (4.3.7)$$

As the sets

$$O(\tau) := \{t : |t - \tau| < \delta(\tau)\} \, (\tau \in [0, 1])$$

form an open covering of $[0, 1]$, there is a finite subcover $\{O(\tau_j)\}_{j \in \{1,\ldots,s\}}$: let $k = 2 \max\{k(\tau_j) : j = 1, \ldots, s\}$. Given $t \in [0, 1]$, there exists $j \in \{1, \ldots, s\}$ such that $t \in O(\tau_j)$: together with (4.3.7) and the definition of k this gives the claimed estimate (4.3.1).

For each $t \in [0, 1]$ let $L_t := (1 - t)\Delta + tL$ and let Γ be the set of all those $t \in [0, 1]$ such that for all $F \in C^\lambda(\overline{\Omega})$ and $G \in C^{2,\lambda}(\partial\Omega)$, there exists $u \in C^{2,\lambda}(\overline{\Omega})$ such that $L_t u = F$ in Ω, $u = G$ on $\partial\Omega$. We know that $0 \in \Gamma$. Let $t_0 \in \Gamma$ and define ε by

$$1/\varepsilon = 2kn(n + 3) \max_{i,j \in \{1,\ldots,n\}} \{1, ||| a_{ij} |||_{\lambda,\Omega}, ||| a_i |||_{\lambda,\Omega}, ||| a |||_{\lambda,\Omega}\}.$$

We claim that all $t \in [0, 1]$ with $|t - t_0| < \varepsilon$ belong to Γ. For given any $u \in C^{2,\lambda}(\overline{\Omega})$, $F(u) := (t - t_0)(\Delta - L)u + f \in C^\lambda(\overline{\Omega})$ and the problem

$$L_{t_0} v = F(u) \text{ in } \Omega, \, v = g \text{ on } \partial\Omega$$

has exactly one solution $v \in C^{2,\lambda}(\overline{\Omega})$ since $t_0 \in \Gamma$. A map $T_{t_0} : C^{2,\lambda}(\overline{\Omega}) \to C^{2,\lambda}(\overline{\Omega})$ is defined by $T_{t_0} u = v$. Given any $u_1, u_2 \in C^{2,\lambda}(\overline{\Omega})$, the function $T_{t_0}(u_1 - u_2)$ is a solution of the Dirichlet problem

$$L_{t_0}(T_{t_0}(u_1 - u_2)) = (t - t_0)(\Delta - L)(u_1 - u_2) \text{ in } \Omega, \, T_{t_0}(u_1 - u_2) = 0 \text{ on } \partial\Omega.$$

Application of (4.3.1) to $T_{t_0}(u_1 - u_2)$ gives

$$||| T_{t_0}(u_1 - u_2) |||_{2,\lambda,\Omega} \leq k |t - t_0| \{||| \Delta(u_1 - u_2) |||_{\lambda,\Omega} + ||| L(u_1 - u_2) |||_{\lambda,\Omega}\},$$

so that $||| T_{t_0}(u_1 - u_2) |||_{2,\lambda,\Omega} \leq \frac{1}{2} ||| u_1 - u_2 |||_{2,\lambda,\Omega}$. Accordingly T_{t_0} is a contraction and thus has a fixed point: there exists $u \in C^{2,\lambda}(\overline{\Omega})$ such that $T_{t_0} u = u$, and so $t \in \Gamma$.

Since ε does not depend on t_0, it follows that $[0, 1] \subset \Gamma$ and the proof of existence is complete. Uniqueness follows from the maximum principle. ∎

In view of (4.3.1) in the proof of Theorem 4.3.7 the following result, usually referred to in the literature as *Schauder's boundary estimate*, is now clear.

Theorem 4.3.8 *Let Ω and L be as in Theorem 4.3.7. suppose that $u, \phi \in C^{2,\lambda}(\overline{\Omega})$, $f \in C^\lambda(\overline{\Omega})$ are such that $Lu = f$ in Ω and $u = \phi$ on $\partial\Omega$. Then there is a constant C, depending only on n, λ, Ω and bounds for the norms in $C^\lambda(\overline{\Omega})$ of a_{ij}, b_i and a, such that*

$$\|u\|_{2,\lambda,\Omega} \leq C \left\{ \|f\|_{\lambda,\Omega} + \|\phi\|_{2,\lambda,\Omega} \right\}.$$

When the hypothesis that $a \leq 0$ in Ω is dropped, the following result of Fredholm type still holds.

Theorem 4.3.9 *Let*

$$L := \sum_{i,j=1}^n a_{ij} D_i D_j + \sum_{i=1}^n a_i D_i + a$$

be uniformly elliptic in $\overline{\Omega}$ with coefficients $a_{ij}, a_i, a \in C^\lambda(\overline{\Omega})$. Then either
(a) the homogeneous problem

$$Lu = 0 \text{ in } \Omega, \ u = 0 \text{ on } \partial\Omega$$

has only the trivial solution $u = 0$, in which case the non-homogeneous problem

$$Lu = f \text{ in } \Omega, \ u = \phi \text{ on } \partial\Omega$$

has a unique solution in $C^{2,\lambda}(\overline{\Omega})$ for all $f \in C^\lambda(\overline{\Omega})$ and all $\phi \in C^{2,\lambda}(\overline{\Omega})$; or
(b) the homogeneous problem has non-trivial solutions which span a finite-dimensional subspace of $C^{2,\lambda}(\overline{\Omega})$.

Proof. The inhomogeneous problem is equivalent to the problem

$$Lv = f - L\phi \text{ in } \Omega, \ v = 0 \text{ on } \partial\Omega.$$

We shall therefore consider the Dirichlet problem with zero boundary data. Put

$$X = \{u \in C^{2,\lambda}(\overline{\Omega}) : u = 0 \text{ on } \partial\Omega\},$$

let $\sigma \geq \sup_\Omega a(x)$ and define $L_\sigma = L - \sigma$. By Theorem 4.3.7, $L_\sigma : X \to C^\lambda(\overline{\Omega})$ is bijective (since $a - \sigma \leq 0$), and $L_\sigma^{-1} : C^\lambda(\overline{\Omega}) \to X$ is bounded (see (4.3.1)). Since $C^{2,\lambda}(\overline{\Omega})$ is compactly embedded in $C^\lambda(\overline{\Omega})$ we may therefore view L_σ^{-1} as a compact linear map of $C^\lambda(\overline{\Omega})$ to itself. Now consider the equation

$$u + \sigma L_\sigma^{-1} u = L_\sigma^{-1} f, \ f \in C^\lambda(\overline{\Omega}). \tag{4.3.8}$$

4.3 More General Elliptic Operators

By the Fredholm-Riesz-Schauder theory of compact linear operators (see Theorem 1.2.9), (4.3.8) has a solution if

$$u + \sigma L_\sigma^{-1} u = 0$$

has only the trivial solution $u = 0$. When this condition is not fulfilled, the theory also tells us that

$$\dim \ker(I + \sigma L_\sigma^{-1}) < \infty,$$

where $I : C^\lambda(\overline{\Omega}) \to C^\lambda(\overline{\Omega})$ is the identity map.

Since L_σ^{-1} maps $C^\lambda(\overline{\Omega})$ onto X, any solution $u \in C^\lambda(\overline{\Omega})$ of (4.3.8) must belong to X. Operate on (4.3.8) with L_σ:

$$Lu = L_\sigma(u + \sigma L_\sigma^{-1} u) = f, u \in X. \quad (4.3.9)$$

This means that solutions of (4.3.8) are in one-to-one correspondence with solutions of (4.3.9), and the result follows. ∎

4.4 Notes

1. The approach to the Schauder estimates given here follows that presented by M. König in [132], [133] and [134]. For other accounts that also rely on estimates of the Newtonian potential we refer to [90], [115] and [141]. Those who are allergic to such procedures may wish to look at (i) the technique based on energy estimates and Campanato–Morrey spaces given in [89], Chapter 3 and [100], Chapter 3; and (ii) the rescaling argument of Leon Simon [193], which culminates in an argument by contradiction.

2. For details of Schauder-type estimates for other types of boundary-value problems, such as the Neumann and oblique derivative problems we refer to [90], [141] and the relevant references given in these books. An interesting discussion of the Neumann problem for the Poisson equation, namely

$$\Delta u = f \text{ in } \Omega, \partial u/\partial \nu = g \text{ on } \partial\Omega,$$

is given in [159].

Chapter 5
Elliptic Operators of Arbitrary Order

5.1 Preliminaries

In what follows we introduce the notion of ellipticity for differential operators of arbitrary order and establish the existence of 'weak' solutions of a Dirichlet problem for such operators. The question of when such weak solutions are classical is investigated for the Laplace operator, together with properties of its eigenvalues and the matter of spectral independence. Unless otherwise stated, the functions appearing are allowed to be either real- or complex-valued.

Let Ω be an open subset of $\mathbb{R}^n$. We consider differential operators of the form

$$A = \sum_{|\alpha| \leq m} a_\alpha D^\alpha, \tag{5.1.1}$$

where m is an arbitrary natural number and $a_\alpha \in C(\overline{\Omega})$ for all $\alpha \in \mathbb{N}_0^n$ with $|\alpha| \leq m$.

Definition 5.1.1 *The* **principal part** *of the operator A is*

$$A_m := \sum_{|\alpha|=m} a_\alpha D^\alpha;$$

the **principal symbol** *of A is the polynomial p_m, where*

$$p_m(x, \xi) = \sum_{|\alpha|=m} a_\alpha(x) \xi^\alpha \ (x \in \overline{\Omega}, \xi \in \mathbb{R}^n).$$

The operator A is called **elliptic at a point** $x \in \overline{\Omega}$ *if for all $\xi \in \mathbb{R}^n \setminus \{0\}$, $p_m(x, \xi) \neq 0$; it is* **elliptic in** $\overline{\Omega}$ *if it is elliptic at each point of $\overline{\Omega}$; it is* **strongly elliptic at** $x \in \overline{\Omega}$ *if there exists $\gamma \in \mathbb{C}$ such that for all $\xi \in \mathbb{R}^n \setminus \{0\}$,*

$$re\ (\gamma p_m(x, \xi)) \neq 0;$$

and it is **strongly elliptic in** $\overline{\Omega}$ if it is strongly elliptic at each point x of $\overline{\Omega}$, the constant γ being independent of x.

Example 5.1.2

(i) $n = 1$, $\Omega = (a, b)$, $A = \sum_{j=0}^{n} a_j D^j$ ($D = \frac{d}{dx}$), $A_m = a_m D^m$, $p_m(x, \xi) = a_m(x) \xi^m$ ($\xi \in \mathbb{R}$). Thus A is elliptic and strongly elliptic at x if $a_m(x) \neq 0$; it is strongly elliptic in $[a, b]$ if there are constants $\gamma_1, \gamma_2 \in \mathbb{R}$ such that for all $x \in [a, b]$,

$$\gamma_1 re\ a_m(x) - \gamma_2 \Im\ a_m(x) \neq 0.$$

(ii) $n = 2$, $\Omega = \mathbb{R}^2$, $A = D_1 + iD_2$ (the Cauchy-Riemann operator). Then A is elliptic in $\mathbb{R}^2$, for

$$p_1(x, \xi) = \xi_1 + i\xi_2 \neq 0 \text{ if } \xi = (\xi_1, \xi_2) \neq 0.$$

A is not strongly elliptic at any point $x \in \mathbb{R}^2$, for given any $\gamma \in \mathbb{C}$, $\gamma = \gamma_1 + i\gamma_2$,

$$re\ (\gamma p_1(x, \xi)) = re\ ((\gamma_1 + i\gamma_2)(\xi_1 + i\xi_2)) = \xi_1 \gamma_1 - \xi_2 \gamma_2,$$

and so $re\ (\gamma p_1(x, \xi))$ may be zero even though $\xi \neq 0$.

(iii) $n = 3$, $\Omega = \mathbb{R}^3$, $A = D_1 + iD_2 + (ix_1 - x_2)D_3$. A is not elliptic at any point of $\mathbb{R}^3$, for

$$p_1(x, \xi) = \xi_1 + i\xi_2 + (ix_1 - x_2)\xi_3 = 0 \text{ if } \xi = (x_2, -x_1, 1).$$

(iv) $n \geq 2$, $m = 2$, $A = \sum_{j,k=1}^{n} a_{jk}(x) D_j D_k + \sum_{j=1}^{n} b_j(x) D_j + c(x)$, $p_2(x, \xi) = \sum_{j,k=1}^{n} a_{jk}(x) \xi_j \xi_k$.

If the a_{jk} are real, then A is strongly elliptic at x if, and only if, $p_2(x, \xi)$ is positive- or negative-definite. In particular, the Laplace operator Δ is strongly elliptic in $\mathbb{R}^n$.

(v) $n \geq 2$, $k \in \mathbb{N}$, $A = \Delta^k$, $p_{2k}(x, \xi) = |\xi|^{2k}$: Δ^k is strongly elliptic in $\mathbb{R}^n$.

Proposition 5.1.3 *Suppose the operator A given by (5.1.1) is strongly elliptic at $x \in \overline{\Omega}$ and that $n \geq 2$. Then m is even.*

Proof. Let $\gamma \in \mathbb{C}$ and $\xi \in \mathbb{R}^n$. Then

$$re\ (\gamma p_m(x, -\xi)) = (-1)^m re\ (\gamma p_m(x, \xi)).$$

Since $n \geq 2$, given $\xi \in \mathbb{R}^n \setminus \{0\}$, there is a continuous function $\tilde{\xi} : [0, 1] \to \mathbb{R}^n$ such that $\tilde{\xi}(0) = \xi$, $\tilde{\xi}(1) = -\xi$ and for all $t \in [0, 1]$, $\xi(t) \neq 0$. As A is strongly elliptic at x, there exists $\gamma \in \mathbb{C}$ such that $re\ (\gamma p_m(x, \xi)) > 0$. But then if m was odd, $re\ (\gamma p_m(x, -\xi)) < 0$, and so there would be a $t_0 \in [0, 1]$ such that $re\ (\gamma p_m(x, \tilde{\xi}(t_0))) = 0$, which contradicts the strong ellipticity of A. Hence m is even. ∎

Remark 5.1.4

This uses the fact that $\mathbb{R}^n \setminus \{0\}$ is connected when $n \geq 2$ and disconnected when $n = 1$.

5.1 Preliminaries

Proposition 5.1.5 *Let A, given by (5.1.1), be elliptic (respectively, strongly elliptic) in $\overline{\Omega}$. Then given any compact set $K \subset \Omega$, there is a constant $c_0 > 0$ such that for all $x \in K$ and all $\xi \in \mathbb{R}^n$,*
$$|p_m(x, \xi)| \geq c_0 |\xi|^m$$

(respectively, $|re \ (\gamma p_m(x, \xi))| \geq c_0 |\xi|^m$, where γ is as in the definition of strong ellipticity).

Proof. Let $S = \{\xi \in \mathbb{R}^n : |\xi| = 1\}$. If A is elliptic in $\overline{\Omega}$, the continuous positive function $|p_m(x, \xi)|$ on the compact set $K \times S$ has a positive minimum c_0. Hence for all $\xi \in \mathbb{R}^n \setminus \{0\}$ and all $x \in K$,
$$\left| p_m(x, \frac{\xi}{|\xi|}) \right| = \frac{1}{|\xi|^m} |p_m(x, \xi)| \geq c_0.$$

If instead A is strongly elliptic in $\overline{\Omega}$, the continuous function $|re \ (\gamma p_m(x, \xi))|$ is used instead. ∎

Corollary 5.1.6 *Let $n \geq 2$ and supppose that A is strongly elliptic in $\overline{\Omega}$; let K be a compact connected subset of Ω. Then either*
$$re \ (\gamma p_m(x, \xi)) \geq c_0 |\xi|^m \text{ for all } x \in K \text{ and all } \xi \in \mathbb{R}^n,$$
or
$$-re \ (\gamma p_m(x, \xi)) \geq c_0 |\xi|^m \text{ for all } x \in K \text{ and all } \xi \in \mathbb{R}^n.$$

Here γ is the constant in the definition of strong ellipticity.

Proof. Since S is connected, so is $K \times S$. Hence the continuous function $re \ (\gamma p_m(x, \xi))$ has a constant sign on the connected set $K \times S$. ∎

Definition 5.1.7 *The operator A given by (5.1.1) is called **uniformly elliptic on $\overline{\Omega}$** if there exists $c_0 > 0$ such that for all $x \in \overline{\Omega}$ and all $\xi \in \mathbb{R}^n$,*
$$|p_m(x, \xi)| \geq c_0 |\xi|^m .$$

*It is called **uniformly strongly elliptic on $\overline{\Omega}$** if there exist $\gamma \in \mathbb{C}$ and $c_0 > 0$ such that for all $x \in \overline{\Omega}$ and all $\xi \in \mathbb{R}^n$,*
$$re \ (\gamma p_m(x, \xi)) \geq c_0 |\xi|^m .$$

Remark 5.1.8

When $n \geq 2$, since we may always multiply the coefficients of A by an appropriate complex number, we shall regard uniform strong ellipticity as characterised by the inequality
$$re \ p_m(x, \xi) \geq c_0 |\xi|^m$$
for all $x \in \overline{\Omega}$ and all $\xi \in \mathbb{R}^n$.

5.2 Gårding's Inequality

Consider a differential operator L of order $2m$ in the so-called 'divergence form':

$$Lu = \sum_{|\alpha|,|\beta| \leq m} (-1)^{|\alpha|} D^\alpha (a_{\alpha\beta} D^\beta u),$$

and suppose that L is uniformly strongly elliptic in a bounded open subset Ω of $\mathbb{R}^n$. Thus there is a constant $c_0 > 0$ such that for all $\xi \in \mathbb{R}^n$ and all $x \in \overline{\Omega}$,

$$re \sum_{|\alpha|=|\beta|=m} a_{\alpha\beta}(x)\xi^\alpha \xi^\beta \geq c_0 |\xi|^{2m}.$$

We make the following additional assumptions about the coefficients $a_{\alpha\beta}$:
(i) there is a constant $c_1 > 0$ such that for all $x \in \Omega$ and all $\alpha, \beta \in \mathbb{N}_0^n$ with $|\alpha|, |\beta| \leq m$,

$$|a_{\alpha\beta}(x)| \leq c_1;$$

(ii) there is a continuous function c_2, with $c_2(t) \downarrow 0$ as $t \downarrow 0$, such that for all $x, y \in \Omega$ and all $\alpha, \beta \in \mathbb{N}_0^n$ with $|\alpha| = |\beta| = m$,

$$|a_{\alpha\beta}(x) - a_{\alpha\beta}(y)| \leq c_2(|x - y|).$$

The function c_2 is called a **modulus of continuity** of the $a_{\alpha\beta}$.

In what follows we shall, for shortness, denote by $\overset{0}{H}{}^m(\Omega)$ the Sobolev space (see 1.3.2) obtained by forming the closure of $C_0^\infty(\Omega)$ in $H^m(\Omega) := W_2^m(\Omega)$.

Theorem 5.2.1 *(Gårding's inequality) Under the above assumptions, there are constants $c > 0$ and $k \geq 0$ (depending only on c_0, c_1, c_2 and Ω) such that for all $\phi \in \overset{0}{H}{}^m(\Omega)$,*

$$re\, a[\phi, \phi] \geq c \|\phi\|_{m,2}^2 - k \|\phi\|_2^2,$$

where

$$a[u, v] = \sum_{|\alpha|,|\beta| \leq m} \int_\Omega D^\alpha u(x).\overline{a_{\alpha\beta}(x) D^\beta v(x)} dx.$$

Proof. First note that it is enough to prove the inequality for functions in $C_0^\infty(\Omega)$. For if this has been done, let $\phi \in \overset{0}{H}{}^m(\Omega)$ and let (ϕ_j) be a sequence in $C_0^\infty(\Omega)$ such that $\|\phi - \phi_j\|_{m,2} \to 0$ as $j \to \infty$. Then

5.2 Gårding's Inequality

$$|a[\phi_j, \phi_j] - a[\phi, \phi]| \leq \sum_{|\alpha|, |\beta| \leq m} \left| \int_\Omega \overline{a_{\alpha\beta}}(D^\alpha \phi_j . \overline{D^\beta \phi_j} - D^\alpha \phi . \overline{D^\beta \phi}) dx \right|$$

$$\leq c_1 \sum_{|\alpha|, |\beta| \leq m} \int_\Omega \left| D^\alpha \phi_j . \overline{D^\beta \phi_j} - D^\alpha \phi . \overline{D^\beta \phi} \right| dx$$

$$\leq c_1 \sum_{|\alpha|, |\beta| \leq m} \int_\Omega \left\{ \left| D^\alpha \phi_j . \overline{D^\beta (\phi_j - \phi)} \right| + \left| \overline{D^\beta \phi} . D^\alpha (\phi_j - \phi) \right| \right\} dx$$

$$\to 0 \text{ as } j \to \infty,$$

the final step following from Schwarz's inequality.

We now deal with successively more complicated situations.

(a) Suppose that L is homogeneous, that is, $a_{\alpha\beta} = 0$ if $|\alpha| + |\beta| < 2m$, and has constant coefficients.

Let $\phi \in C_0^\infty(\Omega)$. Then using the properties of the Fourier transform $\widehat{f}$ of a function f we have

$$a[\phi, \phi] = \sum_{|\alpha|=|\beta|=m} (D^\alpha \phi, a_{\alpha\beta} D^\beta \phi)_2 = \sum_{|\alpha|=|\beta|=m} \overline{a_{\alpha\beta}}((2\pi i \xi)^\alpha \widehat{\phi}, (2\pi i \xi)^\beta \widehat{\phi})_2$$

$$= \int_{\mathbf{R}^n} |\widehat{\phi}(\xi)|^2 \sum_{|\alpha|=|\beta|=m} \overline{a_{\alpha\beta}}(2\pi \xi)^\alpha (2\pi \xi)^\beta d\xi.$$

Hence

$$\text{re } a[\phi, \phi] \geq c_0 \int_{\mathbf{R}^n} |\widehat{\phi}(\xi)|^2 \sum_{|\alpha|=m} (2\pi \xi)^\alpha (2\pi \xi)^\alpha d\xi.$$

Since

$$\sum_{|\alpha|=m} (D^\alpha \phi, D^\alpha \phi)_2 = \int_{\mathbf{R}^n} |\widehat{\phi}(\xi)|^2 \sum_{|\alpha|=m} (2\pi \xi)^\alpha (2\pi \xi)^\alpha d\xi,$$

it follows that

$$\text{re } a[\phi, \phi] \geq c_0 \sum_{|\alpha|=m} \|D^\alpha \phi\|_2^2 = c_0 \|\phi\|_{m,2}^2 \,, \text{ say}.$$

(b) Now suppose that L is homogeneous but may have variable coefficients, and that $\phi \in C_0^\infty(\Omega)$ is such that there exist $x_0 \in \Omega$ and a neighbourhood V of x_0 with supp $\phi \subset V$; we shall assume that V is 'small' in a sense to be made precise in a moment.

Then
$$a[\phi,\phi] = \sum_{|\alpha|=|\beta|=m} (D^\alpha \phi, a_{\alpha\beta}(x_0) D^\beta \phi)_2$$
$$+ \sum_{|\alpha|=|\beta|=m} (D^\alpha \phi, (a_{\alpha\beta}(\cdot) - a_{\alpha\beta}(x_0)) D^\beta \phi)_2$$
$$= I_1 + I_2, \text{ say.}$$

By (a),
$$re\ I_1 \geq c_0 \|\phi\|_{m,2}^2.$$

By hypothesis (ii),
$$|I_2| \leq \sum_{|\alpha|=|\beta|=m} \int_{\mathbf{R}^n} c_2(|x-x_0|) |D^\alpha \phi| |D^\beta \phi| dx.$$

Thus if diam V is small enough, say diam $V < \delta$, where δ depends only on c_0 and c_2, use of Schwarz's inequality shows that
$$|I_2| \leq \frac{1}{2} c_0 \|\phi\|_{m,2}^2.$$

Hence
$$re\ a[\phi,\phi] \geq \frac{1}{2} c_0 \|\phi\|_{m,2}^2.$$

(c) Finally suppose merely that L satisfies the hypotheses of the theorem. Let (V_j) be a finite open covering of $\overline{\Omega}$, where diam $V_j < \delta$ for all j, and let (ψ_j) be a partition of unity subordinate to this covering. Set $\delta_j = \sqrt{\psi_j}$; we may and shall assume that each $\delta_j \in C^\infty$, for if not, we could replace ψ_j by $\psi_j^2 / \sum_k \psi_k^2$. Then
$$a[\phi,\phi] = \sum_{|\alpha|=|\beta|=m} \int_\Omega D^\alpha \phi \cdot \overline{a_{\alpha\beta}} D^\beta \overline{\phi} dx$$
$$+ \sum_{|\alpha|,|\beta| \leq m; |\alpha|+|\beta|<2m} \int_\Omega D^\alpha \phi \cdot \overline{a_{\alpha\beta}} D^\beta \overline{\phi} dx$$
$$= J_1 + J_2, \text{ say.}$$

Using hypothesis (i) and Schwarz's inequality, we see that

5.2 Gårding's Inequality

$$|J_2| \le C \, \|\phi\|_{m,2} \, \|\phi\|_{m-1,2}.$$

As for J_1, using Leibniz's formula we obtain

$$J_1 = \sum_j \sum_{|\alpha|=|\beta|=m} \int_\Omega (\delta_j D^\alpha \phi)(\delta_j \overline{a_{\alpha\beta} D^\beta \phi}) dx$$

$$= \sum_j \sum_{|\alpha|=|\beta|=m} \int_\Omega D^\alpha(\delta_j \phi) \overline{a_{\alpha\beta}} D^\beta (\delta_j \overline{\phi}) dx$$

$$- \sum_j \sum_{|\alpha|=|\beta|=m, |\alpha'+\gamma'|>0, |\alpha''+\gamma''|<2m} \int_\Omega c_{\alpha',\alpha'',\gamma',\gamma''} D^{\alpha'} \delta_j . D^{\alpha''} \phi . \overline{a_{\alpha\beta}}$$

$$\times D^{\gamma'} \delta_j . D^{\gamma''} \overline{\phi} \, dx$$

$$= J_3 - J_4, \text{ say.}$$

Yet another application of Schwarz's inequality shows that

$$|J_4| \le C_0 \, \|\phi\|_{m,2} \, \|\phi\|_{m-1,2}.$$

Since $\delta_j \phi \in C_0^\infty(V_j)$, we may use (b) to conclude that

$$\text{re } J_4 \ge \frac{1}{2} c_0 \sum_j \|\delta_j \phi\|_{m,2}^2$$

$$= \frac{1}{2} c_0 \sum_j \sum_{|\alpha|=m} \int_\Omega D^\alpha(\delta_j \phi) . \overline{D^\alpha(\delta_j \phi)} dx$$

$$\ge \frac{1}{2} c_0 \sum_j \sum_{|\alpha|=m} \int_\Omega \delta_j D^\alpha \phi . \delta_j \overline{D^\alpha \phi} \, dx + J_5,$$

where J_5 is the value of J_4 when $a_{\alpha\beta} = \delta_{\alpha\beta}$ (the Kronecker delta), so that

$$|J_5| \le C_1 \, \|\phi\|_{m,2} \, \|\phi\|_{m-1,2}.$$

Thus

$$\text{re } a[\phi, \phi] \ge \frac{1}{2} c_0 \|\phi\|_{m,2}^2 - C_2 \|\phi\|_{m,2} \|\phi\|_{m-1,2}. \tag{5.2.1}$$

The second term on the right-hand side of (5.2.1) is estimated as follows:

$$C_2 \|\phi\|_{m,2} \|\phi\|_{m-1,2} = \left(\sqrt{2\eta}\, \|\phi\|_{m,2}\right)\left(\frac{C_2}{\sqrt{2\eta}}\, \|\phi\|_{m-1,2}\right)$$

$$\leq \eta \|\phi\|_{m,2}^2 + \frac{C_2^2}{4\eta} \|\phi\|_{m-1,2}^2,$$

for any $\eta > 0$. We now use the result that

$$\|\phi\|_{j-1,2} \leq \varepsilon \|\phi\|_{j,2} + C \|\phi\|_2, \tag{5.2.2}$$

for all $\varepsilon \in (0,1)$ and all $j \in \mathbb{N}$, where $C = C(\Omega, j, \varepsilon)$. This follows from Ehrling's inequality (Lemma 4.3.5), but we give an independent proof below. Application of (5.2.2) shows that for small enough $\varepsilon > 0$,

$$C_2 \|\phi\|_{m,2} \|\phi\|_{m-1,2} \leq \eta \|\phi\|_{m,2}^2 + \frac{C_2^2}{4\eta}\left(\varepsilon \|\phi\|_{m,2}^2 + C_3 \|\phi\|_2^2\right),$$

where C_3 depends only on ε, m and Ω. A second application of (5.2.2) shows that

$$\|\phi\|_{m,2} \leq \frac{1}{1-\varepsilon} \|\phi\|_{m,2} + c \|\phi\|_2.$$

Thus an appropriate choice of ε gives

$$\frac{1}{2} c_0 \|\phi\|_{m,2}^2 \geq \frac{1}{3} c_0 \|\phi\|_{m,2}^2 - C_4 \|\phi\|_2^2,$$

where $C_4 = C_4(\Omega, m, c_0)$. Hence from (5.2.1) we obtain

$$\mathrm{re}\, a[\phi, \phi] \geq \frac{1}{3} c_0 \|\phi\|_{m,2}^2 - C_4 \|\phi\|_2^2 - \eta \|\phi\|_{m,2}^2$$

$$- \frac{C_2^2}{4\eta}\left(\varepsilon \|\phi\|_{m,2}^2 + C_3 \|\phi\|_2^2\right)$$

$$= \left(\frac{1}{3} c_0 - \eta - \frac{C_2^2 \varepsilon}{4\eta}\right) \|\phi\|_{m,2}^2 + \left(C_4 + \frac{C_2^2 C_3}{4\eta}\right) \|\phi\|_2^2.$$

Now choose $\eta = c_0/9$, $\varepsilon \leq \frac{4\eta c_0}{9 C_2^2}$.

To give another proof of (5.2.2), observe that it is enough to show that given $\varepsilon \in (0,1)$, there exists $\lambda(\varepsilon) \geq 0$ such that

5.2 Gårding's Inequality

$$\|\phi\|_{j-1,2} \leq \varepsilon \|\phi\|_{j,2} + \lambda \|\phi\|_2. \tag{5.2.3}$$

If (5.2.3) were false, there would be an $\varepsilon > 0$ and a sequence (ϕ_k) in $C_0^\infty(\Omega)$ such that

$$\|\phi_k\|_{j-1,2} > \varepsilon \|\phi_k\|_{j,2} + k \|\phi_k\|_2. \tag{5.2.4}$$

We may suppose that $\|\phi_k\|_{j,2} = 1$ for all k. Since H^j is compactly embedded in H^{j-1}, there are a subsequence of (ϕ_k), again denoted by (ϕ_k) for convenience, and an element ϕ of H^{j-1} such that $\|\phi_k - \phi\|_{j-1,2} \to 0$ as $k \to \infty$. By (5.2.4), $\|\phi_k\|_2 \to 0$ as $k \to \infty$: thus $\phi = 0$. This contradicts the fact that, again by (5.2.4), $\|\phi\|_{j-1,2} \geq \varepsilon$. ∎

5.3 The Dirichlet Problem

Let L be as in the last section and suppose that Ω is a bounded open subset of $\mathbb{R}^n$ with boundary $\partial\Omega \in C^\infty$. Let $f, g_1, ..., g_{m-1}$ be given functions, with f defined on Ω and the g_j defined on $\partial\Omega$. The classical Dirichlet problem for L is that of the existence of a function $u \in C^{2m}(\Omega) \cap C^{m-1}(\overline{\Omega})$ such that

$$Lu = f \text{ in } \Omega, \quad \frac{\partial^j u}{\partial \nu^j} = g_j \ (j = 0, ..., m-1) \text{ on } \partial\Omega, \tag{5.3.1}$$

where $\frac{\partial}{\partial \nu}$ means differentiation along the outward normal to $\partial\Omega$.

This problem can be reduced to that in which the boundary data are zero, if the g_j are smooth enough. In fact, it can be shown (see [85], Lemma 1.13.1) that if $g_j \in C^k(\partial\Omega)$ $(j = 0, ..., m-1)$, then there exists $\Phi \in C^k(\overline{\Omega})$ such that $\frac{\partial^j \Phi}{\partial \nu^j} = g_j$ $(j = 0, ..., m-1)$ on $\partial\Omega$. Taking this for granted, put $v = u - \Phi$: then problem (5.3.1) reduces to

$$Lv = f - L\Phi \text{ in } \Omega, \quad \frac{\partial^j u}{\partial \nu^j} = 0 \ (j = 0, ..., m-1) \text{ on } \partial\Omega,$$

that is, a problem with zero boundary data. A direct attack on this problem which depends on estimates of singular integrals is given in [3]. Here we describe another method of approach, one that is familiar in mathematics: if a problem is difficult, replace it by a related problem that is easier to solve and then try to show that the solution of the related problem is actually a solution of the original problem. The related problem in our case is called the generalised Dirichlet problem and it is easy to show that it has a solution since the famous inequality due to Gårding (Theorem 5.2.1) has been established. To get back to the classical Dirichlet problem requires what is called regularity theory, and it is here that the technicalities (which cannot be avoided indefinitely) become apparent.

We shall deal with the problem with zero boundary data in what follows. The first step is to formulate a generalised Dirichlet problem. Observe that if u is a solution of the classical Dirichlet problem (with zero boundary data), then on multiplication of both sides of the equation $\overline{Lu} = \overline{f}$ by $\phi \in C_0^\infty(\Omega)$ and integrating over Ω we obtain

$$a[\phi, u] = (\phi, f)_{2,\Omega},$$

where a is as in the last section. Various integrations by parts are needed to obtain this. This leads us to introduce the following **generalised Dirichlet problem** (GDP): does there exist $u \in \overset{0}{W}{}_2^m(\Omega) = \overset{0}{H}{}^m(\Omega)$ such that for all $\phi \in \overset{0}{H}{}^m(\Omega)$,

$$a[\phi, u] = (\phi, f)_{2,\Omega} \;? \tag{5.3.2}$$

Note that the requirement that $u \in \overset{0}{H}{}^m(\Omega)$ is natural, given the boundary conditions imposed in the original Dirichlet problem: membership of $\overset{0}{H}{}^m(\Omega)$ rather than $H^m(\Omega)$ amounts to asking for the boundary conditions to be satisfied in some weak sense. If there is a solution of the GDP, it is usually referred to as a **weak solution** of the classical Dirichlet problem.

To establish the existence of a solution of the GDP for L under the assumptions on L and its coefficients listed in the last section it is convenient to use the Lax-Milgram lemma (Lemma 1.2.7). First we consider the special case in which Gårding's inequality (see Theorem 5.2.1) holds in a strong form.

Theorem 5.3.1 *Suppose that there is a positive constant c such that for all $\phi \in \overset{0}{H}{}^m(\Omega)$,*

$$re\; a[\phi, \phi] \geq c \, \|\phi\|_{m,2,\Omega}^2.$$

Then given any $f \in L_2(\Omega)$, there is a unique solution of the GDP (5.3.2).

Proof. The bilinear function a satisfies the hypotheses imposed on B in the Lax-Milgram lemma. Moreover, f gives rise to a unique functional F on $\overset{0}{H}{}^m(\Omega)$ according to the rule

$$F(\psi) = (\psi, f)_{2,\Omega}.$$

Thus by the Lax-Milgram lemma, there is a unique $u \in \overset{0}{H}{}^m(\Omega)$ such that

$$(\psi, f)_{2,\Omega} = a[\psi, u] \text{ for all } \psi \in \overset{0}{H}{}^m(\Omega).$$

This proves the theorem. ∎

Now suppose that we merely have the ordinary form of Gårding's inequality available, that is,

5.3 The Dirichlet Problem

$$re\ a[\phi, \phi] \geq c\ \|\phi\|_{m,2}^2 - k_0\ \|\phi\|_2^2$$

for all $\phi \in \overset{0}{H}{}^m(\Omega)$. Here $c > 0$ and $k_0 \geq 0$ are constants independent of ϕ. We write $\|\cdot\|_{m,2}$ for $\|\cdot\|_{m,2,\Omega}$ and $(\cdot, \cdot)_2$ for $(\cdot, \cdot)_{2,\Omega}$ in the remainder of this section.

Theorem 5.3.2 *Let $f \in L_2(\Omega)$. Then if $k \geq k_0$, the GDP for $L + k$ has a unique solution.*

Proof. The form corresponding to a for $L + k$ is a_k, where

$$a_k[u, v] = a[u, v] + k(u, v)_2.$$

Hence

$$re\ a_k[\phi, \phi] = re\ a[\phi, \phi] + k\ \|\phi\|_2^2$$
$$\geq c\ \|\phi\|_{m,2}^2 + (k - k_0)\ \|\phi\|_2^2$$
$$\geq c\ \|\phi\|_{m,2}^2.$$

Now use Theorem 5.3.1. ∎

Theorem 5.3.3 *Let $f \in L_2(\Omega)$ be given. Then either*
(i) there is a unique solution of the GDP for L;
or
(ii) there are functions $v_1, ..., v_l$ in $\overset{0}{H}{}^m(\Omega)$ such that $a(v_i, \phi) = 0$ for $i = 1, ..., l$ and all $\phi \in \overset{0}{H}{}^m(\Omega)$, and there is a solution of the GDP for L if and only if $(f, v_i)_{L_2(\Omega)} = 0$ for $i = 1, ..., l$.
In case (ii) the solution (if it exists) is not unique.

Proof. For simplicity we write H instead of $\overset{0}{H}{}^m(\Omega)$. Let $k \geq k_0$, $k \neq 0$, $L_k = L + k$, $a_k[u, v] = a[u, v] + k(u, v)_{L_2(\Omega)}$. By Theorem 5.3.2, given $g \in L_2(\Omega)$, there is a unique solution $w \in H$ of

$$a_k[\phi, w] = (\phi, g)_2 \text{ for all } \phi \in H. \tag{5.3.3}$$

Hence there is a well-defined map $T_k : L_2(\Omega) \to H$ given by $T_k g = w$. Note that $u \in H$ is a solution of the GDP for L if and only if

$$a[\phi, u] = (\phi, f)_2 \text{ for all } \phi \in H,$$

and this holds if and only if

$$a_k[\phi, u] = (\phi, ku + f)_2 \text{ for all } \phi \in H.$$

In turn this holds if and only if

$$u = T_k(ku + f) = kT_k u + T_k f$$

(note that T_k is plainly linear), which is true if and only if

$$u - Tu = f_1, \text{ where } T = kT_k \text{ and } f_1 = T_k f. \tag{5.3.4}$$

From (5.3.2),

$$c\,\|w\|_{m,2}^2 \leq |a_k[w, w]| = |(w, g)_2| \leq \|w\|_{m,2}\,\|g\|_2,$$

and so

$$c\,\|w\|_{m,2} \leq \|g\|_2,$$

which implies that

$$c\,\|T_k g\|_{m,2} \leq \|g\|_2.$$

Hence $T = kT_k : L_2(\Omega) \to H$ is bounded. Since H is compactly embedded in $L_2(\Omega)$, we may regard T as a compact linear map from $L_2(\Omega)$ to $L_2(\Omega)$. With this understanding, (5.3.4) is an equation to which we may apply the Fredholm-Riesz-Schauder theory of compact linear operators (see Theorem 1.2.9). We conclude that either
(i) given any $f_1 \in L_2(\Omega)$, there is a unique solution of (5.3.4);
or
(ii) there are non-trivial solutions of $u - Tu = 0$, and (5.3.4) has a solution if and only if $(f_1, v_j)_2 = 0$ for a finite number of functions v_j ($j = 1, ..., h$) which form a basis of the space of solutions of $v - T^*v = 0$.

Suppose that (i) holds. Then given any $f \in L_2(\Omega)$, there is a solution of the GDP for L, and the solution is unique since $f = 0$ implies that $f_1 = 0$, which implies that $u = 0$.

Now suppose that (ii) holds. We use the formal adjoint L^* of L: this is the differential operator given by

$$L^* v = \sum_{|\alpha|, |\beta| \leq m} (-1)^{|\alpha|} D^\alpha (\overline{a_{\alpha\beta}} D^\beta v).$$

The terminology is justified by the fact that for all $u, v \in C_0^\infty(\Omega)$ (and hence for all $u, v \in H$),

$$a[v, u] = (v, Lu)_2 = (L^* v, u)_2.$$

Since the coefficients of L^* satisfy conditions exactly analogous to those satisfied by the coefficients of L, Gårding's inequality will hold for L^* as will Theorem 5.3.2, for large enough k.

5.3 The Dirichlet Problem

Regarding T as a bounded linear map from $L_2(\Omega)$ to itself, its adjoint T^* is defined by
$$(T^*v, g)_2 = (v, Tg)_2 \text{ for all } v, g \in L_2(\Omega).$$

Let $v, g \in L_2(\Omega)$. Put $Tg = h$, $T^*v = w$: then
$$a_k[\phi, h] = a_k[\phi, Tg] = a_k[\phi, kT_kg] = k(\phi, g)_2$$
for all $\phi \in H$ and all large enough k. Hence, with $\phi = w$,
$$a_k[w, h] = k(w, g)_2 :$$
this is justified by showing, just as for T (that is, using the analogue of (5.3.2)), that T^* maps $L_2(\Omega)$ into H.

We now prove that $T^* = k(L^* + k)^{-1}$. To do this, let $k(L^* + k)^{-1}v = w_1$. Then
$$a_k[w_1, \psi] = k(v, \psi)_2$$
for all $\psi \in \overset{0}{H}{}^m(\Omega)$. Thus
$$a_k[w_1, h] = k(v, h)_2 = k(v, Tg)_2 = k(T^*v, g)_2 = k(w, g)_2,$$
and so
$$a_k[w_1, h] = a_k[w, h],$$
that is,
$$0 = a_k[w - w_1, h] = k(w - w_1, g)_2.$$

Since g is an arbitrary element of $L_2(\Omega)$, we must have $w = w_1$, and so $T^*v = k(L^* + k)^{-1}v$.

It follows that $T^*v = w$ if and only if $k(L^* + k)^{-1}v - v = 0$, which holds if and only if $a_k[v, \phi] = k(v, \phi)_2$. In turn this is equivalent to $v \in H$ and $a[v, \phi] = 0$ for all $\phi \in H$. The conditions $(f_1, v_j)_2 = 0$ are equivalent to $(f, v_j)_2 = 0$, for
$$(f_1, v_j)_2 = (T_k f, v_j)_2 = k^{-1}(Tf, v_j)_2 = k^{-1}(f, T^*v_j)_2 = k^{-1}(f, v_j)_2.$$

∎

5.4 A Little Regularity Theory

Under what circumstances is a solution of the GDP for L a solution of the classical Dirichlet problem for L? In general this is a highly technical matter, but to cast some light on it, and also to show how the notion of a weak solution arises naturally, we consider the simple case when L is the Laplace operator Δ, and assume that all the functions appearing are real-valued. By way of preparation for this, recall that if X is a Banach space with norm $\|\cdot\|$ and $E : X \to \mathbb{R} \cup \{\infty\}$, then E is called Fréchet-differentiable at $x \in X$ if there exists an element $E'(x)$ of X^* such that

$$\lim_{\|h\| \to 0} \left| E(x+h) - E(x) - \langle h, E'(x) \rangle_X \right| / \|h\| = 0.$$

The functional E is said to be of class C^1 if it is Fréchet-differentiable at every point of X and the map $x \longmapsto E'(x)$ of X to X^* is continuous. We say that E is Gâteaux-differentiable at $x \in X$ if there exists $G(x) \in X^*$ such that for all $h \in X$,

$$\langle h, G(x) \rangle_X = \lim_{t \to 0} \frac{E(x+th) - E(x)}{t}.$$

Note that $\langle h, G(x) \rangle_X = \frac{d}{dt} E(x+th)|_{t=0}$: this may be thought of as a directional derivative of E. If E is Fréchet-differentiable at x, it is Gâteaux-differentiable at x and $G(x) = E'(x)$. If E has a minimum at $x \in X$ and E is Gâteaux-differentiable at x, then $G(x) = 0$.

Theorem 5.4.1 *Let M be a weakly closed subset of a reflexive Banach space X and let $E : M \to \mathbb{R} \cup \{\infty\}$ be coercive in the sense that $E(u) \to \infty$ as $\|u\| \to \infty$, $u \in M$; suppose that E is not identically equal to ∞ on M. Assume additionally that E is sequentially weakly lower-semi-continuous on M with respect to X; that is, for all $u \in M$ and all sequences (u_m) in M with $u_m \rightharpoonup u$,*

$$E(u) \leq \liminf_{m \to \infty} E(u_m).$$

Then E is bounded from below on M and attains its infimum on M.

Proof. Let $a = \inf \{E(u) : u \in M\}$ and let (u_m) be a minimising sequence: $E(u_m) \to a$. Since E is coercive, (u_m) is bounded; as X is reflexive, we may assume that $u_m \rightharpoonup u$ for some $u \in X$. Because M is weakly closed, $u \in M$; by weak lower-semi-continuity, $E(u) \leq \liminf_{m \to \infty} E(u_m) = a$. ∎

Now suppose that Ω is a bounded open subset of $\mathbb{R}^n$ and that the Sobolev space $X := \overset{0}{W}{}^1_2(\Omega)$ is endowed with the norm $\|\cdot\| : u \longmapsto \left(\int_\Omega |\nabla u(x)|^2 \, dx \right)^{1/2}$ (equivalent to the usual norm, by the Friedrichs inequality (Theorem 1.3.7)). Let $f \in L_2(\Omega)$ and consider the functional E defined on X by

$$E(u) = \frac{1}{2} \int_\Omega |\nabla u(x)|^2 \, dx - \int_\Omega f(x) u(x) \, dx.$$

Use of the Hölder and Friedrichs inequalities shows that, for some constant $c > 0$,

$$E(u) \geq \frac{1}{2} \|u\|^2 - c \|u\|,$$

5.4 A Little Regularity Theory

so that E is coercive on X. Now let (u_m) be a sequence in M with $u_m \rightharpoonup u$. Then $\int_\Omega f(x) u_m(x) dx \to \int_\Omega f(x) u(x) dx$ as $m \to \infty$ since $f \in L_2(\Omega)$ and so $f \in X^*$; also $\|u\| \le \liminf\limits_{m \to \infty} \|u_m\|$. Thus E is sequentially weakly lower-semi-continuous: by Theorem 5.4.1, there is a minimiser $u \in X$ of E. Since

$$\frac{d}{dt} E(u+th)|_{t=0} = \int_\Omega (\nabla u . \nabla h - fh)\, dx,$$

it follows that

$$\int_\Omega (\nabla u . \nabla h - fh)\, dx = 0 \text{ for all } h \in C_0^\infty(\Omega).$$

This means that u is a weak solution of the GDP for the Poisson equation $\Delta u = f$. Note that if $f = 0$, then by Theorem 2.2.10, u is equivalent to a function that is harmonic and hence analytic in Ω. This is a simple example of interior regularity, that is, smoothness of the variational or weak solution away from the boundary. Even in this special case it still remains to show that the classical boundary values are satisfied. To establish interior regularity of the weak solution when f is not zero it is convenient to begin with the case when Ω is the whole of $\mathbb{R}^n$.

Theorem 5.4.2 *Let $f \in W_2^m(\mathbb{R}^n)$ for some $m \in \mathbb{N}_0$ and let $u \in W_2^1(\mathbb{R}^n)$ be a weak solution of the equation $-\Delta u = f$ in $\mathbb{R}^n$. Then $u \in W_2^{m+2}(\mathbb{R}^n)$.*

Proof. First suppose that $f \in L_2(\mathbb{R}^n)$. Since u and the distributional Laplacian Δu belong to $L_2(\mathbb{R}^n)$ we may apply the Fourier transform, denoted by $\wedge$. Then $\xi \longmapsto (1 + |\xi|^2) \hat{u}(\xi) \in L_2(\mathbb{R}^n)$, which means that $u \in W_2^2(\mathbb{R}^n)$. If $f \in W_2^1(\mathbb{R}^n)$, then (in the sense of distributions) $-\Delta(D_i u) = D_i f \in L_2(\mathbb{R}^n)$ ($i = 1, ..., n$) and so $u \in W_2^3(\mathbb{R}^n)$. The general case follows by induction. ∎

Armed with this result we can now deal with the case in which Ω is bounded.

Theorem 5.4.3 *Let Ω be a bounded open subset of $\mathbb{R}^n$, suppose that $f \in W_2^m(\Omega)$ for some $m \in \mathbb{N}_0$ and let $u \in \overset{\circ}{W}_2^1(\Omega)$ be a weak solution of the Dirichlet problem*

$$-\Delta u = f \text{ in } \Omega, \ u = 0 \text{ on } \partial\Omega.$$

Let $x_0 \in \Omega$ and suppose that $B(x_0, 2r) \subset \Omega$. Then $u \in W_2^{m+2}(B(x_0, r))$.

Proof. Let $\psi \in C_0^\infty(\mathbb{R}^n)$ be such that

$$\psi(x) = 1 \text{ if } x \in B(x_0, r),\ \psi(x) = 0 \text{ if } x \in \mathbb{R}^n \setminus B(x_0, 2r).$$

Put $u_0 = \psi u$, extend u_0 to the whole of $\mathbb{R}^n$ by setting it equal to zero outside Ω and, for simplicity, denote this extension by u_0. Then (see [54], p. 253) $u_0 \in W_2^2(\mathbb{R}^n)$. We claim that in the weak sense

$$-\Delta u_0 = \psi f - 2\nabla \psi \cdot \nabla u - (\Delta \psi) u,$$

where $g := \psi f - 2\nabla\psi \cdot \nabla u - (\Delta\psi)u \in W_2^m(\mathbb{R}^n)$. To verify this, note that for all $\phi \in C_0^\infty(\mathbb{R}^n)$,

$$\int_{\mathbb{R}^n} g\phi dx = \int_{\mathbb{R}^n} \{f\psi - 2\nabla\psi \cdot \nabla u - (\Delta\psi)u\} \phi dx$$

$$= \int_\Omega \{f(\psi\phi) - 2(\nabla\psi \cdot \nabla u)\phi + (\nabla\psi \cdot \nabla u)\phi + (\nabla\psi \cdot \nabla\phi)u\} dx$$

$$= \int_\Omega \{\nabla u \cdot \nabla(\psi\phi) - (\nabla\psi \cdot \nabla u)\phi + (\nabla\psi \cdot \nabla\phi)u\} dx$$

$$= \int_\Omega \{(\nabla u \cdot \nabla\phi)\psi + (\nabla\psi \cdot \nabla\phi)u\} dx = \int_{\mathbb{R}^n} \nabla(u\psi) \cdot \nabla\phi dx,$$

as required. By Theorem 5.4.2 it follows that $u_0 \in W_2^{m+2}(\mathbb{R}^n)$, and so $u \in W_2^{m+2}(B(x_0, r))$. ∎

Corollary 5.4.4 *Let Ω be a bounded open subset of $\mathbb{R}^n$, suppose that $m > n/2$ and that $f \in W_2^m(\Omega)$; let $u \in \overset{0}{W}_2^1(\Omega)$ be a weak solution of the Dirichlet problem*

$$-\Delta u = f \text{ in } \Omega, \ u = 0 \text{ on } \partial\Omega.$$

Then $u \in C^2(\Omega)$.

Proof. By Theorem 5.4.3, for every small enough ball centred at $x_0 \in \Omega$ we have that $u \in W_2^{m+2}(B(x_0, r)) \subset C^2\left(\overline{B(x_0, r)}\right)$ (see Theorem 1.3.4). The result follows. ∎

Under the conditions of the Corollary, the weak solution of the problem actually satisfies the equation $-\Delta u = f$ in a classical sense. For given any $x_0 \in \Omega$ and any $r > 0$ such that $B(x_0, 2r) \subset \Omega$, the function $-\Delta u - f$ belongs to $L_2(B(x_0, r))$. Since

$$\int_{B(x_0,r)} (-\Delta u - f) h dx = \int_{B(x_0,r)} (\nabla u \cdot \nabla h - fh) dx = 0$$

for all $h \in C_0^\infty(B(x_0, r))$, and $C_0^\infty(B(x_0, r))$ is dense in $L_2(B(x_0, r))$, it follows that

$$\int_{B(x_0,r)} |\Delta u + f|^2 dx = 0.$$

Thus $-\Delta u = fu$ at all points of $B(x_0, r)$ and consequently throughout Ω. Given some smoothness of the boundary of Ω, information about the attainment of the boundary values of u can be obtained by somewhat similar techniques. As this is a rather technical matter, we do not go into this here but simply refer to [27], Chapter 9 for a thorough description of the procedure.

For details of the relationship between classical and generalised Dirichlet problems for more general elliptic operators we refer to [27], Chapter 9 and [85], 17. As we shall see later, this variational approach (essentially Dirichlet's principle) does not lead to the Laplace operator when we move outside a Hilbert space framework.

5.5 Eigenvalues of the Laplacian

Let Ω be a bounded domain in $\mathbb{R}^n$: all functions appearing in this section are assumed to be real-valued. We consider the Dirichlet eigenvalue problem

$$-\Delta u = \lambda u \text{ in } \Omega, \ u \upharpoonright_{\partial\Omega} = 0 \qquad (5.5.1)$$

in the weak sense discussed in the last section: a real number λ will be said to be an eigenvalue of this problem (in the weak sense) if there is a function $u \in \overset{0}{W}{}^1_2(\Omega) \setminus \{0\}$ such that

$$\int_\Omega \nabla u(x) \cdot \nabla \phi(x)\, dx = \lambda \int_\Omega u(x)\phi(x)\, dx \qquad (5.5.2)$$

for all $\phi \in \overset{0}{W}{}^1_2(\Omega)$. Such a function u is called an eigenvector corresponding to λ. For simplicity we shall write $H(\Omega)$ or even H instead of $\overset{0}{W}{}^1_2(\Omega)$. To analyse this problem we begin by noting that, by Theorem 5.3.1, given any $f \in L_2(\Omega)$, there is a unique weak solution of the Dirichlet problem

$$-\Delta u = f \text{ in } \Omega, \ u = 0 \text{ on } \partial\Omega;$$

that is, there is a unique $u \in H$ such that

$$\int_\Omega \nabla u(x) \cdot \nabla \phi(x)\, dx = \int_\Omega f(x)\phi(x)\, dx \text{ for all } \phi \in H.$$

Denote this unique function by Gf: thus G is a bounded linear map of $L_2(\Omega)$ into H. The natural embedding id: $H \to L_2(\Omega)$ is compact (see Theorem 1.3.5); thus $T :=$ id $\circ\ G$ is a compact linear map from $L_2(\Omega)$ to $L_2(\Omega)$. Moreover, T is self-adjoint and positive: for given any $f, g \in L_2(\Omega)$,

$$\int_\Omega gTf\,dx = \int_\Omega gGf\,dx = \int_\Omega \nabla(Gf) \cdot \nabla(Gg)\, dx = \int_\Omega fTg\,dx$$

and

$$\int_\Omega fTf\,dx = \int_\Omega \nabla(Gf) \cdot \nabla(Gf)\, dx > 0 \text{ if } f \ne 0.$$

Hence from the theory of compact, positive self-adjoint operators (see, for example, [54], II.5), there exists an orthonormal basis of eigenfunctions $\{w_m\}$ of $L_2(\Omega)$ and a corresponding sequence $\{\mu_m\}$ of positive eigenvalues of T with $\mu_m \to 0$:

$$Tw_m = \mu_m w_m \ (m \in \mathbb{N}).$$

Put $\lambda_m = 1/\mu_m$: then $w_m = T(\lambda_m w_m)$, $w_m \in H$ and

$$\int_\Omega \nabla w_m(x) \cdot \nabla \phi(x)\, dx = \lambda_m \int_\Omega w_m(x)\phi(x)\, dx \text{ for all } \phi \in H,$$

so that each w_m is an eigenvector of the Laplacian with correponding eigenvalue λ_m. By proceeding as in the proof of Theorem 5.4.3 it follows that $w_m \in H \cap C^\infty(\Omega)$; moreover, $-\Delta w_m = \lambda_m w_m$ in a classical sense at all points of Ω. We summarise these arguments as follows.

Theorem 5.5.1 *There is an orthonormal basis $\{w_m\}$ of $L_2(\Omega)$ and a corresponding non-decreasing sequence $\{\lambda_m\}$ of positive real numbers, with $\lambda_m \to \infty$ as $m \to \infty$, such that for all $m \in \mathbb{N}$ and all $\phi \in H$,*

$$\int_\Omega \nabla w_m(x) \cdot \nabla \phi(x)\, dx = \lambda_m \int_\Omega w_m(x)\phi(x) dx.$$

Each w_m belongs to $H \cap C^\infty(\Omega)$ and $-\Delta w_m = \lambda_m w_m$ at all points of Ω.

Note in addition that if Ω has smooth enough boundary, then each w_m belongs to $C(\overline{\Omega})$ and is zero on $\partial\Omega$: see [27], Chapter 9 and [85], 17.

We thus have countably many eigenvector-eigenvalue pairs (w_m, λ_m) corresponding to the weak form of the Dirichlet eigenvalue problem. By arguing as in [54], Corollary II.5.4 it can be seen that these are the only such pairs corresponding to this problem. We also claim that when H is provided with the inner product

$$(u, v)_H := \int_\Omega \nabla u \cdot \nabla v\, dx,$$

the functions $\lambda_m^{-1/2} w_m$ form an orthonormal basis of H. To see this, observe that

$$\left(\lambda_k^{-1/2} w_k, \lambda_m^{-1/2} w_m\right)_H = \lambda_k^{-1/2} \lambda_m^{-1/2} \int_\Omega \nabla w_k \cdot \nabla w_m\, dx = \lambda_k^{1/2} \lambda_m^{-1/2} \int_\Omega w_k w_m\, dx$$
$$= \delta_{km}.$$

Moreover, if $u \in H$ is such that $(u, w_m)_H = 0$ all $m \in \mathbb{N}$, then

$$0 = \int_\Omega \nabla u \cdot \nabla w_m\, dx = \lambda_m \int_\Omega u w_m\, dx.$$

Since the w_m form a basis of $L_2(\Omega)$ it follows that $u = 0$ and the claim is substantiated.

Next, with an eye to future work on the p-Laplacian, we give a variational characterisation of the pairs (w_m, λ_m). This involves the **Rayleigh quotient,** by which is meant

$$R(v) := \frac{\int_\Omega \nabla v \cdot \nabla v\, dx}{\int_\Omega |v|^2 dx} \quad (v \in H \backslash \{0\}). \tag{5.5.3}$$

5.5 Eigenvalues of the Laplacian

Theorem 5.5.2 *(The Courant min-max principle) For each $k \in \mathbb{N}$ let $W_k := sp\{w_1, ..., w_k\}$. Then*

$$\lambda_1 = \min_{v \in H \setminus \{0\}} R(v) = R(w_1), \tag{5.5.4}$$

and for $m \geq 2$,

$$\lambda_m = R(w_m) = \max_{v \in W_m \setminus \{0\}} R(v), \tag{5.5.5}$$

$$\lambda_m = \min_{v \perp W_{m-1}, v \neq 0} R(v) \tag{5.5.6}$$

and

$$\lambda_m = \min_{W \subset H, \dim W = m} R(v). \tag{5.5.7}$$

Proof. For each $k \in \mathbb{N}$, let $W_k = sp\{w_1, ..., w_k\}$; note that $\lambda_k = R(w_k)$. Suppose that $v \perp W_{m-1}$ (in the sense of $L_2(\Omega)$). Then the 'Fourier' expansion of v in $L_2(\Omega)$ is

$$v = \sum_{k=m}^{\infty} \alpha_k w_k, \quad \alpha_k = \int_{\Omega} v w_k \, dx.$$

Thus $v_l := \sum_{k=m}^{l} \alpha_k w_k \to v$ in $L_2(\Omega)$. In fact, $v_l \to v$ also in H: for the Fourier expansion of v in H is

$$v = \sum_{k=m}^{\infty} \lambda_k^{-1/2} \beta_k w_k,$$

where

$$\beta_k = \lambda_k^{-1/2} \int_{\Omega} \nabla v \cdot \nabla w_k \, dx = \lambda_k^{1/2} \int_{\Omega} v w_k \, dx = \lambda_k^{1/2} \alpha_k,$$

and so $v = \sum_{k=m}^{\infty} \alpha_k w_k$ in the sense of H. It follows that $R(v_l) \to R(v)$. Since

$$R(v_l) = \frac{\sum_{k=m}^{l} \alpha_k^2 \int_{\Omega} \nabla w_k \cdot \nabla w_k \, dx}{\sum_{k=m}^{l} \alpha_k^2} = \frac{\sum_{k=m}^{l} \alpha_k^2 \lambda_k}{\sum_{k=m}^{l} \alpha_k^2} \geq \lambda_m,$$

we see that $R(v) \geq \lambda_m$, which gives (5.5.6) as the minimum is attained at w_m; putting $m = 1$ we obtain (5.5.4).

To derive (5.5.5), observe that if $v \in W_m$ then $v = \sum_{k=1}^{m} \alpha_k w_k$ and

$$R(v) = \frac{\sum_{k=1}^{m} \alpha_k^2 \lambda_k}{\sum_{k=1}^{m} \alpha_k^2} \leq \lambda_m,$$

which shows that the maximum of $R(v)$ over W_m is at most λ_m; as this upper bound is attained at $w_m \in W_m$, the desired result follows. Lastly, let W be a linear subspace of H with dimension m. Then there exists $w \in W$ such that $w \perp w_j$ for $j = 1, ..., m-1$. Hence $R(w) \geq \lambda_m$ and so $\max_{v \in W} R(v) \geq \lambda_m$. Thus

$$\min_{\dim W=m} \max_{v \in W} R(v) \geq \lambda_m$$

and the maximum of $R(v)$ over the m−dimensional subspace W_m is exactly λ_m. Now (5.5.7) is immediate. ∎

Lemma 5.5.3 *Let $w \in H\backslash\{0\}$ be such that $R(w) = \lambda_1$. Then w is an eigenvector corresponding to λ_1.*

Proof. Without loss of generality we may assume that $\int_\Omega |w|^2 dx = 1$. Let $v \in H$ and $t > 0$. Then $w + tv \in H$ and $R(w + tv) \geq R(w)$; that is,

$$\frac{\int_\Omega \nabla(w+tv) \cdot \nabla(w+tv)\, dx}{\int_\Omega |(w+tv)|^2 dx} \geq \int_\Omega \nabla w \cdot \nabla w\, dx = \lambda_1.$$

Hence

$$\frac{t}{2}\int_\Omega \nabla v \cdot \nabla v\, dx + \int_\Omega \nabla w \cdot \nabla v\, dx \geq \lambda_1 \left(\int_\Omega wv\, dx + \frac{t}{2}\int_\Omega |v|^2 dx \right).$$

Let $t \to 0$: then $\int_\Omega \nabla w \cdot \nabla v\, dx \geq \lambda_1 \int_\Omega wv\, dx$. Repetition of this argument with $t < 0$ gives the opposite inequality and we conclude that $\int_\Omega \nabla w \cdot \nabla v\, dx = \lambda_1 \int_\Omega vw\, dx$. As this holds for all $v \in H$ the result follows. ∎

Theorem 5.5.4 *The first eigenvalue λ_1 is simple and the corresponding eigenvector does not change sign in Ω.*

We shall not prove this result here as it will be established in the more general setting of the p−Laplacian in Theorem 9.4.6, with a proof essentially identical to that for the case of the Laplacian.

As we have seen above, eigenvectors of the Dirichlet Laplacian in Ω belong to $C^\infty(\Omega)$. More can be said, for they are in fact analytic in Ω. This follows from the next result.

Theorem 5.5.5 *Let Ω be an open subset of $\mathbb{R}^n$, let $\lambda \in \mathbb{R}$ and suppose that $u \in C^\infty(\Omega)$ satisfies $\Delta u = \lambda u$ in Ω. Then u is (real) analytic in Ω.*

Proof. Let B, B_1 be concentric open balls, with centre x, such that $\overline{B} \subset B_1 \subset \Omega$; let $m \in \mathbb{N}$, $m > n/2$, and let $r \in C_0^\infty(B_1)$ be such that $0 \leq r \leq 1$ and $r = 1$ in a neighbourhood of B. We claim that there are positive constants A (≥ 1) and C such that for all $\alpha \in \mathbb{N}_0^n$,

$$\left\| r^{|\alpha|} D^\alpha u \right\|_{m,2,B_1} \leq C A^{|\alpha|} |\alpha|!. \tag{5.5.8}$$

Accepting this for the moment, it follows from the embedding of $W_2^m(B)$ in $C\left(\overline{B}\right)$ that there is a constant C' such that for all $\alpha \in \mathbb{N}_0^n$,

$$\sup_{y \in B} |D^\alpha u(y)| \leq C' A^{|\alpha|} |\alpha|!.$$

5.5 Eigenvalues of the Laplacian

If we can show that the Taylor series

$$\sum_{k=0}^{\infty} \sum_{|\alpha|=k} \frac{(y-x)^\alpha}{\alpha!} D^\alpha u(x)$$

is absolutely and uniformly convergent on B, when B has a sufficiently small radius δ, then it will follow that u is analytic in Ω. To do this we use the multinomial theorem

$$\sum_{|\alpha|=k} \frac{k!}{\alpha!} x^\alpha = \left(\sum_{i=1}^n x_i\right)^k$$

with each $x_i = 1$ to obtain

$$\sum_{|\alpha|=k} \frac{k!}{\alpha!} = n^k.$$

Thus for all $x \in B$,

$$\left| \sum_{k=K}^{\infty} \sum_{|\alpha|=k} \frac{(y-x)^\alpha}{\alpha!} D^\alpha u(x) \right| \leq C' \sum_{k=K}^{\infty} (\delta A n)^k,$$

from which the claim of analyticity follows, δ being taken so that $\delta A n < 1$.

It remains to prove (5.5.8). We use induction on $|\alpha|$ together with the fact that for all $v \in \overset{0}{W}_2^{m+2}(B_1)$ and $\alpha \in \mathbb{N}_0^n$, $|\alpha| = 2$,

$$\|D^\alpha v\|_{m,2,B_1} \leq \|\Delta v\|_{m,2,B_1}, \tag{5.5.9}$$

which follows since

$$\|D^\alpha v\|_{m,2,B_1} = \|D^\alpha v\|_{m,2,\mathbb{R}^n} = \left\|(1+|\xi|^2)^{m/2} \xi^\alpha \widehat{v}(\xi)\right\|_{2,\mathbb{R}^n}$$
$$\leq \left\|(1+|\xi|^2)^{m/2} |\xi|^2 \widehat{v}(\xi)\right\|_{2,\mathbb{R}^n} = \|\Delta v\|_{m,\mathbb{R}^n} = \|\Delta v\|_{m,\Omega}.$$

For the induction process we note that (5.5.8) is plainly true when $|\alpha| \leq 1$, if C is large enough: henceforth we suppose that such a choice of C has been made. Now assume that (5.5.8) holds when $|\alpha| \leq N$, where $N \geq 1$. For simplicity we shall denote the norm on $W_2^m(B_1)$ by $\|\cdot\|$. Use of (5.5.9) shows that if $|\alpha| = N-1$ and $|\beta| = 2$, then

$$\begin{aligned}
\|r^{N+1} D^{\alpha+\beta} u\| &\leq \|D^\beta (r^{N+1} D^\alpha u)\| + \|D^\beta (r^{N+1} D^\alpha u) - r^{N+1} D^{\alpha+\beta} u\| \\
&\leq \|\Delta (r^{N+1} D^\alpha u)\| + \|D^\beta (r^{N+1} D^\alpha u) - r^{N+1} D^{\alpha+\beta} u\| \\
&\leq \|r^{N+1} D^\alpha \Delta u\| + \|\Delta (r^{N+1} D^\alpha u) - r^{N+1} (\Delta D^\alpha u)\| \\
&\quad + \|D^\beta (r^{N+1} D^\alpha u) - r^{N+1} D^{\alpha+\beta} u\| \\
&= I_1 + I_2 + I_3.
\end{aligned}$$

To handle these terms we note that, using Leibniz's theorem, there is a constant C_1 such that for all $f, g \in W_2^m(B_1)$,

$$\|fg\| \leq C_1 \|f\| \|g\|. \tag{5.5.10}$$

For I_3, we put $D^\beta = D_j D_k$ and observe that

$$\begin{aligned} D^\beta \left(r^{N+1} D^\alpha u\right) - r^{N+1} D^{\alpha+\beta} u &= (N+1) \left\{r^N (D_j r) D_k + r^N (D_k r) D_j\right\} D^\alpha u \\ &\quad + (N+1) r^N (D^\beta r) D^\alpha u \\ &\quad + (N+1) N r^{N-1} (D_j r)(D_k r) D^\alpha u. \end{aligned}$$

Thus with the aid of (5.5.10) and the inductive hypothesis,

$$\begin{aligned} I_3 &\leq (N+1) C_1 \left\{\|D_j r\| \|r^N D_k D^\alpha u\| + \|D_k r\| \|r^N D_j D^\alpha u\|\right\} \\ &\quad + (N+1) C_1 \|r D^\beta r\| \|r^{N-1} D^\alpha u\| + (N+1) N C_1 \|(D_j r) D_k r\| \|r^{N-1} D^\alpha u\| \\ &\leq 4 C_1 C_2 C A^N (N+1)!, \end{aligned}$$

where C_2 is a constant depending only on norms involving r and its derivatives. The choice $A \geq 12 C_1 C_2$ gives $I_3 \leq C A^{N+1}(N+1)!/3$. A similar procedure establishes the same estimate for I_2, this time requiring $A \geq 12n C_1 C_2$. Finally, for I_1, we use (5.5.10) and the inductive hypothesis to obtain

$$I_1 \leq |\lambda| \|r^{N+1} D^\alpha u\| \leq |\lambda| \|r^2\| \|r^{N-1} D^\alpha u\| \leq C A^{N+1}(N+1)!/3,$$

choosing A suitably large. Thus

$$\|r^{N+1} D^{\alpha+\beta} u\| \leq C A^{N+1}(N+1)!$$

when A is appropriately chosen, and the proof is complete. ∎

We conclude this chapter with a celebrated result, the Faber-Krahn inequality, concerning the first eigenvalue $\lambda_1(\Omega)$ of the Dirichlet Laplacian on a bounded open set Ω. To do this it is convenient to recall some definitions and results. Given any $u \in W_2^1(\Omega)$, its *distribution function* $\mu_u : [0, \infty) \to [0, \infty]$ is defined by

$$\mu_u(\lambda) = |\{x \in \Omega : |u(x)| > \lambda\}|_n; \tag{5.5.11}$$

and we know from 1.3 that its *non-increasing rearrangement* is the function $u^* : [0, \infty) \to [0, \infty]$ given by

$$u^*(t) = \inf \{\lambda \in [0, \infty) : \mu_u(\lambda) \leq t\}. \tag{5.5.12}$$

Its *symmetric rearrangement* $u^\star$ is defined by

5.5 Eigenvalues of the Laplacian

$$u^\star(x) = u^*\left(\omega_n |x|^n\right) \ (x \in \Omega^\star), \tag{5.5.13}$$

where $\Omega^\star$ is the ball in $\mathbb{R}^n$ centred at 0 and with the same $n-$measure as Ω. For further details see, for example, [55], 3.2. Note in particular that u and $u^\star$ have the same distribution function and that $\int_{\Omega^\star} |u^\star|^2 dx = \int_{\Omega} |\nabla u|^2 dx$. The *Pólya-Szegö principle* (see [55], p. 84) asserts that for all $u \in \overset{0}{W}{}_2^1(\Omega)$, we have $u^\star \in \overset{0}{W}{}_2^1(\Omega^\star)$ and

$$\int_{\Omega^\star} |\nabla u^\star|^2 dx \leq \int_{\Omega} |\nabla u|^2 dx. \tag{5.5.14}$$

A well-known result of Brothers and Ziemer [28] implies that if $u \in \overset{0}{W}{}_2^1(\Omega)$,

$$E_t := \{x \in \mathbb{R}^n : |u(x)| > t\} \ (t \geq 0)$$

and equality holds in (5.5.14), then

$$E_t \text{ is equivalent to an open ball for each } t \geq 0 \tag{5.5.15}$$

(that is, it coincides with a ball except for a set of zero $n-$measure).

Theorem 5.5.6 *Let Ω be a bounded, open, connected subset of $\mathbb{R}^n$. Then $\lambda_1(\Omega) \geq \lambda_1(\Omega^\star)$; in the case of equality, Ω is a ball.*

Proof. The first eigenvalue $\lambda_1(\Omega)$ is given by

$$\lambda_1(\Omega) = \inf \left\{ \frac{\int_\Omega |\nabla u|^2 dx}{\int_\Omega |u|^2 dx} : u \in \overset{0}{W}{}_2^1(\Omega) \right\} \tag{5.5.16}$$

and corresponds to an element u of $\overset{0}{W}{}_2^1(\Omega)$ which, without loss of generality, we shall assume to be positive in Ω. That $\lambda_1(\Omega) \geq \lambda_1(\Omega^\star)$ is an immediate consequence of the Pólya-Szegö principle.

Now suppose that $\lambda_1(\Omega) = \lambda_1(\Omega^\star)$. By (5.5.15), $\Omega = \{x \in \Omega : u(x) > 0\}$ coincides with an open ball apart from a set of zero measure; let K be this ball. Since Ω is open and connected and $|K|_n = |\Omega|_n$, we must have $\Omega \subset K$: thus $\Omega \cup S = K$ for some set S with zero $n-$measure. Since the eigenvector u corresponding to $\lambda_1(\Omega)$ belongs to $\overset{0}{W}{}_2^1(\Omega)$, its extension by zero to K must belong to $\overset{0}{W}{}_2^1(K)$ and so $\lambda_1(\Omega) \geq \lambda_1(K) = \lambda_1(\Omega^\star) = \lambda_1(\Omega)$. Hence u is the first eigenfunction of the Dirichlet Laplacian on the ball K, and therefore is simple and radial. It follows that all the level sets of u are balls and thus $\Omega = K$. ∎

We observe that scaling arguments show that Theorem 5.5.6 can be equivalently rewritten as

$$|\Omega|_n^{2/n} \lambda_1(\Omega) \geq |B|_n^{2/n} \lambda_1(B)$$

for all open balls B, with equality if and only if Ω is a ball.

5.6 Spectral Independence

We give a brief account of the way in which semigroup theory can be used to show that various realisations of suitable differential operators have the same spectrum. Proofs of some of the more elementary assertions are given largely to acclimatise the reader to the style of argument; results requiring more sophisticated proof are simply stated and complete references given. First, some basic definitions and facts.

Let X be a Banach space. A C_0−*semigroup* on X is a family of operators $T = \{T(t) : t \geq 0\} \subset B(X)$ such that

- (i) $T(s)T(t) = T(s+t)$ for all $s, t \geq 0$,
- (ii) $T(0) = I$, the identity map,
- (iii) for each $f \in X$, $T(\cdot)f : [0, \infty) \to X$ is continuous.

If in addition $\|T(t)\| \leq 1$ for all $t \geq 0$, then T is said to be a C_0− *contraction semigroup*.

The *infinitesimal generator* of a C_0−semigroup on X is the map $A : D(A) \to X$ defined by

$$A(f) = \lim_{t \to 0} \frac{T(t)f - f}{t} = \frac{d}{dt} T(t) f \mid_{t=0} \quad (f \in D(A)),$$

where $D(A)$ is the set of all $f \in X$ for which the above limit exists. Given a map $A \in B(X)$, the definition of e^{tA} as $\sum_{n=0}^{\infty} (tA)^n / n!$ makes sense and it is easy to see that

$$T := \{T(t) = e^{tA} : t \geq 0\}$$

is a C_0−semigroup on X that satisfies (iii) in the stronger form

(iii′) $\|T(t) - I\| \to 0$ as $t \to 0$.

It can be shown that, conversely, if T is a C_0−semigroup on X that satisfies (iii′), then its generator A belongs to $B(X)$ and $T(t) = e^{tA}$ for all $t \geq 0$: see [91], p. 15.

Theorem 5.6.1 *Let T be a C_0− semigroup on X. Then there exist $M \geq 1$ and $\omega \in \mathbb{R}$ such that*

$$\|T(t)\| \leq M e^{\omega t} \text{ for all } t \geq 0.$$

Proof. We first claim that there exist $M \geq 1$ and $\delta > 0$ such that $\|T(t)\| \leq M$ if $0 \leq t \leq \delta$. For if not, then there is a sequence $\{t_n\}$, with $t_n \to 0$, such that $\|T(t_n)\| \to \infty$.

5.6 Spectral Independence

However, since $T(t_n)f \to f$ for all $f \in X$, it follows that $\{T(t_n)f\}$ is bounded for every $f \in X$, so that by the Banach-Steinhaus theorem, $\{\|T(t_n)\|\}$ is bounded: contradiction. Set $\omega := \delta^{-1} \log M \geq 0$ and let $t \geq 0$. Then there exist $n \in \mathbb{N}_0$ and $\eta \in [0, \delta)$ such that $t = n\delta + \eta$. As $T(t) = (T(\delta))^n T(\eta)$, we have

$$\|T(t)\| \leq \|T(\delta)\|^n \|T(\eta)\| \leq M^{n+1} \leq Me^{\omega t}$$

since $\log(M^n) = n \log M = n\omega\delta \leq \omega t$. ∎

Corollary 5.6.2 *For every $f \in X$ the mapping $t \mapsto T(t)f : [0, \infty) \to X$ is continuous.*

Proof. Let $h \geq 0$. Then

$$\|T(t+h)f - T(t)f\| \leq \|T(t)\|\|T(h)f - f\| \leq Me^{\omega t}\|T(h)f - f\| \to 0$$

as $h \to 0^+$. Similarly,

$$\|T(t)f - T(t-h)f\| \leq \|T(t-h)\|\|T(h)f - f\| \leq Me^{\omega(t-h)}\|T(h)f - f\| \to 0$$

as $h \to 0^+$. ∎

The integrals of Banach space-valued functions in what follows are Bochner integrals, defined in the same way as the Riemann integrals of real-valued functions: see, for example, [222], pp. 132-136.

Lemma 5.6.3 *For every $f \in X$ and all $t \geq 0$,*

$$\lim_{h \to 0^+} \frac{1}{h} \int_t^{t+h} T(s)f\,ds = T(t)f.$$

Proof. Clearly

$$\left\| \frac{1}{h} \int_t^{t+h} T(s)f\,ds - T(t)f \right\| = \left\| \frac{1}{h} \int_t^{t+h} (T(s) - T(t))f\,ds \right\|$$

$$\leq \frac{1}{h} \int_t^{t+h} \|T(s)f - T(t)f\|\,ds.$$

Given $\varepsilon > 0$, let h be so small that $\|T(s)f - T(t)f\| < \varepsilon$ when $|t - s| < h$. Then

$$\left\| \frac{1}{h} \int_t^{t+h} T(s)f\,ds - T(t)f \right\| < \varepsilon,$$

and the proof is complete. ∎

Theorem 5.6.4 *Let T be a C_0- semigroup on X with infinitesimal generator A, and let $f \in D(A)$. Then*

$$\frac{d}{dt}T(t)f = AT(t)f = T(t)Af.$$

Proof. By the definition of A,

$$\left(\frac{T(h)-I}{h}\right)T(t)f = T(t)\left(\frac{T(h)-I}{h}\right)f \to T(t)Af \text{ as } t \to 0^+.$$

Hence $T(t)f \in D(A)$ and

$$AT(t)f = T(t)Af = D^+T(t)f,$$

where

$$D^+T(t)f = \lim_{h \to 0^+}\left(\frac{T(h)-I}{h}\right)T(t)f;$$

by $D^-T(t)f$ we shall later mean the corresponding limit as $h \to 0^-$. Note that

$$\frac{T(t)f - T(t-h)f}{h} - T(t)Af = T(t-h)\left(\frac{T(h)f - f}{h} - Af\right) + (T(t-h) - T(t))Af.$$

Moreover,

$$\left\|T(t-h)\left(\frac{T(h)f-f}{h} - Af\right)\right\| \le Me^{\omega t}\left\|\frac{T(h)f-f}{h} - Af\right\| \to 0 \text{ as } h \to 0^+,$$

and

$$\|(T(t-h) - T(t))Af\| \to 0 \text{ as } h \to 0^+.$$

Hence

$$D^-T(t)f = T(t)Af = D^+T(t)f,$$

which completes the proof. ∎

Theorem 5.6.5 *Let T be a C_0- semigroup on X with infinitesimal generator A and let $f \in X$. Then for all $t \ge 0$, $\int_0^t T(s)f ds \in D(A)$ and*

$$A\left(\int_0^t T(s)f ds\right) = T(t)f - f.$$

Proof. Let $h > 0$. Then by Lemma 5.6.3,

5.6 Spectral Independence

$$\left(\frac{T(h) - I}{h}\right) \int_0^t T(s)f\,ds = \frac{1}{h} \int_0^t (T(s+h)f - T(s)f)\,ds$$

$$= \frac{1}{h}\left\{\int_t^{t+h} T(s)f\,ds - \int_0^h T(s)f\,ds\right\}$$

$$\to T(t)f - f \text{ as } h \to 0^+.$$

Hence $\int_0^t T(s)f\,ds \in D(A)$ and the result follows. ∎

Corollary 5.6.6 *Let T be a C_0– semigroup on X with infinitesimal generator A. Then A is closed and $D(A)$ is dense in X.*

Proof. Let $f \in X$. By Theorem 5.6.5 and Lemma 5.6.3, $\int_0^t T(s)f\,ds \in D(A)$ and $\frac{1}{h}\int_0^h T(s)f\,ds \to T(0)f = f$: thus $D(A)$ is dense in X. Now suppose that $\{f_n\} \subset D(A)$ is such that $f_n \to f$ and $Af_n \to g$ in X. Then by Theorem 5.6.4,

$$\left(\frac{T(h)-I}{h}\right)f = \lim_{n\to\infty} \left(\frac{T(h)-I}{h}\right)f_n = \lim_{n\to\infty} \frac{1}{h}\int_0^h T(s)Af_n\,ds$$

$$= \frac{1}{h}\int_0^h T(s)g\,ds \to g$$

as $h \to 0^+$. Hence $f \in D(A)$ and $Af = g$: A is closed. ∎

Theorem 5.6.7 *Let T_1 and T_2 be C_0–semigroups on X with the same infinitesimal generator A. Then $T_1 = T_2$.*

Proof. Let $f \in D(A)$. Then

$$\frac{d}{ds}\{T_1(t-s)T_2(s)f\} = -AT_1(t-s)T_2(s)f + T_1(t-s)AT_2(s)f = 0.$$

Thus $T_1(t-s)T_2(s)f$ must be constant, as a function of s, and so $T_2(t)f = T_1(t)f$ for all $t \geq 0$. Since $D(A)$ is dense in X the result follows. ∎

Now we give the celebrated Hille-Yosida theorem.

Theorem 5.6.8 *(Hille-Yosida). A map $A : D(A) (\subset X) \to X$ is the generator of a C_0– contraction semigroup if and only if A is closed, densely defined and*

$$\|(\lambda I - A)^{-1}\| \leq 1/\lambda \text{ for all } \lambda > 0.$$

If the hypothesis of contractivity is omitted, so that we merely have $\|T(t)\| \leq Me^{\omega t}$ for all $t \geq 0$, then the theorem holds in the following modified form.

Theorem 5.6.9 *A map $A : D(A) \to X$ is the generator of a C_0-semigroup T if and only if A is closed, densely defined and constants $M \geq 1, \omega \in \mathbb{R}$ exist such that $\lambda \in \rho(A)$ for each $\lambda > \omega$ and*

$$\|(\lambda I - A)^{-n}\| \leq M(\lambda - \omega)^{-n} \text{ for all } \lambda > \omega \text{ and all } n \in \mathbb{N}_0.$$

In this case, $\|T(t)\| \leq Me^{\omega t}$ for all $t \geq 0$.

For proofs of these results see, for example, [91] and [127].

Now let T be a contraction C_0-semigroup with infinitesimal generator A and set $R(\lambda) = (\lambda I - A)^{-1}$ and $A_\lambda = \lambda AR(\lambda) = \lambda^2 R(\lambda) - \lambda I, \lambda > 0$. It is straightforward to show that for all $f \in X$,

$$\lim_{\lambda \to \infty} \lambda R(\lambda) f = f \text{ and } \lim_{\lambda \to \infty} A_\lambda f = Af.$$

Using the Hille-Yosida theorem it can be shown that for all $f \in X$ and all $t \geq 0$,

$$T(t)f = \lim_{n \to \infty} \left(I - \frac{t}{n} A\right)^{-n} f = \lim_{n \to \infty} \left(\frac{n}{t} R\left(\frac{n}{t}\right)\right)^n f.$$

Because of the first of these equalities, this result is called an exponential formula for $T(t)$ and is often written as $T(t)f = e^{tA} f$ just as in the case when A is bounded.

Suppose that X, Y are Banach spaces, both continuously embedded in some topological vector space Z. We say that maps $B_X \in B(X)$ and $B_Y \in B(Y)$ are **consistent** if

$$B_X x = B_Y x \text{ for all } x \in X \cap Y.$$

Let T_X, T_Y be C_0-semigroups on X, Y respectively with generators A_X, A_Y respectively. Then T_X, T_Y are said to be consistent if $T_X(t)$ and $T_Y(t)$ are consistent for all $t \geq 0$.

Let Ω be an open subset of $\mathbb{R}^n$, and let T be a C_0-semigroup on $L_2(\Omega)$ with generator A. We identify $L_2(\Omega)$ with a subspace of $L_2(\mathbb{R}^n)$ by extending functions by 0 outside Ω. For each $p \in [1, \infty)$, let G_p be the *Gaussian semigroup* on $L_p(\mathbb{R}^n)$: this is defined by $G_p(t)f = k_t * f$, where

$$k_t(x) = \frac{1}{(4\pi t)^{n/2}} \exp\left(-\frac{x^2}{4t}\right) \quad (x \in \mathbb{R}^n, t \geq 0).$$

Verification that G_p really is a C_0-semigroup is best achieved by using the Fourier transform.

Definition 5.6.10 *The semigroup T is said to satisfy an **upper Gaussian estimate** if there exist constants $b > 0$ and $c \geq 1$ such that for all $f \in L_2(\Omega)$ and all $t \in [0, 1]$,*

$$|T(t)f| \leq cG_2(bt)|f|.$$

5.6 Spectral Independence

Lemma 5.6.11 *Suppose that T is a C_0-semigroup on $L_2(\Omega)$ that satisfies an upper Gaussian estimate with constants b, c as in the definition. Then for all $t \geq 0$ and all $f \in L_2(\Omega)$,*
$$|T(t)f| \leq ce^{wt}G_2(bt)|f|, \text{ where } w = \log c.$$

Proof. Plainly
$$|T(n)f| = |T(1)^n f| \leq c^n G_2(bn)|f| \quad (n \in \mathbb{N}).$$

Given $t > 1$, let $n \in \mathbb{N}$ be such that $s := t - n \in [0, 1)$, and observe that

$$|T(t)f| = |T(n)T(s)f| \leq c^n G_2(bn)cG_2(bs)|f|$$
$$= c^{n+1}G_2(bt)|f| = ce^{nw}G_2(bt)|f|$$
$$\leq ce^{wt}G_2(bt)|f|.$$

∎

Lemma 5.6.12 *Let T be as in the last lemma. Then there are consistent semigroups T_p on $L_p(\Omega)$ ($p \in [1, \infty)$) such that $T = T_2$ and*
$$|T_p(t)f| \leq ce^{wt}G_p(bt)|f| \; (f \in L_p(\Omega), t \geq 0).$$

Proof. Since $G_2(t)$ has bounded extensions to each $L_p(\mathbb{R}^n)$, there exist consistent operators $T_p(t) \in B(L_p(\Omega))$ such that $T_2(t) = T(t)$ ($t \geq 0$). The semigroup property follows from a density argument; it remains to establish strong continuity. Let $p \in [1, \infty)$. It is enough to prove that for each $f \in L_p(\Omega) \cap L_2(\Omega)$, $T(t)f \to f$ in $L_p(\Omega)$ as $t \to 0^+$. Put $f_n = T(t_n)f$ and $g_n = ce^{wt_n}G(gt_n)|f|$, where $t_n \to 0^+$. As $f_n \to f$ in $L_2(\Omega)$, we may suppose that $f_n \to f$ a.e.; and by passage to a subsequence, we can assume that $\|g_n - g_{n-1}\|_p \leq 2^{-n}$. Define
$$h = \sum_{n \geq 2}|g_n - g_{n-1}| + |g_1|,$$
and observe that $h \in L_p(\mathbb{R}^n)$ and $|f_n| \leq g_n \leq h$ ($n \in \mathbb{N}$). Use of the dominated convergence theorem now shows that $f_n \to f$ in $L_p(\Omega)$. ∎

Theorem 5.6.13 *Let T be a C_0-semigroup on $L_2(\Omega)$ that satisfies an upper Gaussian estimate, let T_p be the consistent semigroups on $L_p(\Omega)$ ($p \in [1, \infty)$) whose existence is guaranteed by the last lemma, and let A_p be the generator of T_p. Let $\rho_\infty(A_p)$ be the connected component of $\rho(A_p)$ (the resolvent set of A_p) that contains $\{\lambda \in \mathbb{C} : \text{re }\lambda > w\}$ for some $w \in \mathbb{R}$. Then $\rho_\infty(A_p)$ is independent of the particular $p \in [1, \infty)$.*

The proof of this result is given in [7].

Corollary 5.6.14 *Suppose that in addition to the hypotheses of the last theorem, the generator A of T is self-adjoint. Then the spectrum $\sigma(A_p)$ of A_p is independent of $p \in [1, \infty)$.*

Proof. Since A is self-adjoint, $\sigma(A_2) \subset \mathbb{R}$; together with the Hille-Yosida theorem this implies that for some $w \in \mathbb{R}$, $\{\lambda \in \mathbb{C} : \text{re } \lambda > w\}$ lies in the resolvent set of A_2, and hence $\rho_\infty(A_2) = \rho(A_2)$. Thus by Theorem 5.6.13, $\rho_\infty(A_p) = \rho(A_2)$ for all $p \in [1, \infty)$, which implies that $\sigma(A_p) \subset \mathbb{R}$: hence $\rho(A_p) = \rho_\infty(A_p)$ and the result follows. ∎

The question of spectral independence is one that has attracted great interest: for a comprehensive account, with many references, see [163].

One of the most interesting applications of this result is to the Laplacian. Let Ω be a bounded open subset of $\mathbb{R}^n$ $(n \geq 2))$. The Dirichlet Laplacian Δ^D is defined in $L_2(\Omega)$ as the map with domain $\mathcal{D}(\Delta^D) := \left\{ u \in \overset{0}{W}{}_2^1(\Omega) : \Delta u \in L_2(\Omega) \right\}$ (where Δu is to be understood in the distributional sense) given by $\Delta^D u = \Delta u$ for all $u \in \mathcal{D}(\Delta^D)$; if $\partial\Omega$ is sufficiently smooth, $\mathcal{D}(\Delta^D) = W_2^2(\Omega) \cap \overset{0}{W}{}_2^1(\Omega)$. Such results are discussed in more detail in Chapters 6 and 7. Then Δ^D is self-adjoint and generates a C_0-semigroup T on $L_2(\Omega)$ that satisfies an upper Gaussian estimate: in fact (see, for example, [8], [44], [163]), $0 \leq T(t) \leq G_2(t)$ $(t \geq 0)$. Let T_p be the consistent semigroups on $L_p(\Omega)$ $(p \in [1, \infty))$ such that $T_2 = T$, and let $\Delta^{(p,D)}$ be the generator of T_p: we refer to $\Delta^{(p,D)}$ as the p-Dirichlet Laplacian. Then by Corollary 5.6.14, $\sigma(\Delta^{(p,D)}) = \sigma(\Delta^D)$ for all $p \in [1, \infty)$. Naturally it is of interest to know what is the domain of $\Delta^{(p,D)}$. This is discussed, for smooth $\partial\Omega$, in [47]: it turns out that if $\partial\Omega \in C^2$, then $\mathcal{D}(\Delta^{(p,D)}) = W_p^2(\Omega) \cap \overset{0}{W}{}_p^1(\Omega)$ for all $p \in (1, \infty)$ and $\Delta^{(p,D)} u = \Delta u$ for all $u \in \mathcal{D}(\Delta^{(p,D)})$. We return to this matter in 8.3. Results are also available when $\partial\Omega$ is less smooth. In [216], it is assumed that Ω is a bounded Lipschitz domain that satisfies the uniform outer ball condition in the sense that there exists $R > 0$ such that given any $x \in \partial\Omega$, there is an open ball $B \subset \mathbb{R}^n \setminus \Omega$ with radius R such that $x \in \partial B$. It is shown that

$$\mathcal{D}(\Delta^{(p,D)}) = \left\{ u \in \overset{0}{W}{}_p^1(\Omega) : \Delta u \in L_p(\Omega) \right\}$$

and $\Delta^{(p,D)} u = \Delta u$ for all $u \in \mathcal{D}(\Delta^{(p,D)})$. Moreover, if in addition $p \in (1, 2]$, then

$$\mathcal{D}(\Delta^{(p,D)}) = W_p^2(\Omega) \cap \overset{0}{W}{}_p^1(\Omega).$$

For questions of spectral independence involving fairly general second-order uniformly elliptic operators we refer to [163] and the references contained in it.

5.7 Notes

1. Unlike the three previous chapters, Chapter 5 is silent on the subject of maximum principles, the reason being that for elliptic equations of order greater than 2, results of the sharpness of the classical maximum principle of Chapter 4 are not available. For an account of what can be obtained for higher-order equations (and indeed systems of such equations) we refer to [138]. Note in particular the so-called Miranda-Agmon principle, which for solutions of the biharmonic equation $\Delta^2 u = 0$ in a domain Ω asserts that

$$\max_{\overline{\Omega}} |\nabla u| \leq C(\Omega) \max_{\partial \Omega} |\nabla u|.$$

2. As in the case of the Poisson equation, smoothness on $\overline{\Omega}$ of the eigenvectors of the Dirichlet Laplacian can be established, given some smoothness of $\partial \Omega$, by the procedures of [27], Chapter 9.

3. While we have concentrated here on the Dirichlet problem for the Laplacian, a similar analysis can be carried out for other boundary-value problems, such as the Neumann problem. The behaviour of the eigenvalues of the Laplacian, with Dirichlet or Neumann boundary conditions, has attracted a vast amount of attention ever since the pioneering work of Weyl over a hundred years ago. For each $\lambda > 0$ let

$$N(\lambda) := \sum_{\lambda_k < \lambda} 1,$$

where λ_k denotes the kth eigenvalue (allowing for multiplicities) of the Dirichlet or Neumann problem for the Laplacian. A famous conjecture of Weyl [215] suggests that, when $n = 2$,

$$N(\lambda) = \frac{|\Omega|}{4\pi} \lambda \mp \frac{|\partial \Omega|}{4\pi} \lambda^{1/2} + o\left(\lambda^{1/2}\right) \text{ as } \lambda \to \infty,$$

where the - sign refers to Dirichlet boundary conditions and + to Neumann conditions. This was established under certain conditions on Ω by Ivrii [109] and Melrose [155] in 1980. In fact they gave the corresponding result for arbitrary values of $n \subset \mathbb{N}$. The book [181] by Safarov and Vassiliev gives a detailed account of such two-term asymptotic formulae, not only for the Laplacian but for general elliptic operators of arbitrary order acting on manifolds. Of course the assertion concerning the counting function N can be translated into one about the behaviour of the λ_k : for $\Omega \subset \mathbb{R}^n$ the statements

$$N(\lambda) = c_0 \lambda^{n/2} + c_1 \lambda^{(n-1)/2} + o\left(\lambda^{(n-1)/2}\right)$$

and

$$\lambda_k = c_0^{-2/n} k^{2/n} - \frac{2c_1}{nc_0^{1+1/n}} k^{1/n} + o\left(k^{1/n}\right)$$

are equivalent.

Chapter 6
Operators and Quadratic Forms in Hilbert Space

6.1 Self-Adjoint Extensions of Symmetric Operators

Before proceeding to the main theme of this section, we give a brief resumé of the background and introduce the terminology; details may be found in [54].

Let X and Y be complex Banach spaces and $X \times Y := \{\{x, y\} : x \in X, y \in Y\}$ the product space with norm

$$\|\{x, y\}\|_{X \times Y} := \left(\|x\|_X^2 + \|y\|_Y^2\right)^{1/2}.$$

A linear operator T with domain $\mathcal{D}(T)$ in X and range $\mathcal{R}(T)$ in Y is *closed* if its graph $G(T) := \{\{x, Tx\} : x \in \mathcal{D}(T)\}$ is a closed subset of $X \times Y$; we denote the set of all such maps by $\mathcal{C}(X, Y)$ and by $\mathcal{C}(X)$ if $Y = X$. Equivalently, T is closed if, given a sequence $(x_n) \in \mathcal{D}(T)$ which is such that $x_n \to x$ in X and $Tx_n \to y$ in Y, then it follows that $x \in \mathcal{D}(T)$ and $y = Tx$. Hence the domain of a bounded closed operator is closed. Moreover, T is closed if and only if $\mathcal{D}(T)$ is complete with respect to the norm

$$\|x\|_T := \left(\|x\|_X^2 + \|Tx\|_Y^2\right)^{1/2}; \qquad (6.1.1)$$

this is called the *graph norm* of T. If T is closed and its inverse T^{-1} exists, then T^{-1} is closed. The operator T is said to be *closable* if the closure $\overline{G(T)}$ of $G(T)$ in $X \times Y$ is a graph: this means that any $\{x, y\} \in \overline{G(T)}$ must be such that y is uniquely determined by x, or, equivalently, $\{0, y\} \in \overline{G(T)}$ must imply that $y = 0$. If T is closable, $\overline{G(T)}$ is the graph of an operator called the *closure* of T, and denoted by $\overline{T}$; it is the minimal closed extension of T in the following sense. An operator R is an *extension* of T if $\mathcal{D}(R) \supset \mathcal{D}(T)$ and $Rx = Tx$ for all $x \in \mathcal{D}(T)$; we write $R \supset T$ or $T \subset R$. If R is a closed extension of T, then $\overline{T} \subset R$.

We write $\mathcal{N}(T)$ for the *kernel*, or *null space*, of T, i.e., $\{x \in \mathcal{D}(T) : Tx = 0\}$, and call nul $T := \dim \mathcal{N}(T)$ the *nullity* of T. Also, def $T := \operatorname{codim} \mathcal{R}(T)$ is called the *deficiency* of T and the *index*, ind T, of T is defined by ind $T =$ nul $T -$ def T. An operator $T \in \mathcal{C}(X, Y)$ is said to be *semi-Fredholm* if $\mathcal{R}(T)$ is closed and at least one of nul T and def T is finite; if $\mathcal{R}(T)$ is closed and both nul T and def T are finite, then T is called a *Fredholm* operator.

Let $X = Y = H$, a Hilbert space, and denote its inner-product and norm by $(\cdot, \cdot)$, $\|\cdot\|$, respectively; we write $(\cdot, \cdot)_H$, $\|\cdot\|_H$ if necessary to avoid confusion. The *resolvent set* $\rho(T)$ of an operator T in H is $\{\lambda \in \mathbb{C} : (T - \lambda I)^{-1} \in \mathcal{B}(H)\}$, where I the identity operator on H and $\mathcal{B}(H)$ is the set of bounded linear operators mapping H into itself. The *spectrum* of T, $\sigma(T)$, is the complement of $\rho(T)$ in $\mathbb{C}$. The *numerical range* of a linear operator T in H is the set of complex numbers

$$\Theta(T) := \{(Tu, u) : u \in \mathcal{D}(T), \|u\| = 1\};$$

it is a convex subset of $\mathbb{C}$. The closure $\Gamma(T)$ of $\Theta(T)$ in $\mathbb{C}$ is therefore a closed convex set. If $\Gamma(T)$ does not fill out the whole of $\mathbb{C}$, its complement $\tilde{\Delta}(T)$ is a connected open set, except in the special case in which $\Gamma(T)$ is an infinite strip bounded by two parallel lines, the limiting case of coincident lines being included. In this exceptional case $\tilde{\Delta}(T)$ consists of two components $\tilde{\Delta}_1(T)$, $\tilde{\Delta}_2(T)$ which are half-planes. If T is bounded, $\Gamma(T)$ is bounded and $\tilde{\Delta}(T)$ is connected.

Let $\delta(\lambda) = \text{dist}(\lambda, \Gamma(T)) > 0$. Then, for $u \in \mathcal{D}(T)$, $\|u\| = 1$,

$$\delta \leq |(Tu, u) - \lambda| = |([T - \lambda I]u, u)| \leq \|(T - \lambda I)u\|.$$

Hence nul $(T - \lambda I) = 0$, and $(T - \lambda I)^{-1}$ exists; thus $(T - \lambda I)^{-1}$ is a bounded closed operator on $\mathcal{R}(T - \lambda I)$ which implies that $\mathcal{R}(T - \lambda I)$ is a closed subspace of H. Consequently $T - \lambda I$ is semi-Fredholm and

$$\text{ind } (T - \lambda I) = -\text{def } (T - \lambda I); \tag{6.1.2}$$

see [54]. A closed operator has the important property that its index is constant in any connected component of its so-called *semi-Fredholm domain* $\Phi_+(T) := \{\lambda \in \mathbb{C} : T - \lambda I \text{ semi-Fredholm}\}$. Hence from (6.1.2), def $(T - \lambda I)$ is constant in any connected component of $\tilde{\Delta}(T)$. If we also have that def$(T - \lambda I) = 0$ for $\lambda \in \tilde{\Delta}(T)$, then $\mathcal{R}(T - \lambda I) = H$ and so $(T - \lambda I)^{-1}$ is bounded on H; hence $\tilde{\Delta}(T)$ lies in the resolvent set $\rho(T)$ of T and

$$\|(T - \lambda I)^{-1}\| \leq 1/\text{dist}\{\lambda, \Gamma(T)\}.$$

The *field of regularity* $\Pi(T)$ is the set of $\lambda \in \mathbb{C}$ for which there exists a positive constant $k(\lambda)$ such that

$$\|(T - \lambda I)u\| \geq k(\lambda)\|u\|, \quad u \in \mathcal{D}(T).$$

It is readily shown that $\Pi(T)$ is open; also for any $\lambda \in \Pi(T)$, nul $T = 0$ and $(T - \lambda I)^{-1}$ is bounded. By the *closed graph theorem*, $\mathcal{R}(T - \lambda I)$ is therefore closed and so $\Pi(T)$ is a subset of the semi-Fredholm domain $\Phi_+(T)$.

6.1 Self-Adjoint Extensions of Symmetric Operators

Definition 6.1.1 *If $T \in \mathcal{C}(H)$ and $\tilde{\Delta}(T)$ is connected, the constant $m(T) := \text{def}(T - \lambda I)$, with $\lambda \in \tilde{\Delta}(T)$, is called the **deficiency index** of T. If $\tilde{\Delta}(T)$ has two components, the constants $m_i(T) = \text{def}(T - \lambda I)$, $\lambda \in \tilde{\Delta}_i(T)$, ($i = 1, 2$), are called the **deficiency indices** of T.*

Let

$$\mathcal{D}^* := \{y \in H : \exists z \in H \text{ such that } (Tx, y) = (x, z) \;\forall\, x \in \mathcal{D}(T)\}.$$

The element z in $\mathcal{D}^*$ is determined uniquely by y if and only if $\mathcal{D}(T)$ is dense in H, and then the *adjoint* T^* of T is defined by $T^* y = z$ for $y \in \mathcal{D}^*$; thus

$$(Tx, y) = (x, T^* y) \quad \forall x \in \mathcal{D}(T), \; y \in \mathcal{D}(T^*) = \mathcal{D}^*.$$

The adjoint is always a closed operator. A linear operator T with domain and range in H is said to be *symmetric* if its domain $\mathcal{D}(T)$ is *dense* in H and

$$(Tx, y) = (x, Ty) \quad \forall\, x, y \in \mathcal{D}(T).$$

Since $\mathcal{D}(T)$ is a dense subspace of H, the adjoint T^* of T exists and $T \subset T^*$; the converse is clearly true. The closure $\overline{T}$ of T is also symmetric and $T \subset \overline{T} \subset T^* = \overline{T}^*$. If $T = T^*$ then T is said to be *self-adjoint*. A self-adjoint operator has no proper symmetric extensions, for if S is symmetric and $T \subset S$, then $T \subset S \subset S^* \subset T^* = T$ and so $T = S$. A symmetric operator T is said to be *essentially self-adjoint* if its closure $\overline{T}$ is self-adjoint. It is readily established that the following statements are equivalent:

1. T is essentially self-adjoint,
2. $\overline{T}$ is the unique self-adjoint extension of T,
3. T^* is symmetric,
4. T^* is self-adjoint.

The numerical range of a symmetric operator T is a subset of the real line since $\overline{(Tx, x)} = (x, Tx)$; hence the upper and lower open half-planes $\mathbb{C}_\pm$ are connected subsets of $\tilde{\Delta}(T) = \mathbb{C} \setminus \Gamma(T)$. The deficiency indices of T are therefore

$$m_+(T) \equiv m_1(T) = \text{def}\,(T + iI) = \dim \mathcal{R}(T + iI)^\perp = \text{nul}\,(T^* - iI),$$
$$m_-(T) \equiv m_2(T) = \text{def}\,(T - iI) = \dim \mathcal{R}(T - iI)^\perp = \text{nul}\,(T^* + iI).$$
(6.1.3)

If $\tilde{\Delta}(T)$ is connected, $m_+(T) = m_-(T)$ is the deficiency index of T.

A symmetric operator T is *bounded below*, or *lower semi-bounded*, if there exists $c > -\infty$ such that

$$(Tu, u) \geq c\|u\|^2, \quad \forall u \in \mathcal{D}(T);$$

we write $T \geq cI$ and $\inf\{(Tu, u) : u \in \mathcal{D}(T), \|u\| = 1\}$ is called the *lower bound* of T. T is said to be *non-negative* if $c \geq 0$ and *positive* if $c > 0$. A lower semi-bounded symmetric operator T has equal deficiency indices, since $\tilde{\Delta}(T)$ is connected.

We are now ready to discuss the von Neumann theory of extensions of symmetric operators. As it has been extensively covered in the literature, we give only a brief survey of the main features, as a reminder to the reader and for later ease of reference. We refer to [54], Section III.4 for missing details.

Let T be a closed symmetric operator with dense domain $\mathcal{D}(T)$ in a Hilbert space H. The following are the main features of the von Neumann theory.

1. For $\lambda \in \mathbb{C}_\pm$, the upper and lower open half-planes, $T - \lambda I$ has trivial kernel $\mathcal{N}(T - \lambda I) = \ker(T - \lambda I) = \{0\}$ and closed range $\mathcal{R}(T - \lambda I)$.
2. The deficiency indices $m_+(T)$, $m_-(T)$ are equal (with the same cardinality if infinite) if and only if T has a self-adjoint extension. T is self-adjoint if and only if $m_+(T) = m_-(T) = 0$.
3. If one, and only one, of the deficiency indices $m_+(T)$, $m_-(T)$ is zero, then T is *maximal symmetric*, i.e., it has no proper symmetric extensions.
4. The domain of T and its deficiency subspaces $\mathcal{N}_+ = \ker(T^* - iI)$, $\mathcal{N}_- = \ker(T^* + iI)$, are closed subspaces of $\mathcal{D}(T^*)$ endowed with the graph norm induced by the inner product

$$(u, v)_{T^*} := (T^*u, T^*v) + (u, v), \quad u, v \in \mathcal{D}(T^*),$$

and we have the direct sum

$$\mathcal{D}(T^*) = \mathcal{D}(T) \dotplus \mathcal{N}_+ \dotplus \mathcal{N}_-. \tag{6.1.4}$$

In fact, the direct sum is an orthogonal sum with respect to the graph inner-product on $\mathcal{D}(T^*)$.

5.
$$\dim \mathcal{D}(T^*)/\mathcal{D}(T) = m_+(T) + m_-(T). \tag{6.1.5}$$

If T has a self-adjoint extension,

$$\dim \mathcal{D}(T^*)/\mathcal{D}(T) = 2m_+(T) \tag{6.1.6}$$

since then $m_+(T) = m_-(T)$.

6. There is a one-one correspondence between the set of all self-adjoint extensions of T and the set of all unitary maps from $\mathcal{N}_+$ to $\mathcal{N}_-$. If V is one such unitary map and T_V the corresponding self-adjoint extension of T, then T_V is the restriction of T^* to $\mathcal{D}(T_V)$, where

$$\mathcal{D}(T_V) := \{\varphi + \varphi_+ + V\varphi_+, \ \varphi \in \mathcal{D}(T), \ \varphi_+ \in \mathcal{N}_+\}. \tag{6.1.7}$$

6.1 Self-Adjoint Extensions of Symmetric Operators

We also have

$$\mathcal{D}(T_V) = \{u \in \mathcal{D}(T^*) : \beta[u, \varphi_+ + V\varphi_+] = 0, \ \forall \varphi_+ \in \mathcal{N}_+\}. \quad (6.1.8)$$

where

$$\beta[u, v] = (T^*u, v) - (u, T^*v), \quad u, v \in \mathcal{D}(T^*). \quad (6.1.9)$$

Remark 6.1.2

Calkin proved in [34] (see also [54], Theorem III.4.6) that for any closed symmetric operator T, given any $\lambda \in \mathbb{C} \setminus \mathbb{R}$, there exists a closed operator S such that $T \subset S \subset T^*$ and $\lambda \in \rho(S)$. If S is symmetric and $\lambda_0 \in \mathbb{C}_+$ say, then $\tilde{\Delta}_1(S) \subset \rho(S)$ and S is either maximal symmetric or self-adjoint; S is self-adjoint if λ_0 is real since then $\tilde{\Delta}(T)$ is connected. If there exists a real $\lambda \in \Pi(T)$, then Calkin proved that there exists a self-adjoint operator S such that $T \subset S \subset T^*$. This result will have a role to play in the determination of all the lower semi-bounded self-adjoint extensions of a lower semi-bounded symmetric operator T later. The direct sum decompositions in the next proposition, which complement (6.1.4), will also be important in the aforementioned problem; they are established in [190], Proposition 14.11.

Proposition 6.1.3 *Let T be a densely defined closed symmetric operator in H with equal deficiency indices, and set $\mathcal{N}_\lambda := \mathcal{N}(T^* - \lambda I)$ for $\lambda \in \mathbb{C}$. Let A be a self-adjoint extension of T with $\mu \in \rho(A)$. Then we have the direct sum decompositions*

$$\mathcal{D}(T^*) = \mathcal{D}(T) \dotplus (A - \mu I)^{-1}\mathcal{N}_{\bar{\mu}} \dotplus \mathcal{N}_\mu, \quad (6.1.10)$$

$$\mathcal{D}(A) = \mathcal{D}(T) \dotplus (A - \mu I)^{-1}\mathcal{N}_{\bar{\mu}}, \quad (6.1.11)$$

$$\mathcal{D}(T^*) = \mathcal{D}(A) \dotplus \mathcal{N}_\mu, \quad (6.1.12)$$

$$\mathcal{D}(T^*) = \mathcal{D}(T) \dotplus A(A - \mu I)^{-1}\mathcal{N}_{\bar{\mu}} \dotplus (A - \mu I)^{-1}\mathcal{N}_{\bar{\mu}}. \quad (6.1.13)$$

Proof. Since $T \subset A$ and $\mu \in \rho(A)$, it follows that for any $x \in \mathcal{D}(T)$,

$$\|(T - \mu I)x\| = \|(A - \mu I)x\| \geq c\|x\|,$$

where $1/c = \|(A - \mu I)^{-1}\|$. Hence $\mathcal{R}(T - \mu I)$ is closed and has orthogonal complement $\mathcal{N}_{\bar{\mu}}$. We therefore have the orthogonal decomposition

$$H = \mathcal{R}(T - \mu I) \oplus \mathcal{N}_{\bar{\mu}}. \quad (6.1.14)$$

Then, for any $x \in \mathcal{D}(T^*)$, there exists $x_0 \in \mathcal{D}(T)$ and $z_1 \in \mathcal{N}_{\bar{\mu}}$ such that

$$(T^* - \mu I)x = (T - \mu I)x_0 + z_1.$$

Let $z_2 = x - x_0 - R_\mu(A)z_1$, where $R_\mu(A) := (A - \mu I)^{-1}$. Since $T \subset T^*$ and $R_\mu(A)z_1 \in \mathcal{D}(A) \subset \mathcal{D}(T^*)$, we have

$$(T^* - \mu I)z_2 = (T^* - \mu I)x - (T - \mu I)x_0 - (A - \mu I)R_\mu(A)z_1$$
$$= z_1 - z_1 = 0.$$

Consequently $z_2 \in \mathcal{N}_\mu$ and $x \in \mathcal{D}(T) + R_\mu(A)\mathcal{N}_{\bar{\mu}} + \mathcal{N}_\mu$. We have therefore proved that $\mathcal{D}(T^*) \subset \mathcal{D}(T) + R_\mu(A)\mathcal{N}_{\bar{\mu}} + \mathcal{N}_\mu$. The reverse inclusion is clearly true and so

$$\mathcal{D}(T^*) = \mathcal{D}(T) + R_\mu(A)\mathcal{N}_{\bar{\mu}} + \mathcal{N}_\mu. \tag{6.1.15}$$

It remains to prove that the sum is direct. Suppose that $x_0 + R_\mu(A)z_1 + z_2 = 0$, where $x_0 \in \mathcal{D}(T)$, $z_1 \in \mathcal{N}_{\bar{\mu}}$ and $z_2 \in \mathcal{N}_\mu$. Then

$$0 = (T^* - \mu I)\left(x_0 + R_\mu(A)z_1 + z_2\right) = (T - \mu I)x_0 + z_1.$$

But the two terms on the right-hand side are orthogonal by (6.1.14) and hence $(T - \mu I)x_0 = z_1 = 0$. Thus $R_\mu(A)(T - \mu I)x_0 = x_0 = 0$ and $z_2 = -x_0 - R_\mu(A)z_1 = 0$. The proof of (6.1.10) is therefore complete.

Let $x \in \mathcal{D}(A)$. Then from (6.1.10),

$$x = x_0 + R_\mu(A)z_1 + z_2, \quad x_0 \in \mathcal{D}(T), \ z_1 \in \mathcal{N}_{\bar{\mu}}, \ z_2 \in \mathcal{N}_\mu.$$

Hence $z_2 \in \mathcal{D}(A)$, $(A - \mu I)z_2 = (T^* - \mu I)z_2 = 0$ and so $z_2 = 0$ since $\mu \in \rho(A)$. We have therefore proved that

$$\mathcal{D}(A) \subset \mathcal{D}(T) + R_\mu(A)\mathcal{N}_{\bar{\mu}}.$$

The reverse inequality is obvious giving the sum in (6.1.11). The fact that the sum in (6.1.10) is direct implies the same in (6.1.11). The identity (6.1.12) follows from (6.1.10) and (6.1.11).

To prove (6.1.13), we start by replacing μ by $\bar{\mu}$ in (6.1.12), which we may since A is self-adjoint. We then have, on using (6.1.12),

$$\mathcal{D}(T^*) = \mathcal{D}(T) \dotplus R_\mu(A)\mathcal{N}_{\bar{\mu}} \dotplus \mathcal{N}_{\bar{\mu}}. \tag{6.1.16}$$

Since

$$u + R_\mu(A)v = AR_\mu(A)u + R_\mu(A)(v - \mu u)$$

and

$$AR_\mu(A)u + R_\mu(A)v = u + R_\mu(A)(v + \mu u),$$

it follows that

$$R_\mu(A)\mathcal{N}_{\bar{\mu}} \dotplus \mathcal{N}_{\bar{\mu}} = AR_\mu(A)\mathcal{N}_{\bar{\mu}} \dotplus R_\mu(A)\mathcal{N}_{\bar{\mu}}.$$

On inserting this in (6.1.16), (6.1.13) follows. ∎

6.2 Characterisations of Self-Adjoint Extensions

In this section, we shall establish a one-one correspondence between the set of self-adjoint extensions of a closed symmetric operator T acting in H and two other sets of self-adjoint operators, the first defined by linear subspaces of the Cartesian product space $H \times H$, and the second determined by boundary triplets associated with T^*. A fuller treatment may be found in [190]. We first provide some background information.

6.2.1 Linear Relations

Let H_1 and H_2 be Hilbert spaces, and let G_1, G_2, be linear subspaces of $H_1 \times H_2$, whose topology is determined by the inner-product

$$(\{x_1, y_1\}, \{x_2, y_2\})_{H_1 \times H_2} = (x_1, x_2)_{H_1} + (y_1, y_2)_{H_2},$$

and norm

$$\|\{x, y\}\|_{H_1 \times H_2} := \{\|x\|_{H_1}^2 + \|y\|_{H_2}^2\}^{1/2},$$

where $(\cdot, \cdot)_{H_i}$, $\|\cdot\|_{H_i}$ are the inner-product and norm on H_i, $i = 1, 2$. The *sum, scalar multiple, product, adjoint* G_i^* and *inverse* G_i^{-1} are defined by

$$G_1 + G_2 = \{\{x, y_1 + y_2\} : \{x, y_i\} \in G_i, i = 1, 2\}$$
$$\alpha G_i = \{\{x, \alpha y\} : \alpha \in \mathbb{C}, \{x, y\} \in G_i\}$$
$$G_1 G_2 = \{\{x, y\} : \{x, v\} \in G_2 \text{ and } \{v, y\} \in G_1 \text{ for some } v \in H_2\}$$
$$G_i^* = \{\{y, x\} : (v, y)_2 = (u, x)_1, \text{ for all } \{u, v\} \in G_i\},$$
$$G_i^{-1} = \{\{y, x\} : \{x, y\} \in G_i\}.$$

A linear subspace G of $H_1 \times H_2$ is also called a *linear relation* from H_1 into H_2, with *domain* $\mathcal{D}(G)$, *range* $\mathcal{R}(G)$, *null space* $\mathcal{N}(G)$ and *multi-valued part* $\mathcal{M}(G)$ defined by

$$\mathcal{D}(G) = \{x \in H_1 : \{x, y\} \in G \text{ for some } y \in H_2\},$$
$$\mathcal{R}(G) = \{y \in H_2 : \{x, y\} \in G \text{ for some } x \in H_1\},$$
$$\mathcal{N}(G) = \{x \in H_1 : \{x, 0\} \in G\}$$
$$\mathcal{M}(G) = \{y \in H_2 : \{0, y\} \in G\}.$$

If $\mathcal{M}(G) \neq \{0\}$, the linear relation may be viewed as a multi-valued linear map from $\mathcal{D}(G)$ onto $\mathcal{R}(G)$. The closure $\overline{G}$ is the closure of G as a linear subspace of $H_1 \times H_2$, and G is said to be closed if $\overline{G} = G$. If $H_1 = H_2 = H$, with inner-product $(\cdot, \cdot)$, G

is said to be *symmetric* if $G \subset G^*$, that is

$$(x, v) = (y, u), \quad \text{for all } \{x, y\}, \{u, v\} \in G,$$

and *self-adjoint* if $G = G^*$.

A linear relation G determines an operator T acting between H_1 and H_2 if and only if its multi-valued part $\mathcal{M}(G) = \{0\}$, in which case G is the *graph* $G(T)$ of T, namely,

$$G(T) = \{\{x, Tx\} : x \in \mathcal{D}(T)\}.$$

Since $\{x, y\} \in G(T)$ if and only if $x \in \mathcal{D}(T)$ and $y = Tx$, T can be identified with its graph. The domain, range and null space of T are the corresponding subspaces of H_1 and H_2 of its graph $G(T)$, and the multi-valued part $\mathcal{M}(G(T))$ of the graph is now the zero subspace $\{0\}$ of H_2. Furthermore, if T^* exists, $G(T)^* = G(T^*)$ and if T^{-1} exists, $G(T)^{-1} = G(T^{-1})$. The operator T is *closable* if and only if the closure of $G(T)$ is a graph, and in that case, the graph of the closure $\overline{T}$ of T is the closure $\overline{G(T)}$ of $G(T)$; thus T is closed if and only if $G(T)$ is closed. It is worth emphasising that for a linear relation, the inverse, adjoint and closure always exist as linear relations, which is in contrast to the case for operators; for instance, if T is a non-closable operator, the closure $\overline{G(T)}$ is a closed linear subspace, and $\overline{T}$ is only defined as a linear relation. Properties of graphs and linear relations noted above are proved in a similar way to those for operators and are left as exercises. Hereafter in this section $H_1 = H_2 = H$, with inner-product $(\cdot, \cdot)$ and norm $\|\cdot\|$.

If G is a closed linear relation, the subspace $G_\infty = \{\{0, y\} \in G\}$ of $H \times H$ is closed, and so is its orthogonal complement $G \ominus G_\infty$. Since $\{0, y\} \in G \ominus G_\infty$ implies that $y = 0$, $G \ominus G_\infty$ is the graph of a closed operator $T_0(G)$ which is called the *operator part* of G. Thus $G = G(T_0) \oplus G_\infty$; $\oplus$ denotes the orthogonal sum and $\ominus$ the orthogonal complement in $H \times H$.

The following proposition is the basis of the results in this section. We denote by $S(H)$ the set of all self-adjoint operators B which are densely defined in a closed subspace H_B of H and $\mathcal{R}(B) \subset H_B$. Also, for a linear relation β, we write $\beta \geq 0$ if and only if $\{y, x\} \geq 0$ for all $\{x, y\} \in \beta$.

Proposition 6.2.1 *There is a one-to-one correspondence between the set of operators $B \in S(H)$ and the set of self-adjoint linear relations $\mathcal{B}$ on H given by*

$$\mathcal{B} = \{\{x, Bx + y\} : x \in \mathcal{D}(B), y \in H_B^\perp\} \tag{6.2.1}$$

$$= G(B) \oplus \left(\{0\} \times H_B^\perp\right), \tag{6.2.2}$$

where $B = T_0(\mathcal{B})$ is the operator part of $\mathcal{B}$, and $H_B^\perp$ is the multi-valued part $\mathcal{M}(\mathcal{B})$ of $\mathcal{B}$. Moreover $\mathcal{B} \geq 0$ if and only if $B \geq 0$.

Proof.

Suppose $B \in S(H)$ and let $\mathcal{B}$ be defined by (6.2.1). For $\{x, y\}, \{u, v\} \in \mathcal{B}$, $x, u \in \mathcal{D}(B)$ and $y = Bx + y'$, $v = Bu + v'$, where $y', v' \in H_B^\perp$. Thus

6.2 Characterisations of Self-Adjoint Extensions

$$(x, v) = (x, Bu + v') = (x, Bu) = (Bx, u) = (Bx + y', u) = (y, u)$$

and $\mathcal{B}$ is a self-adjoint linear relation with $\mathcal{M}(\mathcal{B}) = H_B^\perp$.

Conversely, assume that $\mathcal{B}$ is a self-adjoint linear relation on H. Let B be the operator part $T_0(\mathcal{B})$ of $\mathcal{B}$ and $H_B = \mathcal{M}(\mathcal{B})^\perp$. For $x \in \mathcal{M}(\mathcal{B}) = \mathcal{M}(\mathcal{B}^*)$, we have $\{0, x\} \in \mathcal{B}^*$, which means that $(v, 0) = (u, x)$ for all $u \in \mathcal{D}(B)$, with $v = Bu$. This implies that $\mathcal{M}(\mathcal{B}) = \mathcal{D}(B)^\perp$ and hence

$$\left(\mathcal{D}(B)^\perp\right)^\perp = \mathcal{M}(\mathcal{B})^\perp = H_B.$$

As $\left(\mathcal{D}(B)^\perp\right)^\perp$ is the closure of $\mathcal{D}(B)$, B is densely defined in H_B. Since $G(B) \perp (\{0\} \times \mathcal{M}(\mathcal{B}))$, we have $\mathcal{R}(B) \subset \mathcal{M}(\mathcal{B})^\perp = H_B$. Hence B is a closed, densely defined linear operator in H_B. The self-adjointness of B is easily seen to follow from that of $\mathcal{B}$.

Let $\{x, Bx + y)\} \in \mathcal{B}$, Then $x \perp y$ and $(Bx + y, x) = (Bx, x)$. Hence $\mathcal{B} \geq 0$ is equivalent to $B \geq 0$. ∎

6.2.2 Boundary Triplets

Let T be a closed symmetric operator in a Hilbert space H and set

$$\mathcal{N}_\lambda := \ker(T^* - \lambda I), \quad \lambda \in \mathbb{C},$$

for the kernel (null space) of $T^* - \lambda I$.

Definition 6.2.2 A *boundary triplet* for T^* is a triplet $(\mathcal{K}, \Gamma_0, \Gamma_1)$, where $\mathcal{K}$ is a Hilbert space with inner product $(\cdot, \cdot)_\mathcal{K}$, Γ_0 and Γ_1 are linear mappings of $\mathcal{D}(T^*)$ into $\mathcal{K}$, and the following conditions are satisfied:

1. $[x, y]_{T^*} := (T^*x, y) - (x, T^*y) = (\Gamma_1 x, \Gamma_0 y)_\mathcal{K} - (\Gamma_0 x, \Gamma_1 y)_\mathcal{K}, \quad x, y \in \mathcal{D}(T^*)$,
2. $x \mapsto \{\Gamma_0 x, \Gamma_1 x\}$ maps $\mathcal{D}(T^*)$ onto $\mathcal{K} \times \mathcal{K}$.

The equation in (1) is an abstract *Green's identity* for T^* in terms of the *boundary values* $\Gamma_0 x, \Gamma_1 x, \Gamma_0 y, \Gamma_1 y$ of x and y in $\mathcal{D}(T^*)$.

On setting $\Gamma_+ := \Gamma_1 + i\Gamma_0$ and $\Gamma_- := \Gamma_1 - i\Gamma_0$, condition (1) becomes

$$2i[x, y]_{T^*} = (\Gamma_- x, \Gamma_- y)_\mathcal{K} - (\Gamma_+ x, \Gamma_+ y)_\mathcal{K}, \quad x, y \in \mathcal{D}(T^*), \tag{6.2.3}$$

and $x \mapsto \{\Gamma_+ x, \Gamma_- x\} : \mathcal{D}(T^*) \to \mathcal{K} \times \mathcal{K}$ is surjective.

Example 6.2.3

Let T be the closure in $L_2(a, b)$ of the symmetric operator T' defined by $T'f = -f''$ for $f \in C_0^2(a, b)$, where $a, b \in \mathbb{R}$. The Sobolev space $H^2(a, b)$ is the domain of its adjoint T^* and $T^*f = -f''$ for $f \in \mathcal{D}(T^*)$. On integration by parts, we have for $f, g \in \mathcal{D}(T^*)$,

$$[f, g]_{T^*} = f(b)\overline{g'(b)} - f'(b)\overline{g(b)} - f(a)\overline{g'(a)} + f'(a)\overline{g(a)}.$$

Therefore the triplet $(\mathcal{K}, \Gamma_0, \Gamma_1)$ is defined for T^* with $\mathcal{K} = \mathbb{C}^2$,

$$(c, d)_{\mathcal{K}} = c_1\overline{d_1} + c_2\overline{d_2}, \quad c = \{c_1, c_2\}, \; d = \{d_1, d_2\},$$

and

$$\Gamma_0 f = \{f(a), f(b)\}, \quad \Gamma_1 f = \{f'(a), -f'(b)\}.$$

For arbitrary complex numbers c_1, c_2, d_1, d_2 there exists $f \in \mathcal{D}(T^*) = H^2(a, b)$ such that $f(a) = c_1$, $f'(a) = c_2$, $f(b) = d_1$, $f'(b) = d_2$. Thus the condition (2) in Definition 6.2.2 is satisfied.

Proposition 6.2.4 *There exists a boundary triplet $(\mathcal{K}, \Gamma_0, \Gamma_1)$ for T^* if T has equal deficiency indices, in which case $\dim \mathcal{K}$ is equal to the deficiency indices.*

Proof. From (6.1.4), each $x \in \mathcal{D}(T^*)$ can be uniquely written as $x = x_0 + x_+ + x_-$, where $x_0 \in \mathcal{D}(T)$, $x_\pm \in \mathcal{N}_\pm$ and $\mathcal{N}_\pm := \mathcal{N}_{\pm i}$ are the deficiency subspaces of T, whose dimensions $m_\pm$ are assumed to be equal. If also $y = y_0 + y_+ + y_- \in \mathcal{D}(T^*)$ then since $T^*x = Tx_0 + ix_+ - ix_-$, $T^*y = Ty_0 + iy_+ - iy_-$ and T is symmetric, it is easily verified that

$$[x, y]_{T^*} = 2i(x_+, y_+) - 2i(x_-, y_-).$$

As the deficiency indices are equal, there exists an isometry V mapping $\mathcal{N}_+$ onto $\mathcal{N}_-$. Let $P_\pm$ be the projections $P_\pm x = x_\pm$ and set

$$\mathcal{K} = \mathcal{N}_-, \quad \Gamma_+ = 2VP_+, \quad \Gamma_1 = 2P_-.$$

Then

$$\begin{aligned}2i[x, y]_{T^*} &= 4(x_-, y_-) - 4(x_+, y_+) = 4(x_-, y_-) - 4(Vx_+, Vy_+) \\&= (\Gamma_-x, \Gamma_-y)_{\mathcal{K}} - (\Gamma_+x, \Gamma_+y)_{\mathcal{K}}\end{aligned}$$

which is (6.2.3). Given $u_1, u_2 \in \mathcal{K}$, put $x = u_1 + V^{-1}u_2$. Then $\Gamma_+x = 2u_2$ and $\Gamma_-x = 2u_1$. Hence the map $x \mapsto \{\Gamma_+x, \Gamma_-x\} : \mathcal{D}(T^*) \to \mathcal{N}_- \times \mathcal{N}_-$ is surjective, and so $(\mathcal{N}_-, \Gamma_+, \Gamma_-)$ is a boundary triplet for T^*. ∎

In [190], Lemma 14.13 and Proposition 14.5, Schmüdgen also proves the converse of Proposition 6.2.4, namely that if there exists a boundary triplet $(\mathcal{K}, \Gamma_0, \Gamma_1)$ for T^*, then T has equal deficiency indices and $\dim \mathcal{K}$ is equal to the deficiency index.

6.2 Characterisations of Self-Adjoint Extensions

Our next objective is to show that, given a boundary triplet $(\mathcal{K}, \Gamma_0, \Gamma_1)$ of T^*, the self-adjoint extensions S of T can be described in terms of self-adjoint linear relations on $\mathcal{K}$ and then by self-adjoint operators acting in $\mathcal{K}$.

Let $\mathcal{B}$ be a linear relation on $\mathcal{K}$, and denote by $T_\mathcal{B}$ the restriction of T^* to the domain

$$\mathcal{D}(T_\mathcal{B}) := \left\{ x \in \mathcal{D}(T^*) : \{\Gamma_0 x, \Gamma_1 x\} \in \mathcal{B} \right\}. \tag{6.2.4}$$

Note that if $\mathcal{B} = G(B)$, the graph of an operator B, then

$$\mathcal{D}(T_\mathcal{B}) = \ker(\Gamma_1 - B\Gamma_0). \tag{6.2.5}$$

Definition 6.2.5 The **boundary space** *of a linear operator S such that $T \subset S \subset T^*$ is defined to be the linear relation*

$$\mathcal{B}(S) = \{\{\Gamma_0 x, \Gamma_1 x\} : x \in \mathcal{D}(S)\} \tag{6.2.6}$$

on $\mathcal{K}$.

It follows from Definition 6.2.2(2) that for any linear relation $\mathcal{B}$ on $\mathcal{K}$, $\mathcal{B}(T_\mathcal{B}) = \mathcal{B}$.

Lemma 6.2.6 *Let $\mathcal{B}$ be a linear relation on $\mathcal{K}$ and let S be a linear operator such that $T \subset S \subset T^*$ and $\mathcal{B}(S) = \mathcal{B}$. Then*

1. $S^* = T_{\mathcal{B}^*}$;
2. $\overline{S} = T_{\overline{\mathcal{B}}}$;
3. *if S is closed, then $\mathcal{B}(S) = \mathcal{B}$ is closed;*
4. $\mathcal{D}(\overline{T}) = \{x \in \mathcal{D}(T^*) : \Gamma_0 x = \Gamma_1 x = 0\}$.

Proof.

1. From $T \subset S \subset T^*$ and since T is closed, it follows that $T \subset S^* \subset T^*$. Hence $y \in \mathcal{D}(T^*)$ lies in $\mathcal{D}(S^*)$ if and only if for all $x \in \mathcal{D}(S)$,

$$(T^*x, y) = (Sx, y) = (x, S^*y) = (x, T^*y),$$

which, in view of Definition 6.2.2, is equivalent to saying that $(\Gamma_0 x, \Gamma_1 y)_\mathcal{K} = (\Gamma_1 x, \Gamma_0 y)_\mathcal{K}$ for $x \in \mathcal{D}(S)$. Thus, for all $\{u, v\} \in \mathcal{B}(S)$, $(u, \Gamma_1 y)_\mathcal{K} = (v, \Gamma_0 y)_\mathcal{K}$, which means that $\{\Gamma_0 y, \Gamma_1 y\} \in \mathcal{B}^*$, and so $y \in \mathcal{D}(T_{\mathcal{B}^*})$ by (6.2.4). We have therefore shown that $\mathcal{D}(S^*) = \mathcal{D}(T_{\mathcal{B}^*})$, and consequently $S^* = T_{\mathcal{B}^*}$ since S^* and $T_{\mathcal{B}^*}$ are restrictions of T^*.
2. From the first part, $\mathcal{B}(S^*) = \mathcal{B}(T_{\mathcal{B}^*}) = \mathcal{B}^*$. This yields $\overline{S} = (S^*)^* = T_{(\mathcal{B}^*)^*} = T_{\overline{\mathcal{B}}}$, on using $(\mathcal{B}^*)^* = \overline{\mathcal{B}}$ which is proved as for operators.
3. If S is closed, we infer from (2) that $S \subset T_\mathcal{B} \subset T_{\overline{\mathcal{B}}} = \overline{S} = S$. Thus $T_\mathcal{B} = T_{\overline{\mathcal{B}}}$ and Definition 6.2.2(2) implies that $\mathcal{B} = \overline{\mathcal{B}}$.
4. Set $\mathcal{B} = \mathcal{K} \times \mathcal{K}$. Then $\mathcal{B}^* = \{\{0, 0\}\}$ and $T_\mathcal{B} = T^*$. Then from (1), $\overline{T} = (T^*)^* = (T_\mathcal{B})^* = T_{\mathcal{B}^*}$ which completes the proof. ∎

Corollary 6.2.7 *Let $(\mathcal{K}, \Gamma_0, \Gamma_1)$ be a boundary triplet for T^*. Then the restrictions T_0 and T_1 of T^* to the domains $\mathcal{D}(T_0) = \mathcal{N}(\Gamma_0)$, $\mathcal{D}(T_1) = \mathcal{N}(\Gamma_1)$ are self-adjoint extensions of T in H, T_0, T_1 being respectively the extensions of T whose boundary spaces are the linear relations $\mathcal{B}_0 = \{0\} \times \mathcal{K}$, $\mathcal{B}_1 = \mathcal{K} \times \{0\}$, in $\mathcal{K}$.*

Proof. The linear relations $\mathcal{B}_0$, $\mathcal{B}_1$ are easily seen to be self-adjoint. Hence the self-adjointness of $T_{\mathcal{B}_0}$ and $T_{\mathcal{B}_1}$ follows from Lemma 6.2.6. Furthermore $T_{\mathcal{B}_0}$ and $T_{\mathcal{B}_1}$ have domains $\mathcal{N}(\Gamma_0)$, $\mathcal{N}(\Gamma_1)$, respectively, by (6.2.4). ∎

We recall the notation $\mathcal{S}(\mathcal{K})$ for the set of self-adjoint operators B which are densely defined in a closed subspace $\mathcal{K}_B$ of $\mathcal{K}$ and have range in $\mathcal{K}_B$. For any $B \in \mathcal{S}(\mathcal{K})$, we denote by T_B the restriction of T^* to the domain

$$\mathcal{D}(T_B) := \{x \in \mathcal{D}(T^*) : \Gamma_0 x \in \mathcal{D}(B) \text{ and } B\Gamma_0 x = P_B \Gamma_1 x\}, \tag{6.2.7}$$

where P_B is the orthogonal projection of $\mathcal{K}$ onto $\mathcal{K}_B$. Thus

$$\mathcal{D}(T_B) = \{x \in \mathcal{D}(T^*) : \Gamma_0 x = u \in \mathcal{D}(B), \Gamma_1 x = Bu + v \text{ for some } v \in \mathcal{K}_B^\perp\}. \tag{6.2.8}$$

If $\mathcal{B}$ is a linear relation in $\mathcal{K}$ with operator part B and multi-valued part $\mathcal{M}(\mathcal{B}) = (\mathcal{K}_B)^\perp$, then we see from (6.2.1) and (6.2.8) that $T_B = T_\mathcal{B}$.

Theorem 6.2.8 *Let $(\mathcal{K}, \Gamma_0, \Gamma_1)$ be a boundary triplet for T^*, and let S be a linear operator acting in H. Then the following are equivalent:*

1. *S is a self-adjoint extension of T;*
2. *there is a self-adjoint linear relation $\mathcal{B}$ on $\mathcal{K}$ such that $S = T_\mathcal{B}$;*
3. *there is an operator $B \in \mathcal{S}(\mathcal{K})$ such that $S = T_B$.*

Proof. If S is self-adjoint, it follows from Lemma 6.2.6 that $T_\mathcal{B} = T_{\mathcal{B}^*}$, where $\mathcal{B}$ is the boundary space $\mathcal{B}(S)$ of S, and $S = T_\mathcal{B}$. Furthermore as the linear relation $\mathcal{B}$ is uniquely determined by the operator $T_\mathcal{B}$, $\mathcal{B}$ is therefore self-adjoint. Conversely, if $\mathcal{B}$ is a self-adjoint linear relation, it is closed and $T_\mathcal{B}$ is a self-adjoint extension of T by parts (1) and (2) of Lemma 6.2.6.

The remainder of the theorem is a consequence of the fact noted after (6.2.8) that if the linear relation $\mathcal{B}$ in $\mathcal{K}$ has operator part B and $\mathcal{M}(\mathcal{B}) = (\mathcal{K}_B)^\perp$, then $T_B = T_\mathcal{B}$. ∎

More generally, there is a one-one correspondence between all closed linear relations $\mathcal{B}$ on $\mathcal{K}$ and closed operators S which satisfy $T \subset S \subset T^*$, given by $\mathcal{B} \leftrightarrow T_\mathcal{B}$; see [190], Proposition 14.7.

To describe the operator $T_B = T_\mathcal{B}$ defined by the boundary space $\mathcal{B}$ of S, we require maps Γ_0, Γ_1 associated with an appropriate boundary triplet $(\mathcal{K}, \Gamma_0, \Gamma_1)$ of T^*. We follow the development in [190] and use the boundary triplet in [190], example 14.6.

The hypothesis of the following lemma requires the existence of a self-adjoint extension A of T with a real point μ in its resolvent set $\rho(A)$. By Calkin's result

6.2 Characterisations of Self-Adjoint Extensions

referred to in Remark 6.1.2, this hypothesis is satisfied if T has a real point μ in its field of regularity $\Pi(T)$, and so in particular, if T is lower semi-bounded.

Lemma 6.2.9 *Let T be a closed symmetric operator and A a self-adjoint extension of T with a real point μ in $\rho(A)$. By (6.1.10), for any $x \in \mathcal{D}(T^*)$, there exist uniquely determined elements $x_T \in \mathcal{D}(T)$ and $x_0, x_1 \in \mathcal{N}_\mu = \ker(T^* - \mu I)$ such that*

$$x = x_T + (A - \mu I)^{-1} x_1 + x_0.$$

Let $\mathcal{K} = \mathcal{N}_\mu$, $\Gamma_0 x = x_0$, $\Gamma_1 x = x_1$. Then $(\mathcal{K}, \Gamma_0, \Gamma_1)$ is a boundary triplet for T^*.

Proof. The second condition in Definition 6.2.2 is clearly met. Let $x, y \in \mathcal{D}(T^*)$. Then

$$x = x_T + (A - \mu I)^{-1} x_1 + x_0, \quad x_T \in \mathcal{D}(T), \; x_0, x_1 \in \mathcal{N}_\mu$$
$$T^* x = T x_T + A(A - \mu I)^{-1} x_1 + \mu x_0$$

and there are similar expressions for y and $T^* y$. These give

$$(T^* x, y) = (T x_T, y) + \left(A(A - \mu I)^{-1} x_1 + \mu x_0, y_T + (A - \mu I)^{-1} y_1 + y_0\right)$$

and

$$(x, T^* y) = (x_T, T^* y) + \left((A - \mu I)^{-1} x_1 + x_0, T y_T + A(A - \mu I)^{-1} y_1 + \mu y_0\right).$$

Hence,

$$\begin{aligned}(T^* x, y) - (x, T^* y) &= \left(A(A - \mu I)^{-1} x_1 + \mu x_0, y_T + (A - \mu I)^{-1} y_1 + y_0\right) \\ &\quad - \left((A - \mu I)^{-1} x_1 + x_0, T y_T + y_T + A(A - \mu I)^{-1} y_1 + \mu y_0\right) \\ &= \left(A(A - \mu I)^{-1} x_1 + \mu x_0, (A - \mu I)^{-1} y_1 + y_0\right) \\ &\quad - \left((A - \mu I)^{-1} x_1 + x_0, A(A - \mu I)^{-1} y_1 + \mu y_0\right).\end{aligned}$$

Since $A \subset T^*$ and $T^* x_0 = \mu x_0$, we have

$$\begin{aligned}(T^* x, y) - (x, T^* y) &= \left(A(A - \mu I)^{-1} x_1, y_0\right) + \mu \left(x_0, (A - \mu I)^{-1} y_1 + y_0\right) \\ &\quad - \mu \left((A - \mu I)^{-1} x_1, y_0\right) - \left(x_0, A(A - \mu I)^{-1} y_1 + \mu y_0\right) \\ &= (x_1, y_0) + \mu \left((A - \mu I)^{-1} x_1, y_0\right) + \mu \left(x_0, (A - \mu I)^{-1} y_1\right) \\ &\quad + \mu (x_0, y_0) - \mu \left((A - \mu I)^{-1} x_1, y_0\right) - (x_0, y_1) \\ &\quad - \mu \left(x_0, (A - \mu I)^{-1} y_1\right) - \mu (x_0, y_0) \\ &= (x_1, y_0) - (x_0, y_1).\end{aligned}$$

It follows that

$$(T^*x, y) - (x, T^*y) = (\Gamma_1 x, \Gamma_0 y) - (\Gamma_0 x, \Gamma_1 y)$$

and the lemma is proved. ∎

Theorem 6.2.10 *Let T be a closed symmetric operator and A a self-adjoint extension of T with a real $\mu \in \rho(A)$. For $B \in \mathcal{S}(\mathcal{N}_\mu)$, let T_B be the restriction of T^* to the domain*

$$\mathcal{D}(T_B) := \left\{ x + R_\mu(A)(Bu + v) + u : x \in \mathcal{D}(T), u \in \mathcal{D}(B), v \in \mathcal{N}_\mu, v \perp \mathcal{D}(B) \right\}, \quad (6.2.9)$$

where $R_\mu(A) = (A - \mu I)^{-1}$, and so

$$T_B(x + R_\mu(A)(Bu + v) + u) = Tx + (I + \mu R_\mu(A))(Bu + v) + \mu u. \quad (6.2.10)$$

Then T_B is a self-adjoint extension of T in H. Conversely, any self-adjoint extension of T in H is of the form T_B with a uniquely determined $B \in \mathcal{S}(\mathcal{N}_\mu)$.

Proof. The theorem follows from Theorem 6.2.8 (3), on inserting the operators Γ_0 and Γ_1 from Lemma 6.2.9 into (6.2.8) to obtain (6.2.9), and (6.2.10) is a consequence of

$$T^* R_\mu(A) = A R_\mu(A) = I + \mu R_\mu(A), \quad T^* x_0 = \mu x_0, \quad T^* x_1 = \mu x_1.$$

∎

6.2.3 Gamma Fields and Weyl Functions

Let $(\mathcal{K}, \Gamma_0, \Gamma_1)$ be a boundary triplet for T^* and let T_0 be the self-adjoint extension of T with domain $\mathcal{D}(T_0) = \mathcal{N}(\Gamma_0)$ in Corollary 6.2.7. It is proved in [190], Section 14.5 that for $z \in \rho(T_0)$,

$$\gamma(z) := (\Gamma_0 \restriction \mathcal{N}_z)^{-1} \in B(\mathcal{K}, H) \quad \text{and} \quad M(z) := \Gamma_1 \gamma(Z) \in B(\mathcal{K}), \quad (6.2.11)$$

where $B(X, Y)$ is the vector space of bounded linear maps from X to Y, $B(X) = B(X, X)$ and $\mathcal{N}_z := \mathcal{N}(T^* - zI)$.

Definition 6.2.11 *The maps*

1. $\rho(T_0) \ni z \mapsto \gamma(z) \in B(\mathcal{K}, H)$,
2. $\rho(T_0) \ni z \mapsto M(z) \in B(\mathcal{K})$

*are called the **gamma field** and **Weyl function** respectively of the operator T_0 associated with the boundary triplet $(\mathcal{K}, \Gamma_0, \Gamma_1)$.*

6.2 Characterisations of Self-Adjoint Extensions

We shall give a brief survey of the properties of the maps in Definition 6.2.11 which are relevant to our needs, but we refer to [190], Section 14.5, for a detailed discussion with proofs.

The following proposition is included in [190], Propositions 14.14 and 14.15.

Proposition 6.2.12 *For $z, w \in \rho(T_0)$,*

1. $\gamma(w) = (T_0 - zI)(T_0 - wI)^{-1}\gamma(z)$;
2. $\frac{d}{dz}\gamma(z) = (T_0 - zI)^{-1}\gamma(z)$;
3. $M(z)\Gamma_0 u = \Gamma_1 u$ for $u \in \mathcal{N}_z$;
4. $M(z)^* = M(\bar{z})$;
5. $M(w) - M(z) = (w - \bar{z})\gamma(\bar{z})^*\gamma(w)$;
6. $\frac{d}{dz}M(z) = \gamma(\bar{z})^*\gamma(z)$.

It follows from properties (2) and (6) that the gamma field $z \mapsto \gamma(z)$ and Weyl function $z \mapsto M(z)$ are operator-valued analytic functions on the resolvent set $\rho(T_0)$. Furthermore, the Weyl function is a $B(\mathcal{K})$-valued analytic function on $\mathbb{C}_+$ and for $z \in \mathbb{C}_+$ and $y = im\, z$,

$$M(z) - M(z)^* = M(z) - M(\bar{z}) = 2i\gamma(z)^*\gamma(z),;$$

hence $im\, M(z) \geq 0$ for all $z \in \mathbb{C}_+$. Therefore, M is a $B(\mathcal{K})$-valued *Nevanlinna function*. A scalar Nevanlinna function m is analytic in $\mathbb{C}_+$ and satisfies $m : \mathbb{C}_+ \to \mathbb{C}_+$. It has the following integral representation:

$$m(z) = a + bz + \int_{\mathbb{R}} \left(\frac{1}{(t-z)} - \frac{t}{(1+t^2)} \right) d\nu(t), \quad z \in \mathbb{C} \setminus \mathbb{R},$$

where $a, b \in \mathbb{R}$, $b \geq 0$, and ν is a regular Borel measure on $\mathbb{R}$ satisfying $\int (1 + t^2)^{-1} d\nu(t) < \infty$. There is such an integral representation for operator-valued Nevanlinna functions, with, in that case, $a = a^*$, $b = b^* \geq 0$ in $B(\mathcal{K})$, and ν a positive operator-valued Borel measure on $\mathbb{R}$.

In [190], Proposition 14.17, relationships are established between the eigenvalues and spectrum of an operator $T_\mathcal{B}$ defined by (6.2.4) with respect to a closed linear relation $\mathcal{B}$ on $\mathcal{K}$, and properties of the Weyl function M of the operator T_0. In particular, z is an eigenvalue of $T_\mathcal{B}$ if and only if $\mathcal{N}(\mathcal{B} - G(M(z))) \neq \{0\}$, and $z \in \rho(T_\mathcal{B})$ if and only if $0 \in \rho(\mathcal{B} - G(M(z)))$, i.e., $(\mathcal{B} - G(M(z)))^{-1}$ is the graph of a linear operator which is bounded on H; note that $G(M(z))$ is the graph of $M(z)$. Furthermore, if T is symmetric, the *Krein-Naimark resolvent formula* concerning the difference between the resolvents of $T_\mathcal{B}$ and T_0, can be expressed as follows: for $z \in \rho(T_\mathcal{B}) \cap \rho(T_0)$, the relation $(\mathcal{B} - G(M(z)))$ has an inverse $(\mathcal{B} - G(M(z)))^{-1}$ in $B(\mathcal{K})$ and

$$(T_\mathcal{B} - zI)^{-1} - (T_0 - zI)^{-1} = \gamma(z)(\mathcal{B} - G(M(z)))^{-1}\gamma(\bar{z})^* \quad (6.2.12)$$

From Lemma 6.2.9 with $A = T_0$, $\mu \in \rho(T_0)$ real, and $\mathcal{N}_\mu = \ker(T^* - \mu I)$, we have for $v \in \mathcal{K}$, $\Gamma_0 v = v$ and $\Gamma_1 v = 0$. Therefore $\gamma(\mu)v = v$ and $M(\mu)v =$

$\Gamma_1 \gamma(\mu) v = 0$, i.e., $\gamma(\mu) = I_H \upharpoonright_{\mathcal{K}}$ and $M(\mu) = 0$. Consequently, $\gamma(\mu)^* = P_{\mathcal{K}}$, the orthogonal projection of H onto $\mathcal{K}$. It follows from parts (1) and (5) of Proposition 6.2.12 that for $z \in \rho(T_0)$, $\gamma(z)$ and $M(z)$ are given by

$$\gamma(z) = (T_0 - \mu I)(T_0 - zI)^{-1} \upharpoonright_{\mathcal{K}}$$
$$M(z) = (z - \mu) P_{\mathcal{K}} (T_0 - \mu I)(T_0 - zI)^{-1} \upharpoonright_{\mathcal{K}} \quad (6.2.13)$$

The Krein-Naimark formula for the difference of the resolvents of two arbitrary self-adjoint extensions A, $\tilde{A}$ of a symmetric operator T, which does not depend on boundary triplets, is also given in [190], Theorem 14.20. The role of the Weyl function M is played by another operator-valued Nevanlinna function $M_{A,\mathcal{K}}$ defined in terms of a closed subspace $\mathcal{K}$ of $\mathcal{N}(T^* - iI)$ and a self-adjoint operator B on $\mathcal{K}$, $\mathcal{K}$ and B being uniquely determined by the two self-adjoint extensions. To be specific, for $z \in \rho(A)$,

$$M_{A,\mathcal{K}}(z)) = P_{\mathcal{K}} (I + zA)(A - zI)^{-1} \upharpoonright_{\mathcal{K}} \in B(\mathcal{K}), \quad (6.2.14)$$

and for $z \in \rho(\tilde{A}) \cap \rho(A)$,

$$(\tilde{A} - zI)^{-1} - (A - zI)^{-1}$$
$$= (A - iI)(A - zI)^{-1}(B - M_{A,\mathcal{K}}(z))^{-1} P_{\mathcal{K}}(A + iI)(A - zI)^{-1} \quad (6.2.15)$$

6.3 The Friedrichs Extension

A lower semi-bounded symmetric operator T has equal deficiency indices, since $\tilde{\Delta}(T) = \mathbb{C} \setminus \Gamma(T)$ is connected, and hence has a self-adjoint extension. A particularly important self-adjoint extension is the *Friedrichs extension*, which is best studied through quadratic forms. We begin with some basic ideas and notation.

Let Q be a linear subspace of H. A *sesquilinear form* (or just *form*) $t[\cdot, \cdot]$ with domain Q, is a map of $Q \times Q$ into $\mathbb{C}$ and is such that $t[u, v]$ is linear in u and conjugate linear in v. We shall write $t[u]$ for $t[u, u]$ and call it the *quadratic form* associated with $t[\cdot, \cdot]$, with domain $\mathcal{D}(t) = Q$. The *adjoint form* t^* of t is defined by

$$t^*[u, v] = \overline{t[v, u]}$$

and t is said to be *symmetric* if $t^* = t$, i.e.,

$$t[u, v] = \overline{t[v, u]}, \quad \forall\ u, v \in \mathcal{D}(t).$$

The quadratic form $t[\cdot]$ determines the sesquilinear form $t[\cdot, \cdot]$ uniquely by the polarisation identity

6.3 The Friedrichs Extension

$$t[u, v] = \frac{1}{4} \left(t[u+v] - t[u-v] + it[u+iv] - it[u-iv] \right).$$

We shall deal mainly with symmetric forms in this section. The inner-product and norm of H will be denoted by $(\cdot, \cdot)$, $\|\cdot\|$. A symmetric form t is said to be *bounded below* (or *lower semi-bounded*) if there exists $c > -\infty$ such that

$$t[u] \geq c\|u\|^2, \quad \forall\, u \in \mathcal{D}(t),$$

and the supremum of all such c is called the *lower bound* of t and will be denoted by c_ℓ. For any $c \leq c_\ell$,

$$t[u, v] + (1 - c)(u, v) \tag{6.3.1}$$

is an inner product on $\mathcal{D}(t)$, and the associated norms with $c < c_\ell$ are equivalent. Let $\mathcal{Q}(t)$ denote $\mathcal{D}(t)$ with the inner-product

$$(u, v)_{\mathcal{Q}(t)} := t[u, v] + (1 - c_l)(u, v) \tag{6.3.2}$$

and norm

$$\|u\|_{\mathcal{Q}(t)} := (t[u] + (1 - c_\ell)\|u\|)^{1/2}.$$

The inclusion map of $\mathcal{D}(t)$ into H is a natural injection, and it is continuous since

$$\|u\| \leq \|u\|_{\mathcal{Q}(t)}. \tag{6.3.3}$$

The form t is said to be *closed* if $\mathcal{Q}(t)$ is a Hilbert space. If t is not closed, the completion $\tilde{\mathcal{Q}}(t)$ of $\mathcal{Q}(t)$ with respect to $\|\cdot\|_{\mathcal{Q}(t)}$ can be identified with a subspace of H. For by (6.3.3), Cauchy sequences in $\mathcal{Q}(t)$ are also Cauchy sequences in H, and to each $u \in \tilde{\mathcal{Q}}(t)$ there exists a unique $v \in H$. The map $E : u \mapsto v$ is a continuous embedding of $\tilde{\mathcal{Q}}(t)$ into H. The form t then has a closed extension and is said to be *closable*. The *closure* of t is the form $\bar{t}$ defined by

$$\bar{t}[u, v] = \lim_{n \to \infty} t[u_n, v_n],$$

where (u_n), (v_n), are sequences in $\mathcal{Q}(t)$ which converge to u, v, respectively, in $\tilde{\mathcal{Q}}(t)$; moreover, $\tilde{\mathcal{Q}}(t) = \mathcal{Q}(\bar{t})$. The closure is the minimal closed extension of t in the sense that any closed extension of t is also an extension of $\bar{t}$. A typical example of a closable form is determined by a symmetric operator, as we now demonstrate.

Proposition 6.3.1 *Let T be a lower semibounded symmetric operator in H and define*

$$t[u, v] = (Tu, v), \quad \mathcal{D}(t) = \mathcal{D}(T). \tag{6.3.4}$$

Then t is symmetric, densely defined, bounded below and closable.

Proof. It is clear that t is symmetric, densely defined and bounded below. Let $(u_n) \subset Q(t)$ be a Cauchy sequence in $Q(t)$ and $u_n \to 0$ in H. We must show that $u_n \to 0$ in $Q(t)$.

$$\|u_n\|_{Q(t)}^2 = (u_n, u_n - u_m)_{Q(t)} + (u_n, u_m)_{Q(t)}$$
$$\leq \|u_n\|_{Q(t)} \|u_n - u_m\|_{Q(t)} + (Tu_n, u_m) + (1 - c_\ell)(u_n, u_m).$$

Hence, given $\varepsilon > 0$, there exists $N \in \mathbb{N}$ such that for $m, n > N$,

$$\|u_n\|_{Q(t)}^2 < \varepsilon + |(Tu_n, u_m)|.$$

On allowing $m \to \infty$ we get $\|u_n\|_{Q(t)}^2 < \varepsilon$ for $n > N$ and the proposition follows. ∎

The form (6.3.4) may not be closed, even if T is self-adjoint. All we can say is that the domain $\mathcal{D}(T)$ of T is a *core* of the closure $\bar{t}$ of t; this is defined to mean that $\mathcal{D}(T)$ is a dense subspace of the Hilbert space $Q(\bar{t})$. The following representation theorem establishes, in particular, that there exists a lower-semibounded, self-adjoint operator $\tilde{T}$ such that $\bar{t}[u, v] = (\tilde{T}u, v)$ for $u, v \in \mathcal{D}(\tilde{T})$, and $\tilde{T}$ is an extension of T. The self-adjoint operator $\tilde{T}$ is called the *Friedrichs* extension of T and was introduced in [86]. More generally, the theorem that follows is a characterisation of all symmetric, densely defined, lower semi-bounded, closed sesquilinear forms. A generalisation of the result, in which the sesquilinear form is sectorial (i.e., takes values in a sector in $\mathbb{C}$ rather than an interval as here) will be given in Section 6.6 below.

Hereafter, $Q(t)$ will denote both the domain $\mathcal{D}(t)$ of a closed form t as a subspace of H, and also $E^{-1}\mathcal{D}(t)$, the Hilbert space defined by $\mathcal{D}(t)$ endowed with the inner-product (6.3.2). The meaning will be clear from the context, making the inclusion of E^{-1} unnecessary.

Theorem 6.3.2 *Let t be a closed, lower semi-bounded symmetric form whose domain $Q(t)$ is a dense subspace of H. Then there exists a lower semi-bounded self-adjoint operator T which has the following properties:*

1. $\mathcal{D}(T) \subset \mathcal{D}(t)$ *and*

$$t[u, v] = (Tu, v), \quad \forall \, u \in \mathcal{D}(T), \, v \in Q(t); \quad (6.3.5)$$

2. $\mathcal{D}(T)$ *is a core of t;*
3. *if $u \in Q(t)$, $w \in H$ and*

$$t[u, v] = (w, v) \quad (6.3.6)$$

for every v in a core of t, then $u \in \mathcal{D}(T)$ and $Tu = w$. The self-adjoint operator T is uniquely defined by the conditions (1) and (2).
4. *T and t have the same lower bound.*

6.3 The Friedrichs Extension

Proof. We may suppose that t is non-negative, for otherwise if t has lower bound γ, we can argue with the non-negative form $t - \gamma$, where $(t - \gamma)[u, v] := t[u, v] - \gamma(u, v)$. We now have that $(t + 1)[u, v] = (u, v)_{Q(t)}$, the inner product on $Q(t)$. The embedding $E : Q(t) \hookrightarrow H$ is injective, has dense range in H and norm ≤ 1. Its adjoint E^* is a linear map of H into $Q(t)$ given by

$$(x, E\varphi) = (E^*x, \varphi)_{Q(t)} \quad x \in H, \; \varphi \in Q(t).$$

We therefore have the triplet of spaces

$$Q(t) \hookrightarrow^E H \hookrightarrow^{E^*} Q(t),$$

with continuous embeddings E, E^* having dense ranges and norms ≤ 1.

In the proof, it helps to be pedantic and distinguish between elements and their images under the embeddings E and E^*; thus (6.3.5) will read

$$t[u, v] = (TEu, Ev), \quad \forall \; u \in E^{-1}\mathcal{D}(T), \; v \in Q(t).$$

The operator $T' := (EE^*)^{-1} = (E^*)^{-1}E^{-1}$ is self-adjoint in H, with dense domain $\mathcal{D}(T') = \mathcal{R}(EE^*)$ in H, and $E^{-1}\mathcal{D}(T') = \mathcal{R}(E^*)$ is a dense subspace of $Q(t)$. Thus $E^{-1}\mathcal{D}(T') \subset Q(t)$ and for $u \in E^{-1}\mathcal{D}(T')$ and $v \in Q(t)$,

$$(t + 1)[u, v] = (u, v)_{Q(t)} = (E^*T'Eu, v)_{Q(t)}$$
$$= (T'Eu, Ev).$$

Hence parts 1 and 2 are satisfied with $T = T' - I$.

Suppose that for $u \in Q(t)$ and $w \in H$,

$$t[u, v] = (w, Ev)$$

for all v in a dense subspace of $Q(t)$. Then

$$(TEu, Ev) = t[u, v] = (w, Ev) = (E^*w, v)_{Q(t)},$$

and this can be extended to all $v \in Q(t)$ by continuity. It follows that $Eu \in \mathcal{D}(T)$ and $TEu = w$.

Suppose that S is another self-adjoint operator with $E^{-1}\mathcal{D}(S) \subset Q(t)$ which satisfies

$$t[u, v] = (SEu, Ev), \quad \forall \; u \in E^{-1}\mathcal{D}(S), \; v \in Q(t),$$

and $E^{-1}\mathcal{D}(S)$ is dense in $Q(t)$. It follows from part 3 that $Eu \in \mathcal{D}(T)$ and $TEu = SEu$. Hence $S \subset T$ and consequently $S = T$ since T and S are self-adjoint.

Let $u \in Q(t)$ with $\|u\|_{Q(t)} = 1$. Then, by part 2, there exists a sequence $(u_n) \in E^{-1}\mathcal{D}(T)$ such that $u_n \to u$ in $Q(t)$; we may suppose that $\|u_n\|_{Q(t)} = 1$, for otherwise, replace u_n by $u_n/\|u_n\|_{Q(t)}$ (if $u_n \neq 0$). Since $t[u_n] = (TEu_n, Eu_n)$, part 4 follows. ∎

The unique self-adjoint operator T determined in Theorem 6.3.2 will be referred to as the *operator associated with* t. Also t will be called the *form of* T and $Q(t)$ the *form domain* of T; it is customary to use $Q(T)$ to denote the form domain of T, and we shall use the lower case t for the form of T. Note that if S is any lower semi-bounded symmetric operator, the form of its Friedrichs extension is the closure of the form

$$s[u, v] = (Su, v) \quad u, v \in \mathcal{D}(S).$$

We shall write S_F for the Friedrichs extension of S, s_F for its form in Theorem 6.3.2, and $Q(S) = Q(S_F) := Q(s_F)$.

A partial ordering is defined on the set of lower semi-bounded self-adjoint operators as follows: we say that $A \leq B$ if $Q(B) \subset Q(A)$ and for all $u \in Q(B)$, $a[u] \leq b[u]$, where a, b are the forms of A, B respectively. In [119], Theorem VI.2.21, it is proved that if A, B have lower bounds γ_1, γ_2 respectively, then in order for $A \leq B$, it is necessary that $\gamma_1 \leq \gamma_2$ and for every $\lambda < \gamma_1$, $(B - \lambda I)^{-1} \leq (A - \lambda I)^{-1}$, thus,

$$((B - \lambda I)^{-1}u, u) \leq ((B - \lambda I)^{-1}u, u) \ \forall \ u \in H;$$

also, it is sufficient that $(B - \lambda I)^{-1} \leq (A - \lambda I)^{-1}$ for some $\lambda < \min\{\gamma_1, \gamma_2\}$. The next two theorems show that the Friedrichs extension A_F is maximal among lower semi-bounded self-adjoint extensions of A, and is the only such self-adjoint extension with domain in $Q(A_F)$.

Theorem 6.3.3 *Let A be a lower semi-bounded symmetric operator with Friedrichs extension A_F. Let B be any lower semi-bounded self-adjoint extension of A. Then $Q(A_F) \subset Q(B)$ and $b[f] = a[f]$ for all $f \in Q(A_F)$, i.e., $A_F \leq B$. Thus the Friedrichs extension has the smallest form domain amongst the self-adjoint extensions.*

Proof. The form b of B is the closure of the form $b'[u] := (Bu, u)$ defined on $\mathcal{D}(B)$. Since B is an extension of A, we have that $(Bu, u) = (Au, u)$ on $\mathcal{D}(A)$ and hence b is an extension of a. This implies that $Q(B) \supset Q(A)$ and $b = a$ on $Q(A)$. ∎

Theorem 6.3.4 *The Friedrichs extension of A is the only self-adjoint extension of A with domain contained in $Q(A)$.*

Proof. Let B be a lower semi-bounded self-adjoint extension of A with $\mathcal{D}(B) \subset Q(A)$. As in the preceding proof, it follows that b is an extension of a. Thus, for $u \in \mathcal{D}(B)$ and $v \in Q(A)$, we have

$$a[u, v] = (Bu, v).$$

Consequently, by Theorem 6.3.2 (3), $B \subset A_F$, and so $B = A_F$ since both operators are self-adjoint. ∎

6.3 The Friedrichs Extension

Theorem 6.3.5 *Let A be a lower semi-bounded symmetric operator, with Friedrichs extension A_F. Then*

$$\mathcal{D}(A_F) = Q(A) \cap \mathcal{D}(A^*). \tag{6.3.7}$$

Proof. Let B be the restriction of A^* to $Q(A) \cap \mathcal{D}(A^*)$. Since $\mathcal{D}(A)$ is dense in $Q(A)$, for any $u \in Q(A)$, there exists a sequence (u_n) in $\mathcal{D}(A)$ which converges to u in $Q(A)$, and hence in H. Thus, for any $u \in \mathcal{D}(B) = Q(A) \cap \mathcal{D}(A^*)$,

$$\begin{aligned}(Bu, u) &= \lim_{n \to \infty} (A^*u, u_n) \\ &= \lim_{n \to \infty} (u, Au_n) \\ &= \lim_{n \to \infty} a_F[u, u_n] \\ &= a_F[u, u].\end{aligned}$$

Consequently B is symmetric and bounded below, and, as in the proof of the previous theorem, $B \subset A_F$. Since $A \subset A_F$ implies that $A_F \subset A^*$, it follows that $A_F \subset B$ and hence $B = A_F$. ∎

The last theorem in this section is called the *second representation theorem* in [119]. We remind the reader that a non-negative self-adjoint operator S has a square root $S^{1/2}$, which is a non-negative self-adjoint operator satisfying $(S^{1/2})^2 = S$. Furthermore, $\mathcal{D}(S)$ is a dense subspace of $\mathcal{D}(S^{1/2})$ with the graph norm of $S^{1/2}$, namely

$$\|u\|_{\mathcal{D}(T^{1/2})} = \left\{\|T^{1/2}u\|^2 + \|u\|^2\right\}^{1/2}.$$

Theorem 6.3.6 *Let t be a closed, densely defined, non-negative symmetric form, and let T be the associated non-negative self-adjoint operator. Then, $Q(t) = Q(T) = \mathcal{D}(T^{1/2})$ and*

$$t[u, v] = (T^{1/2}u, T^{1/2}v), \quad \forall u, v \in \mathcal{D}(T^{1/2}). \tag{6.3.8}$$

Proof. For $u, v \in \mathcal{D}(T) \subset \mathcal{D}(T^{1/2})$,

$$t[u, v] = (Tu, v) = (T^{1/2}u, T^{1/2}v).$$

The theorem follows by continuity, since $\mathcal{D}(T^{1/2})$ is dense in both $Q(t)$ and $\mathcal{D}(T^{1/2})$ with the graph norm of $T^{1/2}$. ∎

Example 6.3.7

Let Ω be a bounded open subset of $\mathbb{R}^n$ and let T be the non-negative symmetric operator defined by $Tu = -\Delta u$ for $u \in C_0^\infty(\Omega)$. Then, the form domain of its Friedrichs extension T_F is the completion of $C_0^\infty(\Omega)$ with respect to $(t+1)^{1/2}[\cdot]$, where

$$(t+1)[u] = \int_\Omega (-\Delta u + u)\bar{u}d\mathbf{x}$$
$$= \int_\Omega \left(|\nabla u|^2 + |u|^2\right) d\mathbf{x}.$$

Thus the form domain is the Sobolev space $\overset{0}{W}{}^1_2(\Omega)$. The Friedrichs extension satisfies $T_F u = -\Delta u$ in the distributional sense, with $\Delta u \in L_2(\Omega)$; also $\mathcal{D}(T_F) \subset Q(A_F)$. Hence $T_F u = -\Delta u$ with domain

$$\mathcal{D}(T_F) = W^2_{2,loc}(\Omega) \cap \overset{0}{W}{}^1_2(\Omega),$$

where $W^2_{2,loc}(\Omega)$ is the set of functions which are in $W^2_2(K)$ for every compact subset K of Ω. From the proof of Theorem 6.3.2, $T_F + I = (EE^*)^{-1}$ where E is the natural embedding $\overset{0}{W}{}^1_2(\Omega) \hookrightarrow L_2(\Omega)$ and E^* is its adjoint $L_2(\Omega) \hookrightarrow \overset{0}{W}{}^1_2(\Omega)$.

6.4 The Krein-Vishik-Birman (KVB) Theory

The von Neumann theory gives all the self-adjoint extensions of any symmetric operator T. If T is non-negative, the non-negative self-adjoint extensions of T are not evident from the general von Neumann theory, but since any $\mu < 0$ lies in the field of regularity of T, the existence of a self-adjoint extension A of T with $\mu \in \rho(A)$ is guaranteed by Calkin's result referred to in Remark 6.1.2, and so by Theorem 6.2.10, the self-adjoint extensions of T are characterised by self-adjoint operators B acting in the kernel of $T^* - \mu I$. If, in particular, T is assumed to be positive, then we can take A to be the Friedrichs extension T_F of T to obtain the following:

Theorem 6.4.1 *Let T be a positive, closed, symmetric operator and T_F its Friedrichs extension. For $B \in \mathcal{S}(\mathcal{N})$, i.e., a self-adjoint operator acting in $\mathcal{N} := \ker T^*$, let T_B be the restriction of T^* to the domain*

$$\mathcal{D}(T_B) := \left\{x + (T_F)^{-1}(Bu + v) + u : x \in \mathcal{D}(T), u \in \mathcal{D}(B), v \in \mathcal{N}, v \perp \mathcal{D}(B)\right\};$$
(6.4.1)

hence

$$T_B(x + T_F^{-1}(Bu + v) + u) = Tx + Bu + v. \qquad (6.4.2)$$

Then T_B is a self-adjoint extension of T in H. Conversely, every self-adjoint extension of T is of the form T_B for some unique $B \in \mathcal{S}(\mathcal{N})$.

In [136] Krein showed that if T is positive (in fact, he only assumed that T is non-negative), then there are two special self-adjoint extensions of T, T_K and T_F. The fundamental theorem of Krein (to be proved later in this section) is

6.4 The Krein-Vishik-Birman (KVB) Theory

Theorem 6.4.2 *Let T be a positive symmetric operator in H. It has two distinguished positive self-adjoint extensions, namely the Krein extension T_K and the Friedrichs extension T_F. The set of positive self-adjoint extensions of T is precisely the set of self-adjoint operators S satisfying*

$$T_K \leq S \leq T_F. \tag{6.4.3}$$

In the sense of (6.4.3), T_K and T_F are respectively the smallest and largest self-adjoint extensions of T, and are referred to as the "soft" and "hard" extensions of T. The extension T_K was first considered by von Neumann in [213], but it was Krein in [136] who uncovered its extremal role in (6.4.3), as well as only supposing that T is non-negative. The extension T_K is now called the *Krein–von Neumann* extension of T; T_F is the Friedrichs extension of T. Both these extensions have important roles in the application of the theory of symmetric operators. The Friedrichs extension has an acknowledged natural part to play in quantum mechanics, while in [97], Grubb demonstrates that the Krein-von Neumann extension has a natural role in elasticity theory by describing an intimate connection between the eigenvalues of the Krein-von Neumann extension of an elliptic differential operator of even order and those of a higher-order problem concerning the buckling of a clamped plate. An abstract version of the latter connection was established in [13], and this will be reproduced in Proposition 6.4.12 below.

The task of characterising all the non-negative self-adjoint extensions of a positive symmetric operator T was initiated by Krein in [136, 137] and further developed by Višik in [212] and then by Birman in [24]. The Krein-Birman-Vishik (KVB) theory is also presented in [96] as a special case of results characterising the closed operators S which are such that $A_1 \subset S \subset A_2^*$, where A_1, A_2 is a given adjoint pair of closed operators, i.e., $A_1 \subset A_2^*$; the abstract theory in [96] is motivated and illustrated by a comprehensive analysis of boundary conditions which define such operators S when A_1 and A_2 are realisations of elliptic differential operators. There is also a review of the KVB theory in [5], and we borrow liberally from this in the present section.

The von Neumann theory exploited ideas from operator theory, with Cayley transforms playing a leading role. The KVB theory is based on quadratic forms, just as the Friedrichs extension was shown to be in the last section. The essence of the KVB theory is that there is a one-to-one correspondence between all closed, positive quadratic forms and positive self-adjoint operators acting in some closed subspace of H. This necessitates consideration of self-adjoint operators, and their associated forms, defined in proper closed subspaces of H, and the extension of their domains to dense subspaces of H.

In [5] a lower semi-bounded symmetric form t takes values in $(-\infty, \infty]$, and its domain $Q(t)$ is the set of elements $u \in H$ which are such that $t[u] < \infty$. Given a non-negative self-adjoint operator B acting in a closed subspace M of H, we can say from Theorem 6.3.6 that its form satisfies

$$t[u] = \begin{cases} \|B^{1/2}u\|^2, & \text{if } u \in \mathcal{D}(B^{1/2}), \\ \infty, & \text{if } u \notin \mathcal{D}(B^{1/2}). \end{cases} \tag{6.4.4}$$

Note that $(\cdot, \cdot)$, $\|\cdot\|$ will denote the inner-product and norm of H throughout the section. If B is strictly positive, we extend B^{-1} to H by setting $B^{-1} = 0$ on $M^\perp$ and then using linearity; thus with any $u \in H$ written as $u = v + w$, $v \in M$, $w \in M^\perp$, we have $(B - \lambda I)^{-1} u = (B - \lambda I)^{-1} v$. The case $B = 0$ corresponds to $Q(t) = M$, and the case $B = \infty$ to $Q(B) = \{0\}$.

Before proceeding to the main theorems, we need some preliminary results and constructions. We shall assume throughout that T is a closed symmetric operator and that, without loss of generality, $T \geq I$, i.e.,

$$(Tu, u) \geq \|u\|^2, \quad \forall\ u \in \mathcal{D}(T). \tag{6.4.5}$$

Also we set $\mathcal{N} := \ker T^* = [\mathcal{R}(T)]^\perp$, where $\mathcal{R}(T)$ denotes the range of T, and recall that the form domain $Q(T_F)$ of the Friedrichs extension T_F is $\mathcal{D}(T_F^{1/2})$. Thus (6.4.4) is satisfied by $t = t_F$ and $B = T_F$.

Lemma 6.4.3

$$\mathcal{N} \cap Q(T_F) = \{0\}.$$

Proof. Let $u \in \mathcal{N} \cap Q(T_F)$. Then, by Theorem 6.3.2 (1), there exist $u_n \in \mathcal{D}(T)$, $n \in \mathbb{N}$, such that

$$t_F[u] = \lim_{n\to\infty} (Tu_n, u) = \lim_{n\to\infty} (u_n, T^* u) = 0.$$

Since $t_F[u] \geq \|u\|^2$, the lemma follows. ∎

Lemma 6.4.4

$$(i)\quad \mathcal{D}(T^*) = \mathcal{D}(T_F) \dotplus \mathcal{N}. \tag{6.4.6}$$

$$(ii)\quad \mathcal{D}(T_F) = \mathcal{D}(T) \dotplus T_F^{-1} \mathcal{N}. \tag{6.4.7}$$

Proof. (i). Let $\psi \in \mathcal{D}(T^*)$. Then since $T_F \subset T^*$, we have $\varphi = T_F^{-1} T^* \psi \in \mathcal{D}(T^*)$ and $T^*(\psi - \varphi) = T^* \psi - T^* \psi = 0$. Therefore $\eta = \psi - \varphi \in \mathcal{N}$ and $\mathcal{D}(T^*) = \mathcal{D}(T_F) \dotplus \mathcal{N}$. Moreover, if $\varphi \in \mathcal{D}(T_F) \cap \mathcal{N}$, then, $\varphi = 0$ by Lemma 6.4.3.

(ii). Since T is closed and we are assuming that $T \geq I$ in (6.4.5), it follows that $\mathcal{R} = \mathcal{R}(T)$ is closed and then $\mathcal{R} = \mathcal{N}^\perp$. For any $\psi \in \mathcal{D}(T_F)$, we can therefore write $T_F \psi = T\varphi + \eta$, where $\varphi \in \mathcal{D}(T)$ and $\eta \in \mathcal{N}$. This implies that $\psi = \varphi + T_F^{-1} \eta$, and so $\mathcal{D}(T_F) = \mathcal{D}(T) \dotplus T_F^{-1} \mathcal{N}$. If $\varphi \in \mathcal{D}(T) \cap T_F^{-1} \mathcal{N}$, then for some $\eta \in T_F^{-1} \mathcal{N}$, $(T\varphi, T\varphi) = (T T_F^{-1} \eta, T\varphi) = (\eta, T\varphi) = 0$ and so $\varphi = 0$. ∎

Let B be a lower semi-bounded, self-adjoint operator acting in the closed subspace $\mathcal{N} = \ker T^*$ of H, with form b, and define the following form t_B with domain $Q(t_B)$:

$$Q(t_B) = Q(T_F) \dotplus Q(B)$$
$$t_B[u + v] = t_F[u] + b[\eta], \tag{6.4.8}$$

for $u \in Q(T_F)$, $\eta \in Q(B)$.

6.4 The Krein-Vishik-Birman (KVB) Theory

Remark 6.4.5

From Lemma 6.4.4, $u \in \mathcal{D}(T^*)$ if and only if $u = u_F + u_N$, where $u_F \in \mathcal{D}(T_F)$, $u_N \in \mathcal{N} = \ker T^*$, and $u_F = u_T + T_F^{-1}\eta_u$, where $u_T \in \mathcal{D}(T)$ and $\eta_u \in \mathcal{N}$. Let $u, v \in \mathcal{D}(T^*)$. Then with obvious notation,

$$\begin{aligned}(T^*u, v) &= (T_F u_F + T^* u_N, v_F + v_N) \\ &= (T_F u_F, v_F) + (T_F u_F, v_N) \\ &= (u_F, T_F v_F) + (T u_T + \eta_u, v_N) \\ &= (u_F, T_F v_F) + (\eta_u, v_N),\end{aligned}$$

since $(T u_T, v_N) = (u_T, T^* v_N) = 0$. Similarly,

$$(u, T^*v) = (u_F, T_F v_F) + (u_N, \eta_v).$$

Hence

$$(T^*u, v) - (u, T^*v) = (\eta_u, v_N) - (u_N, \eta_v).$$

Let $\Gamma_0 : u \mapsto u_N : \mathcal{D}(T^*) \to \mathcal{N}$, $\Gamma_1 : u \mapsto \eta_u : \mathcal{D}(T^*) \to \mathcal{N}$. Then

$$(T^*u, v) - (u, T^*v) = (\Gamma_1 u, \Gamma_0 v)_\mathcal{N} - (\Gamma_0 u, \Gamma_1 v)_\mathcal{N}. \tag{6.4.9}$$

Since $u \mapsto \langle \Gamma_0 u, \Gamma_1 u \rangle : \mathcal{D}(T^*) \to \mathcal{N} \oplus \mathcal{N}$ is surjective, we have shown that $(\mathcal{N}, \Gamma_0, \Gamma_1)$ is a boundary triplet for T^*.

Theorem 6.4.6 *If $B \geq 0$, t_B is a non-negative, closed, symmetric quadratic form in H and its associated self-adjoint operator T_B is an extension of T.*

Proof. Clearly t_B is symmetric and non-negative. To prove that it is closed we first note that

$$\begin{aligned}t_B[\varphi + \eta] + \|\varphi + \eta\|^2 &= a_F[\varphi] + b[\eta] + \|\varphi + \eta\|^2 \\ &\geq \|\varphi\|^2 + \|\varphi + \eta\|^2.\end{aligned}$$

From these inequalities it follows that if $(\varphi_n + \eta_n)$ is Cauchy in $Q(t_B)$, then (φ_n) and (η_n) are Cauchy in $Q(T_F)$ and $Q(B)$ respectively. Since the forms t_F, b are closed, (φ_n), (η_n) therefore converge in $Q(A_F)$, $Q(B)$, respectively, and we then see that $(\varphi_n + \eta_n)$ converges in $Q(t_B)$. Hence t_B is closed.

To prove that the self-adjoint operator T_B associated with t_B is an extension of T, we invoke Theorem 6.3.2(3) and prove that for all $\psi \in Q(T_B)$ and $\varphi \in \mathcal{D}(T)$,

$$t_B[\psi, \varphi] = (\psi, T\varphi),$$

where $t_B[\cdot, \cdot]$ is the sesquilinear form derived from $t_B[\cdot]$ by the polarisation identity. Let $\psi = \varphi' + \eta'$, $\varphi' \in Q(T_F)$, $\eta' \in \mathcal{N}$. Then, for $\varphi \in \mathcal{D}(T)$, $(\eta', T\varphi) = (T^*\eta', \varphi) = 0$, and so, since $\varphi \in \mathcal{D}(T) \subset Q(T_F)$,

$$t_B[\psi, \varphi] = t_F[\varphi', \varphi]$$
$$= (\varphi', T\varphi) = (\psi, T\varphi)$$

This completes the proof. ∎

Theorem 6.4.7 *Let $\tilde{T}$ be a positive self-adjoint extension of T. Then there exists a positive self-adjoint operator B in $\mathcal{N}$ such that $\tilde{T} = T_B$.*

Proof. Let $\psi \in \mathcal{D}(\tilde{T})$. Since $\mathcal{D}(\tilde{T}) \subset \mathcal{D}(T^*)$, we have from (6.4.6) that $\psi = \varphi + \eta$, with $\varphi \in \mathcal{D}(T_F)$ and $\eta \in \mathcal{N}$. Furthermore, as $Q(T_F) \subset Q(\tilde{T})$ by Theorem 6.3.3, it follows that $\eta \in \tilde{N} := \mathcal{N} \cap Q(\tilde{T})$. Since $\tilde{t}$, the form of $\tilde{T}$, is closed and positive, so is its restriction to $\tilde{N}$ and hence this restriction must be the form of some positive self-adjoint operator B acting in $\tilde{N}$; we set $\tilde{b} = \tilde{t} \upharpoonright_{\tilde{N}}$ and consider it extended to $\mathcal{N}$ by setting it equal to ∞ on $\mathcal{N} \setminus \tilde{N}$.

With $\psi = \varphi + \eta \in \mathcal{D}(\tilde{T})$ as above, we choose $\varphi_n \in \mathcal{D}(T)$ such that $\varphi_n \to \varphi$ in $Q(T_F)$. Then

$$(\varphi, \tilde{T}\eta) = \lim_{n \to \infty} (\varphi_n, \tilde{T}\eta) = \lim_{n \to \infty} (T\varphi_n, \eta)$$
$$= \lim_{n \to \infty} (\varphi_n, T^*\eta) = 0.$$

Thus
$$(\psi, \tilde{T}\psi) = (\varphi, \tilde{T}\varphi) + (\eta, \tilde{T}\eta) = (\varphi, T_F\varphi) + (\eta, B\eta).$$

Consequently, the restriction $\tilde{t} \upharpoonright \mathcal{D}(\tilde{T})$ is equal to $t_B | \mathcal{D}(\tilde{T})$. Any $\psi \in Q(t_B)$ can be written as $\psi = \varphi + \eta$, where $\varphi \in Q(T_F) \subset Q(\tilde{T})$ and $\eta \in Q(\tilde{b})$. Since $\mathcal{D}(\tilde{T})$ is dense in $Q(\tilde{T})$, and hence in $Q(T_F)$ by Theorem 6.3.3, and $\mathcal{D}(\tilde{T}) \cap \tilde{N}$ is dense in $Q(\tilde{b})$, it follows that $\mathcal{D}(\tilde{T})$ is dense in $Q(t_B)$. This implies that $\tilde{T} = T_B$ and the theorem is proved. ∎

Remark 6.4.8

Theorem 6.4.6 and Theorem 6.4.7 establish a one-one correspondence between positive self-adjoint extensions T_B of T and positive self-adjoint operators acting in $\mathcal{N} = \ker T^*$. More details and further information may be found in [5], and the original papers [24], [25, 136], [137, 212]. Theorem 6.4.7 is also Theorem 15.3 in [77]. Note that it is proved in [5] that the word "positive" may be deleted from the theorems if $\dim \mathcal{N} < \infty$.

Remark 6.4.9

In (6.4.4), the Friedrich extension T_F of T is that given by the choice $\mathcal{D}(B) = \{0\}$ (i.e. formally, $B = \infty$). Therefore, as already established in (6.4.7),

$$\mathcal{D}(T_F) = \mathcal{D}(T) \dotplus T_F^{-1}\mathcal{N}.$$

6.4 The Krein-Vishik-Birman (KVB) Theory

The Krein-von Neumann extension T_K corresponds to the case $B = 0$; thus $\mathcal{D}(B)^\perp = \{0\}$ in (6.4.4) giving $\mathcal{D}(T_K) = \mathcal{D}(T) \dotplus \mathcal{N}$. If $\varphi \in \mathcal{D}(T) \cap \mathcal{N}$, then since T is symmetric, $T\varphi = T^*\varphi = 0$ and $\varphi = 0$ since T is positive. Hence

$$\mathcal{D}(T_K) = \mathcal{D}(T) \dotplus \mathcal{N}. \tag{6.4.10}$$

This implies in particular that

$$\ker T_K = \mathcal{N}. \tag{6.4.11}$$

We note also that with $B = 0$ and so $Q(B) = \mathcal{N}$ in (6.4.8),

$$Q(T_K) = Q(T_F) \dotplus \mathcal{N}. \tag{6.4.12}$$

In Theorem 6.2.8, we proved that a boundary triplet $(\mathcal{K}, \Gamma_0, \Gamma_1)$ determines a self-adjoint extension T_B of T defined as the restriction of T^* to the domain

$$\mathcal{D}(T_B) = \left\{ u \in \mathcal{D}(T^*) : \Gamma_0 u \in \mathcal{D}(B), \ B\Gamma_0 u = P_B \Gamma_1 u \right\},$$

where P_B denotes a projection on H. Let $\mathcal{K} = \mathcal{N}$ and consider the boundary triplet in Remark 6.4.5. The choice $B = \infty$, $\mathcal{D}(B) = \{0\}$ yields

$$\mathcal{D}(T_\infty) = \left\{ u \in \mathcal{D}(T^*) : u = u_F \right\}.$$

Thus, T_∞ is the Friedrichs extension T_F of T. If $B = 0$, then

$$\mathcal{D}(T_0) = \left\{ u \in \mathcal{D}(T^*) : u = u_T + u_N \right\}.$$

Thus by (6.4.10), T_0 is the Krein-von Neumann extension T_K of T.

Since T is positive, it follows that $0 \in \mathbb{C} \setminus \Gamma(T)$ and the deficiency indices of T are equal to nul T^*, i.e., the dimension of $\mathcal{N}$. Therefore, if T is not self-adjoint, T_K has an eigenvalue at 0 with multiplicity equal to the deficiency index of T, and this is nul T^*.

Remark 6.4.10

A similar result characterising self-adjoint extensions $\tilde{T}$ of a symmetric operator T is given in [96], Theorem II.2.1. In it the positivity of T is replaced by the hypothesis of Theorem 6.2.10 with $\mu = 0$, namely that there exists a self-adjoint extension A with $0 \in \rho(A)$. This determines the decomposition $\mathcal{D}(T^*) = \mathcal{D}(A) \dotplus \mathcal{N}$ of Proposition 6.1.3 and the result features the projection of $\mathcal{D}(T^*)$ onto closed subspaces of $\mathcal{N}$. In fact, there is considered in [96] the more general problem of characterising all the closed, densely defined operators $\tilde{T}, \tilde{T}'$ which satisfy $T \subset \tilde{T} \subset (T')^*$ and $T' \subset \tilde{T}' \subset T^*$, where T, T' is an adjoint pair of closed, densely defined operators; this problem and Grubb's treatment in [96] is discussed in Section 6.5 below. The case of self-adjoint extensions of a symmetric operator T is, of course, when $T = T'$.

Theorem 1 in [12] is also noteworthy in the context of this section: that a non-negative self-adjoint extension $\tilde{T}$ of a non-negative symmetric operator T is such that the form $(\tilde{T}u, v) - t_K[u, v]$, $u, v \in \mathcal{D}(\tilde{T})$, is non-negative and closable in $Q(T_K)$. Moreover, the formula

$$\tilde{t}[u, v] = t_K[u, v] + t[u, v], \quad u, v \in Q(\tilde{T}) = Q(t),$$

gives a one-one correspondence between all closed forms $\tilde{t}$ associated with non-negative self-adjoint extensions $\tilde{T}$ of T and all non-negative forms t which are closed in $Q(T_K)$ and are such that $t[f] = 0$ for all $f \in Q(T)$.

Proof of Krein's Theorem 6.4.2 Let T_B, T_F, B be the operators and t_B, t_F, b the forms with domains $Q(T_B)$, $Q(T_F)$, $Q(B)$ defined in (6.4.8). Thus

$$Q(T_B) = Q(T_F) \dotplus Q(B)$$

and if $\psi = \varphi + \eta \in Q(T_B)$, where $\varphi \in Q(T_F)$, $\eta \in Q(B)$, then

$$t_B[\psi] = t_F[\varphi] + b[\eta].$$

Therefore $T_B \geq T_{B'}$ if and only if $B \geq B'$. Since T_K and T_F correspond to the choices $B = 0$ and $B = \infty$ respectively, we infer from $0 \leq B \leq \infty$ that $T_K \leq T_B \leq T_F$ for all positive self-adjoint operators B acting in $\mathcal{N}$.

Suppose that S is a self-adjoint operator which satisfies $T_K \leq S \leq T_F$. Then

$$Q(T_F) \subset Q(S) \subset Q(T_K) = Q(T_F) \dotplus \mathcal{N},$$

by (6.4.11), and by Theorem 6.3.3,

$$s[u] = t_F[u] \; \forall \; u \in Q(T_F),$$

where $s[\cdot]$ is the form of S. Let $\varphi \in Q(T_F)$, $\eta \in \mathcal{N}$, $\lambda \in \mathbb{C}$. Then

$$\begin{aligned}
t_F[\varphi] &= t_K[\varphi + \lambda \eta] \\
&\leq s[\varphi + \lambda \eta] \\
&= s[\varphi] + 2re\,[\lambda s[\varphi, \eta]] + |\lambda|^2 s[\eta] \\
&= t_F[\varphi] + 2re\,[\lambda s[\varphi, \eta]] + |\lambda|^2 s[\eta].
\end{aligned}$$

Hence

$$2re\,[\lambda s[\varphi, \eta]] + |\lambda|^2 s[\eta] \geq 0,$$

and since $\lambda \in \mathbb{C}$ is arbitrary, we have that

$$s[\eta] \geq 0, \quad s[\varphi, \eta] = 0, \quad \forall \varphi \in Q(T_F), \; \eta \in \mathcal{N}. \tag{6.4.13}$$

6.4 The Krein-Vishik-Birman (KVB) Theory

This gives

$$s[\varphi + \eta] = s[\varphi] + s[\eta], \quad s[\eta] \geq 0,$$

and on $\mathcal{N}$, $s = b$, the form of some non-negative self-adjoint operator B. Thus $s = t_B$, $S = T_B$ and the proof is complete. ∎

Proposition 6.4.11 *Let T be a positive symmetric operator in H. Then*

$$\mathcal{D}(T_F) \cap \mathcal{D}(T_K) = \mathcal{D}(T); \tag{6.4.14}$$

T_F *and* T_K *are said to be relatively prime.*

Proof. By (6.4.7) and (6.4.10)

$$\mathcal{D}(T_F) = \mathcal{D}(T) \dotplus T_F^{-1}\mathcal{N}$$
$$\mathcal{D}(T_K) = \mathcal{D}(T) \dotplus \mathcal{N}.$$

It therefore suffices to prove that

$$\mathcal{N} \cap T_F^{-1}\mathcal{N} = \{0\}.$$

Let $u \in \mathcal{N} \cap T_F^{-1}\mathcal{N}$. Then $T^*u = 0$ and $u = T_F^{-1}v$, where $T^*v = 0$. Since $T_F \subset T^*$ it follows that $v = T_F u = T^* u = 0$. Hence $u = T_F^{-1} v = 0$ and the proof is complete. ∎

The following result, proved in [13], is an abstract version of Proposition 1 in Grubb's paper [97] which establishes that the Krein-von Neumann extension has a natural role to play in elasticity theory.

Proposition 6.4.12 *Let T be a positive symmetric operator and $\lambda \neq 0$. Then there exists $v \in \mathcal{D}(T_K)$ satisfying*

$$T_K v = \lambda v, \quad v \neq 0, \tag{6.4.15}$$

if and only if there exists $u \in \mathcal{D}(T^*T)$ such that

$$T^* T u = \lambda T u, \quad u \neq 0. \tag{6.4.16}$$

The solutions v of (6.4.15) are in one-to-one correspondence with the solutions u of (6.4.16) given by the formulas

$$u = T_F^{-1} T_K v, \quad v = \lambda^{-1} T u. \tag{6.4.17}$$

Proof. Let (6.4.15) be satisfied, where in view of (6.4.10), $v = u + w$, with $u \in \mathcal{D}(T)$ and $w \in \mathcal{N} = \ker T^*$. Then (6.4.15) is equivalent to

$$v = \lambda^{-1} T_K v = \lambda^{-1} T_K u = \lambda^{-1} T u.$$

Moreover $v \neq 0$ if and only if $u \neq 0$, for $u = 0$ implies $v = 0$, while $v = 0$ implies $u = w = 0$ since the sum in (6.4.10) is a direct sum. Furthermore $u = T_F^{-1} T_K v$ since $T_K v = T_F u$. Finally, $\lambda w = \lambda v - \lambda u = T u - \lambda u \in \mathcal{N}$ implies that $u \in \mathcal{D}(T^*T)$ and

$$0 = \lambda T^* w = T^* (Tu - \lambda u) = T^*Tu - \lambda T^* u = T^*Tu - \lambda Tu.$$

Conversely, suppose $u \in \mathcal{D}(T^*T)$ and $T^*Tu = \lambda Tu$. Then $v := \lambda^{-1} Tu \in \mathcal{D}(T^*)$ and

$$T^* v = \lambda^{-1} T^* Tu = Tu = \lambda v.$$

Since $T^*Tu = \lambda Tu = \lambda T^* u$ implies that $T^* (T - \lambda I) u = 0$ and hence $(T - \lambda I)) u \in \ker T^*$, we have on rewriting v as

$$v = u + \lambda^{-1} (T - \lambda I) u$$

that $v \in \mathcal{D}(T_K)$ by (6.4.10). ∎

Remark 6.4.13

In [97], Proposition 1, Grubb determines an asymptotic formula for the distribution of the eigenvalues of the Krein-von Neumann extension of the minimal operator defined on a smooth domain Ω by an elliptic differential operator of order 2m. Grubb's strategy is to first prove the equivalence of the problem with one for an elliptic system of order m and then use known results which apply to the latter problem. The special case $m = 1$ of Grubb's example is the following problem which is associated with the buckling of a clamped plate:

$$(-\Delta)^2 u = \lambda(-\Delta)u \text{ in } \Omega, \ \lambda \neq 0, \ u \in H_0^2(\Omega), \quad (6.4.18)$$

where Ω is a bounded open set in $\mathbb{R}^n$, $n \geq 2$. The eigenvalue problem for the Krein-von Neumann extension of the Dirichlet Laplacian in $L_2(\Omega)$ is shown to be in one-one correspondence with (6.4.18). A similar strategy is adopted in [13] for the problem on domains whose boundaries are not smooth.

6.5 Adjoint Pairs and Closed Extensions

Let A_0, A_0' be closed, densely defined operators in the Hilbert space H. They are said to form an *adjoint pair* if $A_0 \subset (A_0')^* =: A_1$ and $A_0' \subset A_0^* =: A_1'$. Thus if a densely defined operator A satisfies $A_0 \subset A \subset A_1$, it follows that $A_0' \subset A^* \subset A_1'$. In [96], Grubb considers the very general problem of characterising all the closed operators $\tilde{A}$ in H which, for a given adjoint pair A_0, A_0', satisfy $A_0 \subset \tilde{A} \subset A_1$; the problem considered hitherto concerning the self-adjoint extensions of a symmetric operator A_0 is thus the special case $A_0 = A_0'$. As well as the inherent interest of the abstract problem, the outcome of the investigation is applied in [96] to give a comprehensive

6.5 Adjoint Pairs and Closed Extensions

description of operator realisations of even order elliptic differential expressions on a domain with smooth boundary by means of boundary conditions. We shall give some more details in the section below, but for a thorough treatment a study of the original paper is recommended.

To characterise the operators $\tilde{A}$, it is assumed that there exists a closed operator A_β which is such that $0 \in \rho(A_\beta)$, the resolvent set of A_β, and

$$A_0 \subset A_\beta \subset (A_0')^* =: A_1; \qquad (6.5.1)$$

thus

$$A_0' \subset A_\beta^* \subset A_0^* =: A_1'.$$

Note that the assumption that $0 \in \rho(A_\beta)$ implies that A_0 must have a bounded inverse. Hereafter in this section we denote by $\mathcal{M}$ the set of closed operators $\tilde{A}$ satisfying $A_0 \subset \tilde{A} \subset A_1$, where (6.5.1) is satisfied by the adjoint pair A_0, A_0'; the corresponding set for A_0', A_1' will be denoted by $\mathcal{M}'$.

The first step in Grubb's theory is the following lemma reminiscent of Lemma 6.4.4(i). In it $\mathcal{N}$ stands for the kernel of the operator exhibited.

Lemma 6.5.1

$$\mathcal{D}(A_1) = \mathcal{D}(A_\beta) \dotplus \mathcal{N}(A_1) \qquad (6.5.2)$$

and

$$\mathcal{D}(A_1') = \mathcal{D}(A_\beta^*) \dotplus \mathcal{N}(A_1'). \qquad (6.5.3)$$

Proof. Since $\mathcal{D}(A_\beta) \subset \mathcal{D}(A_1)$ we have $\mathcal{D}(A_\beta) + \mathcal{N}(A_1) \subset \mathcal{D}(A_1)$. Let $u \in \mathcal{D}(A_1)$ and set $u_\beta = A_\beta^{-1} A_1 u$. Then $u_\beta \in \mathcal{D}(A_\beta) \subset \mathcal{D}(A_1)$ and $A_1 u_\beta = A_\beta u_\beta = A_1 u$. Thus $A_1 (u_1 - u_\beta) = 0$ and consequently $\mathcal{D}(A_1) \subset \mathcal{D}(A_\beta) + \mathcal{N}(A_1)$. The sum in (6.5.2) is a direct sum since $v \in \mathcal{D}(A_\beta) \cap \mathcal{N}(A_1)$ implies that $A_\beta v = 0$ and hence $v = 0$ since $0 \in \rho(A_\beta)$.

The proof of (6.5.3) is similar. ∎

We shall denote by P_β and P_η, the projections in (6.5.2) of $\mathcal{D}(A_1)$ onto $\mathcal{D}(A_\beta)$ and $\mathcal{N}(A_1)$ respectively, and represent the corresponding projections in (6.5.3) by $P_{\beta'}$, $P_{\eta'}$.

Lemma 6.5.2 For $u \in \mathcal{D}(A_1)$, $v \in \mathcal{D}(A_1')$,

$$(A_1 u, v) - (u, A_1' v) = (A_1 u, v_{\eta'}) - (u_\eta, A_1' v),$$

where $v_{\eta'} := P_{\eta'} v$, $u_\eta := P_\eta u$ and $(\cdot, \cdot)$ denotes the inner-product of H.

Proof. Let $u_\beta := P_\beta u$, $v_{\beta'} := P_{\beta'} v$. Then $A_1 u_\eta = 0$, $A_1' v_{\eta'} = 0$ and since $u_\beta \in \mathcal{D}(A_\beta)$, $v_{\beta'} \in \mathcal{D}(A_\beta^*)$,

$$\begin{aligned}
(A_1 u, v) - (u, A_1' v) &= (A_1 u, v_{\beta'} + v_{\eta'}) - (u_\beta + u_\eta, A_1' v) \\
&= (A_1 u_\beta, v_{\beta'}) - (u_\beta, A_1' v_{\beta'}) + (A_1 u, v_{\eta'}) - (u_\eta, A_1' v) \\
&= (A_\beta u_\beta, u_{\beta'}) - (u_\beta, A_\beta^* v_\beta) + (A_1 u, v_{\eta'}) - (u_\eta, A_1' v) \\
&= (A_1 u, v_{\eta'}) - (u_\eta, A_1' v).
\end{aligned}$$

∎

Lemma 6.5.3 *For $\tilde{A} \in \mathcal{M}$, $\tilde{A}^* \in \mathcal{M}'$, let V, W be the closed subspaces of $\mathcal{N}(A_1)$, $\mathcal{N}(A_1')$, respectively, defined by $V = \overline{P_\eta \mathcal{D}(\tilde{A})}$, $W = \overline{P_{\eta'} \mathcal{D}(\tilde{A}^*)}$. Define linear relations $T : V \to W$ and $T_1 : W \to V$ as follows:*

$$\begin{aligned}
\mathcal{D}(T) &= P_\eta \mathcal{D}(\tilde{A}) \\
T u_\eta &= (A_1 u)_W, \ u \in \mathcal{D}(\tilde{A}),
\end{aligned} \tag{6.5.4}$$

where $(\cdot)_W$ denotes the orthogonal projection onto W in H;

$$\begin{aligned}
\mathcal{D}(T_1) &= P_{\eta'} \mathcal{D}(\tilde{A}^*) \\
T_1 v_{\eta'} &= (A_1' v)_V, \ v \in \mathcal{D}(\tilde{A}^*),
\end{aligned} \tag{6.5.5}$$

where $(\cdot)_V$ is the orthogonal projection onto V. Then T and T_1 are closed linear operators and $T = T_1^*$, $T_1 = T^*$.

Proof. Let $u \in \mathcal{D}(\tilde{A})$, $v \in \mathcal{D}(\tilde{A}^*)$. Then, as $\tilde{A} \subset A_1$ and $\tilde{A}^* \subset A_1'$, we have $(A_1 u, v) = (u, A_1' v)$, and it follows from Lemma 6.5.2 that

$$(A_1 u, v_{\eta'}) = (u_\eta, A_1' v).$$

Since $u_\eta \in V$ and $v_{\eta'} \in W$, this gives

$$((A_1 u)_W, v_{\eta'}) = (u_\eta, (A_1' v)_V). \tag{6.5.6}$$

Suppose $u_\eta = 0$. Then $((A_1 u)_W, v_{\eta'}) = 0$ for all $v \in \mathcal{D}(\tilde{A}^*)$, and as $W = \overline{P_{\eta'} \mathcal{D}(\tilde{A}^*)}$ it follows that $(Au)_W = 0$. Therefore T is single-valued and defines a linear operator with domain in V and range in W. Similarly, T_1 is a linear operator with domain in W and range in V.

By definition, $\mathcal{D}(T_1)$ is a dense subspace of W and so its adjoint $T_1^* : V \to W$ exists. From (6.5.6),

$$(Tu, v)_W = (u, T_1 v)_V, \text{ for all } u \in \mathcal{D}(T), \ v \in \mathcal{D}(T_1)$$

and so $T \subset T_1^*$.

6.5 Adjoint Pairs and Closed Extensions

Let $z \in \mathcal{D}(T_1^*)$, and define $x = z + A_\beta^{-1} T_1^* z$. Then $x \in \mathcal{D}(A_1)$, $x_\beta = A_\beta^{-1} T_1^* z$, $x_\eta = z$ and $A_1 x = A_\beta x_\beta = T_1^* z$. Moreover, for all $v \in \mathcal{D}(\tilde{A}^*)$ we have by Lemma 6.5.2, and since $z \in V$,

$$\begin{aligned}(\tilde{A}x, v) - (x, \tilde{A}^* v) &= (A_1 x, v_{\eta'}) - (x_\eta, A_1' v) \\ &= (T_1^* z, v_{\eta'}) - (z, A_1' v) \\ &= (T_1^* z, v_{\eta'}) - (z, (A_1' v)_V) \\ &= (T_1^* z, v_{\eta'}) - (z, T_1 v_{\eta'}) \\ &= 0\end{aligned}$$

Thus $x \in \mathcal{D}(\tilde{A}^{**}) = \mathcal{D}(\tilde{A})$. Consequently, $x_\eta = z \in \mathcal{D}(T)$ and $Tx_\eta = (A_1 x)_W = (T_1^* z)_W = T_1^* z$. Hence $Tz = T_1^* z$ and $T_1^* \subset T$. We have therefore proved that $T_1^* = T$. Similarly, $T^* = T_1$ and the lemma is proved. ∎

In Lemma 6.5.3, it was shown that a pair of adjoint operators $\tilde{A} \in \mathcal{M}$, $\tilde{A}^* \in \mathcal{M}'$ gives rise to a pair of adjoint operators $T : V \to W$, $T^* : W \to V$ with V and W closed subspaces of $\mathcal{N}(A_1)$ and $\mathcal{N}(A_1')$ respectively. We next show that given closed subspaces V, W of $\mathcal{N}(A_1)$, $\mathcal{N}(A_1')$ respectively, any pair of adjoint operators $T : V \to W$, $T^* : W \to V$ is determined by Lemma 6.5.3 with respect to adjoint operators $\tilde{A}$, $\tilde{A}^*$.

Lemma 6.5.4 *Let V, W be closed subspaces of $\mathcal{N}(A_1)$, $\mathcal{N}(A_1')$ respectively, and let $T : V \to W$, $T^* : W \to V$ be a pair of adjoint operators. Then the operators $\tilde{A}$, $\tilde{A}'$ in H defined by*

$$\mathcal{D}(\tilde{A}) := \left\{u \in \mathcal{D}(A_1) : u_\eta \in \mathcal{D}(T), (A_1 u)_W = T u_\eta\right\}, \quad \tilde{A} \in \mathcal{M} \qquad (6.5.7)$$

and

$$\mathcal{D}(\tilde{A}') := \left\{v \in \mathcal{D}(A_1') : v_\eta \in \mathcal{D}(T^*), (A_1' v)_V = T^* u_\eta\right\}, \quad \tilde{A}' \in \mathcal{M}' \qquad (6.5.8)$$

are adjoints, and the operators derived from $\tilde{A}$, $\tilde{A}^*$ by Lemma 6.5.3 are precisely $T : V \to W$ and $T^* : W \to V$.

Proof.

For $u \in \mathcal{D}(\tilde{A})$, $v \in \mathcal{D}(\tilde{A}')$, Lemma 6.5.2 gives

$$\begin{aligned}(\tilde{A}u, v) - (u, \tilde{A}'v) &= (A_1 u, v_{\eta'}') - (u_\eta, A_1' v) \\ &= ((A_1 u)_W, v_{\eta'}') - (u_\eta, (A_1' v)_V) \text{ (since } u_\eta \in V, v_{\eta'} \in W\text{)} \\ &= (T u_\eta, v_{\eta'}) - (u_\eta, T^* v_{\eta'}) \\ &= 0.\end{aligned}$$

Hence $\tilde{A} \subset (\tilde{A}')^*$, $\tilde{A}' \subset \tilde{A}^*$. It will follow that $\tilde{A}' = \tilde{A}^*$ if we prove that $\tilde{A}^* \subset \tilde{A}'$, which we shall proceed to do.

In (6.5.1), A_0 is closed with adjoint $A_1 = (A_0')^*$ and it was observed that the assumption $0 \in \rho(A_\beta)$ meant that A_0 has a bounded inverse. Thus we have the orthogonal sum decomposition

$$H = \mathcal{R}(A_0) \oplus \mathcal{N}(A_1'), \tag{6.5.9}$$

and so $\mathcal{R}(A_0) \perp W$. Let u be any element in $\mathcal{D}(A_1)$ of the form $u = z + A_\beta^{-1} Tz + w$, where $z \in \mathcal{D}(T)$, and $w \in \mathcal{D}(A_0)$. Then

$$P_\beta u = A_\beta^{-1} Tz + w, \quad P_\eta u = z.$$

Hence $u_\eta = z \in \mathcal{D}(T)$ and $(A_1 u)_W = (Tz + A_0 w)_W = Tz$ since $A_0 w \perp W$ by (6.5.9). This proves that $u \in \mathcal{D}(\tilde{A})$. For all such u and any $v \in \mathcal{D}(\tilde{A}^*)$, we have

$$\begin{aligned}
0 &= (A_1 u, v) - (u, A_1' v) \\
&= (Tz + A_0 w, v_{\eta'}) - (z, A_1' v) \\
&= (Tz, v_{\eta'}) - (z, (A_1' v)_V).
\end{aligned}$$

This implies that $v_{\eta'} \in \mathcal{D}(T^*)$ and $T^* v_{\eta'} = (A_1' v)_V$, whence, by definition, $v \in \mathcal{D}(\tilde{A}')$. Thus $\tilde{A}^* \subset \tilde{A}'$ and the rest of the lemma follows easily. ∎

The final lemma confirms that every pair T, T^* is associated with only one pair $\tilde{A}, \tilde{A}^*$.

Lemma 6.5.5 *Let $\tilde{A} \in \mathcal{M}$, $\tilde{A}^* \in \mathcal{M}'$ be a pair of adjoint operators, and let $T : V \to W$, $T^* : W \to V$ be derived from $\tilde{A}$ and $\tilde{A}^*$ as in Lemma 6.5.3. Then*

$$\begin{aligned}
\mathcal{D}(\tilde{A}) &= \{u \in \mathcal{D}(A_1) : u_\eta \in \mathcal{D}(T), (A_1 u)_W = T u_\eta\} \\
\mathcal{D}(\tilde{A}^*) &= \{v \in \mathcal{D}(A_1') : v_{\eta'} \in \mathcal{D}(T^*), (A_1' v)_V = T^* v_{\eta'}\}
\end{aligned}$$

Proof. Let $\tilde{A}_1$, $\tilde{A}_1^*$ be the operators defined by (6.5.7) and (6.5.8) in Lemma 6.5.4. Then from the definitions of T and T^* in Lemma 6.5.3, we see that

$$\mathcal{D}(\tilde{A}) \subset \mathcal{D}(\tilde{A}_1) \tag{6.5.10}$$

and

$$\mathcal{D}(\tilde{A}^*) \subset \mathcal{D}(\tilde{A}_1^*). \tag{6.5.11}$$

Since $\tilde{A}$ and $\tilde{A}_1$ lie in $\mathcal{M}$, (6.5.10) gives $\tilde{A} \subset \tilde{A}_1$, and so $\tilde{A}^* \supset \tilde{A}_1^*$. This and (6.5.11) yield $\tilde{A}^* = \tilde{A}_1^*$, and consequently $\tilde{A} = \tilde{A}_1$. ∎

6.5 Adjoint Pairs and Closed Extensions

We are now able to state Grubb's Theorem 1.1 in [96], which is a consequence of the preceding lemmas. We preserve the notation P_η, $P_{\eta'}$, u_η, $v_{\eta'}$ for the projections of $\mathcal{D}(A_1)$, $\mathcal{D}(A_1')$ onto $\mathcal{N}(A_1)$, $\mathcal{N}'(A_1')$, and $u_\eta = P_\eta u$, , $v_{\eta'} = P_{\eta'} v$.

Theorem 6.5.6 *There is a one-one correspondence between all pairs of adjoint operators $\tilde{A}$, $\tilde{A}^*$ with $\tilde{A} \in \mathcal{M}$, $\tilde{A}^* \in \mathcal{M}'$, and all pairs of adjoint operators T, T^* with $T : V \to W$, $T^* : W \to V$, where V, W are respectively closed subspaces of $\mathcal{N}(A_1)$, $\mathcal{N}(A_1')$. The correspondence is given by*

$$\mathcal{D}(\tilde{A}) = \{u \in \mathcal{D}(A_1) : u_\eta \in \mathcal{D}(T), (A_1 u)_W = T u_\eta\},$$
$$\mathcal{D}(\tilde{A}^*) = \{v \in \mathcal{D}(A_1') : v_{\eta'} \in \mathcal{D}(T^*), (A_1' v)_V = T v_{\eta'}\},$$

and

$$\mathcal{D}(T) = P_\eta \mathcal{D}(\tilde{A}), \quad V = \overline{P_\eta \mathcal{D}(\tilde{A})},$$
$$\mathcal{D}(T^*) = P_{\eta'} \mathcal{D}(\tilde{A}^*), \quad W = \overline{P_{\eta'} \mathcal{D}(\tilde{A}^*)}.$$

Since the theorem is completely symmetric in $\tilde{A}$ and $\tilde{A}^*$, and also in T and T^*, an immediate consequence is

Corollary 6.5.7 *There is a one-one correspondence between all closed operators $\tilde{A} \in \mathcal{M}$ and all operators $T : V \to W$ satisfying*

1. *V is a closed subspace of $\mathcal{N}(A_1)$ and W a closed subspace of $\mathcal{N}(A_1')$;*
2. *T is densely defined in V and closed.*

The correspondence is given by

$$\mathcal{D}(\tilde{A}) = \{u \in \mathcal{D}(A_1) : u_\eta \in \mathcal{D}(T), (A_1 u)_W = T u_\eta\},$$

where $\mathcal{D}(T) = P_\eta \mathcal{D}(\tilde{A})$ and so $V = \overline{P_\eta \mathcal{D}(\tilde{A})}$.

Furthermore, if $\tilde{A}$ corresponds to T in the above sense, then $\tilde{A}^*$ corresponds to $T^* : W \to V$ by

$$\mathcal{D}(\tilde{A}^*) = \{v \in \mathcal{D}(A_1') : v_{\eta'} \in \mathcal{D}(T^*), (A_1' v)_V = T v_{\eta'}\}$$

with $\mathcal{D}(T^*) = P_{\eta'} \mathcal{D}(\tilde{A}^*)$ and so $W = \overline{P_{\eta'} \mathcal{D}(\tilde{A}^*)}$.

An alternative description of the correspondence between $\tilde{A}$ and T will follow from the next lemma.

Lemma 6.5.8 *Let W be a closed subspace of $\mathcal{N}(A_1')$ and let T be any operator with $\mathcal{D}(T) \subset \mathcal{N}(A_1)$ and $\mathcal{R}(T) \subset W$. Then the the following sets D_1, D_2 are identical:*

$$D_1 := \{u \in \mathcal{D}(A_1) : u_\eta \in \mathcal{D}(T), (A_1 u)_W = T u_\eta\};$$

$$D_2 := \left\{ u = z + A_\beta^{-1}(Tz + f) + v : z \in \mathcal{D}(T), \ f \in \mathcal{N}(A_1') \ominus W, v \in \mathcal{D}(A_0) \right\}. \tag{6.5.12}$$

Moreover the elements z, f and v are uniquely determined by u in (6.5.12).

Proof. We shall use (6.5.9). If $u = z + A_\beta^{-1}(Tz + f) + v \in D_2$, then clearly $u \in \mathcal{D}(A_1)$. Furthermore, $z \in \mathcal{N}(A_1)$ and $A_\beta^{-1}(Tz + f) + v \in \mathcal{D}(A_\beta)$. Hence $u_\eta = z$. Also $A_1 u = Tz + f + A_1 v$, where $Tz \in W$, $f \in \mathcal{N}(A_1') \ominus W$ and $A_1 v \in \mathcal{R}(A_0)$. Then, since

$$H = W \oplus (\mathcal{N}(A_1') \ominus W) \oplus \mathcal{R}(A_0), \tag{6.5.13}$$

we infer that $(A_1 u)_W = Tz$ and this implies that $u \in D_1$.

Conversely, let $u \in D_1$. From (6.5.9), we can write

$$A_1 u = (A_1 u)_W + (A_1 u)_{\mathcal{N}(A_1') \ominus W} + (A_1 u)_{\mathcal{R}(A_0)}. \tag{6.5.14}$$

Set $(A_1 u)_{\mathcal{N}(A_1') \ominus W} = f$ and $A_\beta^{-1}[(A_1 u)_{\mathcal{R}(A_0)}] = v$. Then $v \in \mathcal{D}(A_0)$ since A_0 is injective.

We have $u = u_\eta + u_\beta$, where $u_\beta = A_\beta^{-1} A_1 u$, and as $u \in D_1$, $u_\eta \in \mathcal{D}(T)$ and $(A_1 u)_W = T u_\eta$. Then, on using (6.5.14),

$$\begin{aligned} u &= u_\eta + A_\beta^{-1} A_1 u \\ &= u_\eta + A_\beta^{-1}(T u_\eta + f + A_1 v) \\ &= u_\eta + A_\beta^{-1}(T u_\eta + f) + v, \end{aligned}$$

where $u_\eta \in \mathcal{D}(T)$, $f \in \mathcal{N}(A_1') \ominus W$ and $v \in \mathcal{D}(A_0)$. Therefore $u \in D_2$.

The uniqueness follows from (6.5.13). ∎

Corollary 6.5.7 can now be expressed as follows.

Theorem 6.5.9 *Let $\tilde{A}$ correspond to $T : V \to W$ as in Corollary 6.5.7. Then $u \in \mathcal{D}(\tilde{A})$ if and only if*

$$u = z + A_\beta^{-1}(Tz + f) + v \text{ for some } z \in \mathcal{D}(T), f \in \mathcal{N}(A_1') \ominus W, v \in \mathcal{D}(A_0). \tag{6.5.15}$$

The elements z, f, v are uniquely determined by u.

Other interesting facts about the correspondence between $\tilde{A}$ and T are given in [96]. For instance, it is shown that in Theorem 6.5.9, the map

$$\{z, f, v\} \mapsto u + A_\beta^{-1}(Tz + f) + v$$

is an isomorphism of $\mathcal{D}(T) \times (\mathcal{N}(A_1') \ominus W) \times \mathcal{D}(A_0)$ onto $\mathcal{D}(\tilde{A})$ when the spaces are endowed with the graph topologies.

6.6 Sectorial Operators

A linear operator S with domain and range in a Hilbert space H is said to be *sectorial* if its numerical range

$$\Theta(S) := \{(Su, u) : u \in \mathcal{D}(S), \|u\| = 1\}$$

is a subset of a sector

$$\Sigma(\mu, \theta) := \{z \in \mathbb{C} : re\, z \geq \mu,\ |\arg(z - \mu)| \leq \theta < \pi/2\}$$

for some $\mu \in \mathbb{R}$ and $\theta \in [0, \pi/2)$; μ is called a vertex and θ a semi-angle. The inner-product and norm of H will be denoted by $(\cdot, \cdot)$, $\|\cdot\|$ as has been our custom. We shall consider only those cases for which $\mu \geq 0$ and call the associated sectorial operator θ-sectorial, or just sectorial if we do not need to be so specific. Thus, when $\theta = 0$, S is a non-negative symmetric operator, while $\theta = \pi/2$ implies that $re\,(Su, u) \geq 0$ for all $u \in \mathcal{D}(S)$. An operator satisfying the latter property is said to be *accretive*. If $\mu > 0$, then S is said to be *coercive*.

Let S be a closed, densely defined θ-sectorial operator; note that by Proposition III.6.2 in [54], any densely defined θ-sectorial operator is closable and its closure is θ-sectorial. Then, for $\lambda \in \mathbb{C} \setminus \Gamma(S)$, where $\Gamma(S) := \overline{\Theta(S)}$, $(S - \lambda I)^{-1}$ exists and the deficiency index $m(S) := \mathrm{def}(S - \lambda I)$ is constant; see Section 6.1. Thus if $m(S) = 0$, $\mathbb{C} \setminus \Gamma(S)$ lies in the resolvent set of S and

$$\|(S - \lambda I)^{-1}\| \leq 1/\mathrm{dist}\{\lambda, \Gamma(S)\}. \tag{6.6.1}$$

A sectorial operator S with this property is said to be *m-sectorial*.

In [9–12], Arlinskii and his co-authors have investigated the existence of m-sectorial extensions of a sectorial operator S, and have established a rich vein of results which are analogues of ones in the KVB theory for the symmetric case. The Friedrichs extension in the first representation theorem in [119] is one such m-sectorial extension, but others exist in general, including a Krein-von Neumann type extension. If S is a symmetric operator, its self-adjoint extensions are all restrictions of the adjoint S^* and are distinguished only by their domains. For a sectorial S, the problem is more of a challenge, as an action and a domain have to be specified to determine the extension.

A sesquilinear form s with domain in H is θ-sectorial if its numerical range

$$\Theta(s) := \{s[u] = s[u, u] : u \in \mathcal{D}(S), \|u\| = 1\}$$

is a subset of a sector $\Sigma(0, \theta)$. Then

$$re\,(s[u]) \geq 0,\quad |im\,(s[u])| \leq (\tan \theta) re\,(s[u]) \tag{6.6.2}$$

for all $u \in \mathcal{D}(s)$, the domain of s. With s_1 the *real part* of s, i.e., $s_1[u,v] = (1/2)(s[u,v] + \overline{s[v,u]})$, we have

$$0 \leq s_1[u] \leq |s[u]| \leq (\sec\theta) s_1[u] \tag{6.6.3}$$

and this implies that s is closed if and only if the inner-product space $Q(s)$ determined by $\mathcal{D}(s)$ with inner-product

$$(u,v)_{Q(s)} := s_1[u,v] + (u,v), \quad u, v \in \mathcal{D}(s) \tag{6.6.4}$$

is complete. Therefore, if s is closed and densely defined, there exists a natural embedding $E : Q(s) \hookrightarrow H$ which is densely defined and $\|E\| \leq 1$. Note that $Q(s) = Q(s_1)$. Also as in Section 6.3 for symmetric forms, we use $Q(s)$ to denote $\mathcal{D}(s)$ as a subspace of H and as the Hilbert space with inner-product (6.6.4), since the meaning will be clear from the context.

Theorem 6.6.1 *Let S be sectorial in H and define*

$$s[u,v] ;= (Su, v), \quad \mathcal{D}(s) = \mathcal{D}(S).$$

Then s is sectorial and closable.

Proof. Let $s_2 ;= \text{im } s = \frac{1}{2i}(s - s^*)$, where $s^*[u,v] = \overline{s[v,u]}$. We shall begin by showing that

$$|s_2[u,v]| \leq \tan(\theta)\, s_1[u] s_1[v] \leq \tan(\theta)\, \|u\|_{Q(s)} \|v\|_{Q(s)}. \tag{6.6.5}$$

The form s_2 is symmetric and we may suppose, without loss of generality, that $s_2[u,v]$ is real; for if $\alpha = \arg(s_2[u,v])$, replacing u in $s_2[u,v]$ by $e^{-i\alpha}u$ will leave (6.6.5) unchanged. Thus by the polar identity,

$$s_2[u,v] = \frac{1}{4}\{(s_2[u+v] - s_2[u-v]\})$$

and so by (6.6.2),

$$|s_2[u,v]| \leq \frac{\tan\theta}{4}(s_1[u+v] + s_1[u-v]) = \frac{\tan\theta}{2}(s_1[u] + s_1[v]).$$

It follows that, for any $a > 0$,

$$|s_2[u,v]| = |s_2[au, \frac{1}{a}v]| \leq \frac{\tan\theta}{2}\left(a^2 s_1[u] + \frac{1}{a^2}s_1[v]\right)$$
$$\leq (\tan\theta) s_1[u] s_1[v] \leq (\tan\theta) \|u\|_{Q(s)} \|v\|_{Q(s)},$$

to establish (6.6.5).

6.6 Sectorial Operators

It is clear that s is sectorial and that the closures of $\Theta(s)$ and $\Theta(S)$ coincide. To prove that s is closable, let $(x_n) \subset \mathcal{D}(s)$ be a Cauchy sequence in $Q(s)$ and suppose that $x_n \to 0$ in H. We must show that $x_n \to 0$ in $Q(s)$. We have

$$\|x_n\|^2_{Q(s)} = (s_1 + 1)[x_n] \leq |(s+1)[x_n]|$$
$$\leq |(s+1)[x_n, x_n - x_m]| + |(s+1)[x_n, x_m]|$$
$$\leq (1 + \tan\theta) \|x_n\|_{Q(s)} \|x_n - x_m\|_{Q(s)} + |(Sx_n, x_m)| + \|x_n\|\|x_m\|,$$

by (6.6.2). Given any $\varepsilon > 0$, there exists an $N \in \mathbb{N}$ such that for $n, m > N$,

$$\|x_n\|^2_{Q(s)} < \varepsilon + |(Sx_n, x_m)|.$$

On allowing $m \to \infty$, it follows that $\|x_n\|^2_{Q(s)} < \varepsilon$ for $n > N$, whence the result. ∎

6.6.1 The Friedrichs Extension

The intimate connection between sectorial forms and sectorial operators is exposed in the following theorem (called in [119] the *first representation theorem*), which is a generalisation of Theorem 6.3.2 concerning the Friedrichs extension of lower-semibounded symmetric operators. Recall that a subspace D of H is said to be a *core* of a closed sectorial form s if it is a dense subspace of $Q(s)$.

Theorem 6.6.2 *Let s be a closed, sectorial form whose domain $\mathcal{D}(s)$ is a dense subspace of H. Then there exists an m-sectorial operator S which has the following properties:*

1. $\mathcal{D}(S) \subset Q(s)$ and

$$s[u, v] = (Su, v), \quad \forall\ u \in \mathcal{D}(S),\ v \in Q(s); \tag{6.6.6}$$

2. $\mathcal{D}(S)$ *is a core of s;*
3. *if $u \in Q(s)$, $w \in H$ and*

$$s[u, v] = (w, v) \tag{6.6.7}$$

for every v in a core of s, then $u \in \mathcal{D}(S)$ and $Su = w$. The m-sectorial operator S is uniquely determined by the conditions (1) and (2).
4. $\Theta(T)$ *is a dense subset of* $\Theta(s)$.

Proof. From (6.6.5),

$$|(s+1)[u, v]| \leq (1 + \tan\theta) \|u\|_{Q(s)} \|v\|_{Q(s)}. \tag{6.6.8}$$

Also,
$$|(s+1)[u]| \geq (s_1+1)[u] = \|u\|^2_{Q(s)}. \tag{6.6.9}$$

Thus $s+1$ is bounded on $Q(s) \times Q(s)$ and coercive on $Q(s)$. It follows from the Lax-Milgram lemma (see Lemma 1.2.7) that there exists an isomorphism $\hat{S}$ of $Q(s)$ onto its adjoint $Q(s)^*$ which is such that

$$s[\varphi, f] = \langle \varphi, \hat{S}f \rangle_{Q(s), Q(s)^*} := (\hat{S}f)(\varphi), \text{ for all } \varphi \in H, \tag{6.6.10}$$

and with $M := (1 + \tan\theta)$,

$$1 \leq \|\hat{S}\| \leq M, \quad M^{-1} \leq \|\hat{S}^{-1}\| \leq 1.$$

Let E be the natural embedding of $Q(s)$ into H, with adjoint $E^* : H \to Q(s)^*$. We then have the triplet of spaces

$$Q(s) \hookrightarrow^E H \hookrightarrow^{E^*} Q(s)^*,$$

E, E^*, being continuous embeddings having dense ranges and norms ≤ 1. Set $S := (E^*)^{-1}\hat{S}E^{-1}$. Then, on distinguishing elements of $Q(s)$ from their images in H, we have from (6.6.10), with $f = Eu$,

$$s[\varphi, u] = (\varphi, E^*SEu) = (E\varphi, SEu);$$

we have identified H with its adjoint and used the fact that E^* is the adjoint of E.

From this point, the proof proceeds along similar lines to the proof of Theorem 6.3.2 and is left as an exercise; full details may be found in [54], Theorem IV.2.4. ∎

The sectorial form s is referred to as the form *associated* with the sectorial operator S, and, following convention, we shall denote $Q(s)$ by $Q(S)$ and call it the *form domain* of S. The closure $\bar{s}$ of the form in Theorem 6.6.1 satisfies Theorem 6.6.2 and the associated m-sectorial operator $\tilde{S}$ say, is an extension of S in view of Theorem 6.6.2(3); $\bar{s}$ is now the form associated with $\tilde{S}$. The operator $\tilde{S}$ is called the *Friedrichs extension* of S as it was in the original case when S was symmetric. If S is already m-sectorial then we must have $\tilde{S} = S$. Therefore Theorems 6.6.1 and 6.6.2 establish a one-one correspondence between the set of closed, densely defined sectorial forms and the set of m-sectorial operators.

Theorem 6.6.3 *Let $\tilde{S}$ be the Friedrichs extension of a densely defined sectorial operator S, and let T be any sectorial extension of S with $\mathcal{D}(T) \subset Q(\bar{s})$, where $\bar{s}$ is the form of $\tilde{S}$. Then $T \subset \tilde{S}$. Thus $\tilde{S}$ is the only m-sectorial extension of S with domain in $Q(\bar{s})$.*

The form domain of $\tilde{S}$ is contained in the form domain of any m-sectorial extension of S.

Proof. Define the sectorial forms

6.6 Sectorial Operators

$$t[u, v] := (Tu, v), \quad u, v \in \mathcal{D}(T), \quad s[u, v] := (Su, v), \quad u, v \in \mathcal{D}(S).$$

Since $\mathcal{D}(t) = \mathcal{D}(T) \supset \mathcal{D}(S) = Q(s)$, and $S \subset T$, the closure of t is an extension of s and hence of $\bar{s}$. Therefore, for $u \in \mathcal{D}(T)$ and $v \in \mathcal{D}(\bar{s})$, we have

$$\bar{s}[u, v] = \bar{t}[u, v] = (Tu, v).$$

By Theorem 6.6.2(3), it follows that $T \subset \tilde{S}$ which proves the first part of the theorem. Let A be any m-sectorial extension of S and define

$$a[u, v] := (Au, v), \quad \mathcal{D}(a) = \mathcal{D}(A).$$

We have observed above that the closure $\bar{a}$ of a is the form associated with the Friedrichs extension of A, which is A itself, since an m-sectorial operator has no proper m-sectorial extensions. As $A \supset S$ implies that $a \supset s$, it follows that $\bar{a} \supset \bar{s}$, and so $\mathcal{D}(\bar{a}) \supset \mathcal{D}(\bar{s})$; thus, $Q(A) \supset Q(\tilde{S})$. ∎

If S is m-sectorial with form s, S^* is m-sectorial with form s^* (which is also densely defined, sectorial and closed). Furthermore, the form $re(s) = (1/2)(s + s^*)$ is a closed, densely defined non-negative symmetric form, and is therefore the form of a non-negative, self-adjoint operator S_R say; S_R is called the *real part* of S. While $S_R = (1/2)(S + S^*)$ when S is bounded on H, this is not true in general. Indeed $\mathcal{D}(S) \cap \mathcal{D}(S^*)$ may not be dense in H. We have observed earlier that $Q(re(s)) = Q(s)$ and hence the form domain $Q(S_R)$ coincides with $Q(S)$. It therefore follows from the 2nd representation theorem, Theorem 6.3.6, that

$$Q(S) = Q(S_R) = \mathcal{D}(S_R^{1/2}), \tag{6.6.11}$$

where $\mathcal{D}(S_R^{1/2})$ is endowed with the graph topology.

The analogue of Theorem 6.3.6 for an m-sectorial operator S with form s, whose numerical range lies in a sector with semi-angle θ (and vertex $\mu \geq 0$), is the following result from [119], Theorem VI-3.2:

$$s[u, v] = \left((1 + iB)S_R^{1/2}u, S_R^{1/2}v\right), \quad u, v \in Q(s) = \mathcal{D}(S_R^{1/2}), \tag{6.6.12}$$

where B is a bounded self-adjoint operator acting in $\overline{R(S)}$ with $\|B\| \leq \tan \theta$, and

$$\mathcal{D}(S) = \left\{u \in Q(s) : (1 + iB)S_R^{1/2}u \in Q(s)\right\}, \quad Su = S_R^{1/2}(1 + iB)S_R^{1/2}u. \tag{6.6.13}$$

6.6.2 The Krein-von Neumann Extension

In [9], [10], Arlinskii's definition of the Krein-von Neumann extension of a densely defined sectorial operator S is modelled on that of Ando and Nishio in [6] for non-negative symmetric operators. Set $H_K := \overline{\mathcal{R}(S)}$ and

$$T(Su) = P_K u, \quad u \in \mathcal{D}(S),$$

where P_K is the orthogonal projection of H onto H_K. Then, for $\varphi = Su$, $\psi = Sv$, $u, v \in \mathcal{D}(S)$,

$$(T\varphi, \psi) = (T(Su), Sv) = (P_K u, Sv) = (P_K u, P_K Sv) = (u, Sv),$$

since $(I - P_K)v \in H_K^\perp$. It follows that T is well-defined, with dense domain $\mathcal{D}(T) = \mathcal{R}(S)$ and dense range in H_K. Moreover, T is sectorial with $\Theta(T) = \Theta(S)$, and T is coercive if and only S is coercive; we shall assume that S is coercive, but note that this assumption is not generally made in the works cited below. The Krein-von Neumann extension of S is defined as

$$S_K := T_F^{-1} P_K, \qquad (6.6.14)$$

where T_F is the Friedrichs extension of T; equivalently, we can write

$$S_K = \left((S^{-1})_F\right).$$

The proof of important properties of S_K collected in the following theorem may be found in [9] and [10].

Theorem 6.6.4 *Let S be a densely defined, closed, coercive, sectorial operator in H with $\Theta(S) \subset \Sigma(0, \theta) := \{z \in \mathbb{C} : |\arg z| \leq \theta < \pi/2\}$, and let $\mathcal{N} := \ker S^*$. Then the following hold.*

1.

$$Q(S_K) = \left\{ u \in H : \sup\left[|(u, Sx)|^2 / \mathrm{re}\,(Sx, x); x \in \mathcal{D}(S), x \neq 0\right] < \infty \right\}$$
$$= \mathcal{R}(T_R^{1/2}) \oplus \mathcal{N}. \qquad (6.6.15)$$

2. *For all $u \in \mathcal{D}(S_K)$,*

$$\inf\left\{\|S_K u - Sv\|^2 + \mathrm{re}\,(S_K(u-v), u-v) : v \in \mathcal{D}(S)\right\} = 0. \qquad (6.6.16)$$

3. *For any m-sectorial extension $\tilde{S}$ of S*

$$Q(S) \subset Q(\tilde{S}) \subset Q(S_K).$$

6.6 Sectorial Operators

4. For any $\lambda \in (\mathbb{C} \setminus \Sigma(0, \theta)) \cup \{0\}$, we have $\mathcal{N}_z := \ker(S^* - \lambda I) \subset Q(S_K)$ and, in particular,

$$Q(S_K) = Q(S) \dotplus \mathcal{N}, \tag{6.6.17}$$

$$s_K[u, v] = s_F[\Pi_F u, \Pi_F v], \quad u, v \in Q(S_K), \tag{6.6.18}$$

where Π_F is the projection onto $Q(S)$ with respect to the decomposition (6.6.17); in (6.6.18), s_K and s_F are the forms associated with S_K and S_F respectively.

5.
$$\mathcal{D}(S_K) = \mathcal{D}(S) \dotplus \mathcal{N}, \quad S_K(u + v) = Su, \tag{6.6.19}$$

for $u \in \mathcal{D}(S), v \in \mathcal{N}$.

6. The sectorial operator S has a unique m-sectorial extension S_F if and only if, for some (and hence all) $z \in \rho(S_F^*)$,

$$\sup\left\{|(\varphi_z, x)|^2 / re\,(Sx, x), x \in \mathcal{D}(S)\right\} = \infty, \quad \varphi_z \in \mathcal{N}_z \setminus \{0\}.$$

The inclusions in Theorem 6.6.4(3) mean that the Friedrichs extension S_F and Krein-von Neumann extension S_K of S have respectively, the smallest and largest form domains of all m-sectorial extensions of S; recall that $Q(S) = Q(S_F)$. A fact of additional interest is uncovered in [10] in relation to m-sectorial extensions which are *extremal*; these satisfy the condition

$$\inf\left\{re\left(\tilde{S}(u-v), u-v\right) : v \in \mathcal{D}(S)\right\} = 0, \quad u \in \mathcal{D}(\tilde{S});$$

the Friedrichs extension is extremal by definition, and the Krein-von Neumann extension is extremal by Theorem 6.6.4(2). It is proved in [10] that amongst the extremal m-sectorial extensions, the Krein-von Neumann extension is uniquely the one with the largest form domain.

Arlinsky and his co-authors have made a comprehensive and penetrating study of sectorial operators and their extensions; see, for instance, [9–11] and the references therein. In [11], a characterisation of all the m-sectorial extensions of S was established. It is expressed in terms of a *boundary pair* (H, Γ) of S: this means that Γ is a bounded linear map from $Q(S_K)$ onto H which is such that $\ker \Gamma = Q(S_F)$. It also features the real part S_{KR} of the operator S_K, which the reader will recall is the non-negative self-adjoint operator associated with $s_{KR} = re\,(s_K) = (1/2)(s_K + s_K^*)$.

Theorem 6.6.5 *Let S be a densely defined, closed, coercive sectorial operator in H and let S_{KR} be the real part of the Krein-von Neumann operator S_K. Let (H, Γ) be a boundary pair for S. Then*

$$Q(\tilde{S}) = \Gamma^{-1} Q(\tilde{\omega}) = \{u \in Q(S_K) : \Gamma u \in Q(\tilde{\omega})\},$$
$$\tilde{s}[u, v] = s[u - (Z_0 - 2\tilde{Y})\Gamma u, v - Z_0 \Gamma v] + \tilde{\omega}[\Gamma u, \Gamma v], \quad u, v \in Q(S_K),$$
$$\tag{6.6.20}$$

establish a one-one correspondence between all closed forms $\tilde{s}$ associated with m-sectorial extensions $\tilde{S}$ of S and all pairs $\langle \tilde{\omega}, \tilde{Y} \rangle$, where $\tilde{\omega}$ is a closed sectorial form in H, and $\tilde{Y} : Q(\tilde{\omega}) \to Q(S)$ is a bounded linear operator which is such that, for some $\delta \in [0, 1]$,

$$|(\tilde{Y}h, Su)|^2 \le \delta^2 re\,[(Su, u)] re\,(\tilde{\omega})[h], \quad u \in \mathcal{D}(S), \; h \in Q(\tilde{\omega});$$

also $Z_0 = (\Gamma \restriction_{\ker S^*})^{-1}$

6.7 Notes

1. The theory of boundary triplets associated with symmetric operators has its origins in the work of Gorbachuk and Gorbachuk [92], but the notion of an abstract boundary triplet was first introduced by Kočubeĭ [130] and Bruk [32]; in fact the idea can be traced back to Calkin's paper on abstract boundary conditions [33]. Boundary triplets have been the subject of intense study involving many authors, especially in the former Soviet Union, and they are now widely used tools in research on the extension of symmetric and non-symmetric operators and their spectral properties. Boundary triplets for adjoint pairs of abstract operators were introduced by Vainerman [210] and Lyantze and Storozh [153].

2. Gamma fields and Weyl M-functions associated with boundary triplets were first investigated by Derkach and Malamud [48]. In [30] a Weyl M-function is defined for a non-symmetric operator using the boundary triplet setting of Lyantze and Storozh, and the relationship between the behaviour of M as an operator-valued function on $\mathbb{C}$ and the spectral properties of the operator investigated. An example is given in which the M-function does not contain all the spectral information of the resolvent of the operator, and it is shown that the results can be applied to elliptic PDEs, where the M-function corresponds to the Dirichlet to Neumann map. The extent to which the analytic properties of the M-function defined for an adjoint pair of operators reflect the spectral properties of a certain restriction of the maximal operator is also studied in [31]. A guide to some of the many authors who have contributed to the theory and application of boundary triplets and Weyl M-functions during the last half century, and especially in recent years, may be found in [30] and the references therein.

Chapter 7
Realisations of Second-Order Linear Elliptic Operators

7.1 Sturm–Liouville Operators: Basic Theory

Let τ denote the second-order linear differential expression given by

$$\tau u = \frac{1}{k}\left(-(pu')' + qu\right) \tag{7.1.1}$$

on an interval (a, b), with $-\infty < a < b \leq \infty$, where the coefficients p, q, k, are real-valued and satisfy the following conditions:

$$p(x) \neq 0, \ k(x) > 0 \text{ a.e on } (a, b) \tag{7.1.2}$$

and
$$1/p, \ k, \ q \in L_{1,loc}(a, b), \tag{7.1.3}$$

where $L_{1,loc}(a, b)$ is the space of functions which are integrable on every compact subset of (a, b). These conditions will always be assumed: they guarantee the validity of the following basic existence theorem; see [158], section 16.2. In it τu is defined for functions u which are such that u and pu' are absolutely continuous on compact subsets of (a, b). We use the convenient notation

$$u^{[1]} := pu' \tag{7.1.4}$$

for what is called the *quasi-derivative* of u.

Theorem 7.1.1 *Let $f \in L_{1,loc}(a, b)$ and suppose that (7.1.2) and (7.1.3) are satisfied. Then given any complex numbers c_0, c_1, and any $x_0 \in (a, b)$, there exists a unique solution of $\tau u = f$ which satisfies $u(x_0) = c_0$, $u^{[1]}(x_0) = c_1$.*

A consequence is that the solutions of the equation

$$\tau u = \lambda u, \quad \lambda \in \mathbb{C}, \tag{7.1.5}$$

form a 2-dimensional vector space over $\mathbb{C}$. If (c_0, c_1), (d_0, d_1) are linearly independent vectors in $\mathbb{C}^2$, then the solutions $u(\cdot, \lambda)$, $v(\cdot, \lambda)$ of (7.1.5) which satisfy $u(x_0, \lambda) = c_0$, $u^{[1]}(x_0, \lambda) = c_1$, $v(x_0, \lambda) = d_0$, $v^{[1]}(x_0, \lambda) = d_1$, form a basis for the space of solutions of (7.1.5).

Definition 7.1.2 *The equation (7.1.5) is said to be **regular** at a if*

$$a \in \mathbb{R}; \ \frac{1}{p}, \ k, \ q \in L_{1,loc}[a, b). \tag{7.1.6}$$

*Otherwise it is said to be **singular** at a. Similarly, we define (7.1.5) to be regular or singular at b. If (7.1.5) is regular at a and b, we say that it is regular on $[a, b]$; in that case*

$$a, b \in \mathbb{R}; \ \frac{1}{p}, \ k, \ q \in L_1(a, b). \tag{7.1.7}$$

If (7.1.5) is regular at a, then Theorem 7.1.1 continues to hold for $x_0 = a$.

7.1.1 The Regular Problem

In this section we assume that (7.1.5) is regular on $[a, b]$, so that (7.1.7) holds. Let H be the weighted space $L_2((a, b); k)$ with inner-product and norm

$$(u, v) := \int_a^b u(x)\overline{v(x)}k(x)dx, \quad \|u\| := (u, u)^{1/2}.$$

Define

$$\mathcal{D}(\tau) := \left\{ u : u, \ u^{[1]} \in AC[a, b], \ \tau u \in H \right\}, \tag{7.1.8}$$

where $AC[a, b]$ denotes the space of functions which are absolutely continuous on $[a, b]$; let $T^+(\tau)$ be the operator defined by $T^+(\tau)u = \tau u$ for $u \in \mathcal{D}(\tau)$. Thus $T^+(\tau)$ has the largest domain of any operator generated by τ in H, and is naturally called the *maximal operator*. The restriction of $T^+(\tau)$ to

$$\mathcal{D}_0(\tau) := \left\{ u \in \mathcal{D}(\tau) : u(a) = u^{[1]}(a) = u(b) = u^{[1]}(b) = 0 \right\} \tag{7.1.9}$$

is called the *minimal* operator generated by τ in H and is denoted by $T(\tau)$. It has the following properties; see [158], Section 17.

Theorem 7.1.3 $T(\tau)$ *is a closed, symmetric operator in H with deficiency indices (2,2), and*

7.1 Sturm–Liouville Operators: Basic Theory

$$T^*(\tau) = T^+(\tau). \tag{7.1.10}$$

It is the closure of the operator $T'(\tau)$ which is the restriction of $T^+(\tau)$ to the set of functions in $\mathcal{D}(\tau)$ with compact supports in (a, b).

By (6.1.7), any self-adjoint extension of $T(\tau)$ is determined by a unitary map V from $\mathcal{N}_+ := \ker(T^+(\tau) - iI)$ onto $\mathcal{N}_- := \ker(T^+(\tau) + iI)$ and is the restriction of $T^+(\tau)$ to

$$\mathcal{D}(T_V) := \{\varphi + \varphi_+ + V\varphi_+ : \varphi \in \mathcal{D}_0(\tau), \ \varphi_+ \in \mathcal{N}_+\}. \tag{7.1.11}$$

Since $\mathcal{N}_+$ is of dimension 2, V is a unitary 2×2 matrix $(a_{i,j})$. Thus, if $\{\varphi_1, \varphi_2\}$ is a basis for $\mathcal{N}_+$, the elements of $\mathcal{D}(T_V)$ are of the form

$$u = \varphi + \psi, \tag{7.1.12}$$

where $\varphi \in \mathcal{D}_0(\tau)$ and ψ is a linear combination of the functions

$$\psi_i(x) = \varphi_i(x) + \sum_{j=1}^{2} a_{i,j} \varphi_j(x). \tag{7.1.13}$$

We also have from (6.1.8)

$$\mathcal{D}(T_V) := \{u \in \mathcal{D}(\tau) : \beta[u, \varphi_+ + V\varphi_+] = 0, \ \forall \varphi_+ \in \mathcal{N}_+\}, \tag{7.1.14}$$

where on integration by parts, for $u, v \in \mathcal{D}(\tau)$,

$$\begin{aligned}\beta[u, v] &= (T^+(\tau)u, v) - (u, T^+(\tau)v) \\ &= \int_a^b (\bar{v}\tau u - u\overline{\tau v}) \, dx \\ &= [u, v](b) - [u, v](a), \end{aligned} \tag{7.1.15}$$

with the notation

$$[u, v](t) = u(t)\bar{v}^{[1]}(t) - \bar{v}(t)u^{[1]}(t).$$

Hence $\mathcal{D}(T_V)$ is the set of functions $u \in \mathcal{D}(\tau)$ which satisfy the conditions

$$[u, \psi_i](b) - [u, \psi_i](a) = 0, \quad i = 1, 2. \tag{7.1.16}$$

Since $\psi_i \in \mathcal{D}(\tau)$, we must have

$$[\psi_j, \psi_i](b) - [\psi_j, \psi_i](a) = 0, \quad i, j = 1, 2. \tag{7.1.17}$$

For $u \in \mathcal{D}_0(\tau)$, $[u, v](b) = [u, v](a) = 0$ for all $v \in \mathcal{D}(\tau)$. Hence with u given by (7.1.12), (7.1.14) is equivalent to

$$[\psi, \psi_i](b) - [\psi, \psi_i](a) = 0, \quad i = 1, 2.$$

The converse of the above is also true, i.e., given arbitrary functions ψ_1, ψ_j which form a basis of the quotient space $\mathcal{D}(\tau)/\mathcal{D}_0(\tau)$ (i.e. no non-trivial linear combination lies in $\mathcal{D}_0(\tau)$) and satisfy (7.1.17), then the set of functions $u \in \mathcal{D}(\tau)$ which satisfy (7.1.16) constitute the domain of a self-adjoint extension of $T(\tau)$; see [158], Section 18.

On putting

$$\alpha_{j1} = -\overline{\psi}_j^{[1]}(a), \quad \alpha_{j2} = \overline{\psi}_j(a),$$
$$\beta_{j1} = \overline{\psi}_j^{[1]}(b), \quad \beta_{j2} = -\overline{\psi}_j(b),$$

in (7.1.16) and (7.1.17), we obtain

Theorem 7.1.4 *Every self-adjoint extension of $T(\tau)$ is determined by linearly independent boundary conditions of the form*

$$\alpha_{j1} u(a) + \alpha_{j2} u^{[1]}(a) + \beta_{j1} u(b) + \beta_{j2} u^{[1]}(b) = 0, \quad j = 1, 2, \tag{7.1.18}$$

where

$$\alpha_{j1} \overline{\alpha}_{i2} - \alpha_{j2} \overline{\alpha}_{i1} = \beta_{j1} \overline{\beta}_{i2} - \beta_{j2} \overline{\beta}_{i1}, \quad j = 1, 2. \tag{7.1.19}$$

Conversely, every system of linearly independent boundary conditions (7.1.18) defines a self-adjoint extension of $T(\tau)$ provided that the relations (7.1.19) are satisfied.

For $u, v \in \mathcal{D}(T^*(\tau))$, Green's theorem gives

$$[u, v]_{T^*(\tau)} = (T^*(\tau)u, v) - (u, T^*(\tau))$$
$$= u(b)\overline{v^{[1]}(b)} - u^{[1]}(b)\overline{v(b)} - u(a)\overline{v^{[1]}(a)} + u^{[1]}(a)\overline{v(a)}.$$

It follows from this and Theorem 7.1.1 that $(\mathcal{K}, \Gamma_0, \Gamma_1)$ is a boundary triplet for $T^*(\tau)$ with

$$\mathcal{K} = \mathbb{C}^2, \quad \Gamma_0 u = (u(a), u(b)), \quad \Gamma_1 u = (u^{[1]}(a), -u^{[1]}(b)).$$

We shall now determine the gamma field and Weyl function for this triplet. For $z \in \mathbb{C}$, the solutions $s(\cdot; z), c(\cdot; z)$ of $\tau u = zu$ which satisfy the initial conditions

$$s(a; z) = 0, \; s^{[1]}(a; z) = 1, \; c(a; z) = 1, \; c^{[1]}(a; z) = 0, \tag{7.1.20}$$

7.1 Sturm–Liouville Operators: Basic Theory

form a basis of $\mathcal{N}_z = \ker(T^+(\tau) - zI)$. Let $T_0(\tau)$ be the self-adjoint extension of $T(\tau)$ with domain $\mathcal{N}(\Gamma_0)$, and suppose that $z \in \rho(T_0(\tau))$. Then $s(b, z) \neq 0$, for otherwise we would have $s(a; z) = s(b; z) = 0$ and hence $s(\cdot; z) \in \mathcal{D}(T_0(\tau))$, implying $T_0(\tau)s(\cdot; z) = zs(\cdot; z)$ in contradiction to $z \in \rho(T_0(\tau))$. Any $u \in \mathcal{N}_z$ can be written as

$$u(x) = (u(b) - u(a)c(b; z))\, s(b; z)^{-1} s(x; z) + u(a)c(x; z)$$

The gamma field is the inverse of the map $\Gamma_0 : \mathcal{N}_z \to \mathcal{K}$ and so for $(c_1, c_2) \in \mathcal{K}$,

$$\gamma(z)(c_1, c_2) = (c_2 - c_1 c(b; z))\, s(b; z)^{-1} s(\cdot; z) + c_1 c(\cdot; z).$$

The last two equations give that $\gamma(z)\Gamma_0 u = u$, from which it follows that the Weyl function satisfies $M(z)\Gamma_0 u = \Gamma_1 u$ and hence that $M(z)(u(a), u(b)) = (u^{[1]}(a), -u^{[1]}(b))$. From this one readily verifies that

$$M(z) = -s(b; z)^{-1} \begin{pmatrix} c(b; z) & -1 \\ -1 & s^{[1]}(b; z) \end{pmatrix}. \tag{7.1.21}$$

7.1.2 One Singular Point

Let (7.1.5) be regular at a and singular at b. The maximal operator $T^+(\tau)$ then has domain

$$\mathcal{D}(\tau) := \left\{ u : u, u^{[1]} \in AC_{loc}[a, b),\ \tau u \in H \right\}. \tag{7.1.22}$$

The minimal operator $T(\tau)$ is the closure of the restriction $T(\tau)'$ of $T^+(\tau)$ to the set of those functions in $\mathcal{D}(\tau)$ whose supports are compact subsets of $[a, b)$. The domain of $T(\tau)$ turns out to be

$$\mathcal{D}_0(\tau) := \left\{ u : u \in \mathcal{D}(\tau),\ u(a) = u^{[1]}(a) = 0,\ [u, z](b) = 0 \text{ for all } z \in \mathcal{D}(\tau) \right\}. \tag{7.1.23}$$

Note that in (7.1.23), $[u, z](b) := \lim_{x \to b}[u, v](x)$ exists in view of the *Lagrange identity*

$$\int_a^x \{\tau u \bar{z} - u \overline{\tau z}\}\, dt = [u, z](x) - [u, z](a), \tag{7.1.24}$$

which follows on integration by parts.

Since the coefficients of τ are real, the equations $(\tau \pm i)u = 0$ have the same number of solutions, and the solution space is at most two-dimensional, by Theorem 7.1.1. Furthermore, an important result of H. Weyl is that for any $\lambda \in \mathbb{C} \setminus \mathbb{R}$, there exists at least one solution of

$$\tau u = \lambda u$$

which lies in H. The equation $\tau u = \lambda u$ is said to be in the *limit-point* case at b if there is precisely one solution in H (up to a constant multiple) for $\lambda \notin \mathbb{R}$; otherwise $\tau u = \lambda u$ is said to be in the *limit-circle* case, and then it turns out that all solutions are in H for all $\lambda \in \mathbb{C}$. All this means that $T(\tau)$ has equal deficiency indices, (m, m), say, and $1 \leq m \leq 2$. The case of deficiency indices $(2, 2)$ is treated in the same way as the regular problem was in Section 7.1.1, culminating in an analogue of Theorem 7.1.4. Hereafter in this subsection we shall assume that the deficiency indices are $(1,1)$, and hence, equivalently, that $\tau u = \lambda u$ is in the limit-point case at b.

The following two theorems are the symmetric cases of Theorems III.10.13 and III.10.14 in [54]; see also [158], Section 18.

Theorem 7.1.5 $T(\tau)$ *is a closed, symmetric operator in H and*

$$T^*(\tau) = T^+(\tau). \tag{7.1.25}$$

If $T(\tau)$ has deficiency indices $(1,1)$, then for all $u, v \in \mathcal{D}(\tau)$,

$$[u, v](b) := \lim_{x \to b} [u, v](x) = 0; \tag{7.1.26}$$

hence by (7.1.23)

$$\mathcal{D}_0(\tau) := \left\{ u : u \in \mathcal{D}(\tau),\ u(a) = u^{[1]}(a) = 0 \right\}. \tag{7.1.27}$$

The domain of a self-adjoint extension T_V of $T(\tau)$ is now the set of functions of the form

$$u = \varphi + \psi,$$

where $\varphi \in \mathcal{D}_0(\tau)$ and $\psi \in \mathcal{N}_+ = \ker\left(T^+(\tau) - iI\right)$.

The analogue of Theorem 7.1.4 in this case is

Theorem 7.1.6 *Let $T(\tau)$ have deficiency indices $(1, 1)$. Every self-adjoint extension of $T(\tau)$ is determined by linearly independent boundary conditions of the form*

$$\alpha_1 u(a) + \alpha_2 u^{[1]}(a) = 0, \tag{7.1.28}$$

with

$$\alpha_1 \bar{\alpha}_2 - \alpha_2 \bar{\alpha}_1 = 0. \tag{7.1.29}$$

Conversely, linearly independent boundary conditions of the form (7.1.28) define a self-adjoint extension of $T(\tau)$ with deficiency indices $(1, 1)$, provided (7.1.29) is satisfied.

An important observation to be made from Theorem 7.1.6 is that no boundary condition needs to be imposed at the singular point b to construct a self-adjoint extension in this limit-point case.

7.1 Sturm–Liouville Operators: Basic Theory

For $u, v \in \mathcal{D}(T^*)$, we have from (7.1.26)

$$[u, v]_{T^*(\tau)} = (T^*(\tau)u, v) - (u, T^*(\tau))$$
$$= u^{[1]}(a)\overline{v(a)} - u(a)\overline{v^{[1]}(a)}.$$

It follows that $(\mathcal{K}, \Gamma_0, \Gamma_1)$, with

$$\mathcal{K} = \mathbb{C}, \quad \Gamma_0 u = u(a), \quad \Gamma_1 u = u^{[1]}(a),$$

is a boundary triplet for $T^*(\tau)$. Let $\{s(\cdot; z), c(\cdot; z)\}$ be the solutions of $\tau u = z u$ for $z \in \mathbb{C}$ determined by (7.1.20), and let $T_0(\tau)$ be the self-adjoint extension of $T(\tau)$ with domain $\mathcal{N}(\Gamma_0)$. If $s(\cdot; z) \in H$, then $s(\cdot; z) \in \mathcal{D}(T_0(\tau))$, which would not be possible if $z \in \mathbb{C} \setminus \mathbb{R} \subseteq \rho(T_0(\tau))$. Therefore, for $z \in \mathbb{C} \setminus \mathbb{R}$, $s(\cdot; z) \notin H$. Since $T(\tau)$ has deficiency indices $(1, 1)$, there must therefore exist a function $m(z)$ such that

$$\psi(x; z) := c(x; z) + m(z)s(x; z) \in H \quad \text{for } z \in \mathbb{C} \setminus \mathbb{R}. \tag{7.1.30}$$

This function m is the celebrated *Titchmarsh–Weyl* function. Since $\mathcal{N}_z = \mathcal{N}(T^*(\tau) - zI)$ is spanned by $\psi(\cdot; z)$, we have that $\gamma(z) = \psi(\cdot; z)$. Also, $M(z) = \Gamma_1 \gamma(z) = \psi^{[1]}(a; z) = m(z)$ by (7.1.20) and (7.1.30).

7.1.3 Two Singular End-Points

This case when both the end-points a, b are singular can be treated by reducing it to the cases of operators $T(\tau, a)$, $T(\tau, b)$ defined with respect to intervals $(a, c]$, $[c, b)$, where $a < c < b$, and each interval has only one singular end-point. The orthogonal sum

$$\tilde{T}'(\tau) = T'(\tau, a) \oplus T'(\tau, b)$$

in

$$L_2((a, b); k) = L_2((a, c); k) \oplus L_2((c, b); k)$$

is densely defined and closable in H, and its closure is given by

$$\tilde{T}(\tau) = T(\tau, a) \oplus T(\tau, b).$$

Furthermore, it can be shown that

$$\mathcal{D}(\tilde{T}(\tau)) = \left\{ u \in \mathcal{D}(\tau) : u(c) = u^{[1]}(c) = 0 \right\}.$$

and the quotient space $\mathcal{D}(T(\tau))/\mathcal{D}(\tilde{T}(\tau))$ has dimension 2; see [54], Theorem III.10.20. It follows that the deficiency index of $T(\tau)$ is given by the formula

$$\operatorname{def} T(\tau) = \operatorname{def} T(\tau, a) + \operatorname{def} T(\tau, b) - 2 \qquad (7.1.31)$$

and hence can take any value between 0 and 2.

We refer to [190], Section 15.3, for an analysis of boundary triplets for T and a description of all the self-adjoint extensions of T in terms of boundary conditions in the cases when both end-points a, b are singular.

7.1.4 The Titchmarsh–Weyl Function and Spectrum

Let T_α be the self-adjoint extension of $T(\tau)$ in the case when (7.1.5) is regular at a and limit-point at b, with domain

$$\mathcal{D}(T_\alpha) = \left\{ u : u \in AC_{loc}[a, b), u(a) \cos \alpha + u^{[1]}(a) \sin \alpha = 0,\ u \in H \right\},$$

where $\alpha \in [0, \pi)$. Let $\varphi(\cdot, \lambda)$, $\theta(\cdot, \lambda)$ be the linearly independent solutions of $\tau u = \lambda u$ which satisfy the initial conditions

$$\varphi(a, \lambda) = -\sin \alpha, \quad \varphi^{[1]}(a, \lambda) = \cos \alpha,$$

$$\theta(a, \lambda) = \cos \alpha, \quad \theta^{[1]}(a, \lambda) = \sin \alpha,$$

for $\alpha \in [0, \pi)$. Since (7.1.5) is assumed to be in the limit-point case at b, there is a function m_α (the Titchmarsh–Weyl function) which is such that $\psi(x, \lambda) = \theta(x, \lambda) + m_\alpha(\lambda) \varphi(x, \lambda)$ is the unique (up to constant multiples) solution of (7.1.5) in H. It follows that in terms of the Titchmarsh–Weyl function $m = m_0$ for the case $\alpha = 0$, already introduced in (7.1.30),

$$m_\alpha(\lambda) = \frac{m(\lambda) \cos \alpha - \sin \alpha}{m(\lambda) \sin \alpha + \cos \alpha}. \qquad (7.1.32)$$

By a fundamental theorem of Weyl (see [201], Section II.2.1), m_α is analytic in the upper and lower open half-planes $\mathbb{C}_\pm$.

In [35], Chaudhuri and Everitt determine the links between the spectral properties of T_α and properties of the associated Titchmarsh–Weyl function m_α in $\mathbb{C}$. They consider the special case in which $a = 0$, $b = \infty$ and $k = 1$ in τ, but their analysis is essentially unchanged for our interval and τ. We shall describe their findings, referring to [35] for details.

Chaudhuri and Everitt use the following classification of points in the spectrum $\sigma(T_\alpha)$ and resolvent set $\rho(T_\alpha)$; $R_\lambda(T_\alpha) := (T_\alpha - \lambda I)^{-1}$ will denote the resolvent operator.

1. $\lambda \in P\sigma(T_\alpha)$ (point spectrum) if $\mathcal{R}(T_\alpha - \lambda I)$ is a closed, non-dense subset of H.
2. $\lambda \in C\sigma(T_\alpha)$ (continuous spectrum) if $\mathcal{R}(T_\alpha - \lambda I) \subsetneq \overline{\mathcal{R}(T_\alpha - \lambda I)} = H$.

7.1 Sturm–Liouville Operators: Basic Theory

3. $\lambda \in PC\sigma(T_\alpha)$ (point continuous spectrum) if $\mathcal{R}(T_\alpha - \lambda I) \subsetneq \overline{\mathcal{R}(T_\alpha - \lambda I)} \subsetneq H$.

Elements of $P\sigma(T_\alpha) \cup PC\sigma(T_\alpha)$ are the eigenvalues of T_α. The spectrum of T_α is said to be discrete if $PC\sigma(T_\alpha) = C\sigma(T_\alpha) = \varnothing$. The main theorem in [35] is

Theorem 7.1.7

1. $\lambda \in \rho(T_\alpha)$ if and only if m_α is analytic at λ. In this case

$$(R_\lambda(T_\alpha)f)(x) = \psi(x,\lambda) \int_a^x \varphi(t,\lambda) f(t) k(t) dt + \varphi(x,\lambda) \int_x^b \psi(t,\lambda) f(t) k(t) dt \tag{7.1.33}$$

   for all $f \in H$ and $\psi(x,\lambda) = \theta(x,\lambda) + m_\alpha(\lambda)\varphi(x,\lambda)$.

2. $\lambda \in P\sigma(T_\alpha)$ if and only if m_α has a pole (simple) at λ. In this case $\varphi(\cdot,\lambda) \in H$ and if r is the residue of m_α at λ, then $\theta(\cdot,\lambda) + r\varphi_\lambda(\cdot,\lambda) \in H$, where $\varphi_\lambda(x,\lambda) = [\partial\varphi(x,\lambda')/\partial\lambda']_{\lambda'=\lambda}$. Also

$$(R_\lambda(T_\alpha)f)(x) = \theta(x,\lambda) \int_a^x \varphi(t,\lambda) f(t) k(t) dt$$
$$+ r\varphi(x,\lambda) \int_a^x \varphi_\lambda(t,\lambda) f(t) k(t) dt$$
$$+ \varphi(x,\lambda) \int_x^b \{\theta(t,\lambda) + r\varphi_\lambda(t,\lambda)\} f(t) k(t) dt \tag{7.1.34}$$

   for all $f \in H \ominus \{\varphi(\cdot,\lambda)\}$, where $\{\varphi(\cdot,\lambda)\}$ is the eigenspace at λ.

3. $\lambda \in C\sigma(T_\alpha)$ if and only if m_α is analytic at λ and $\lim_{\nu \to 0} \nu m_\alpha(\lambda + i\nu) = 0$. If $\gamma\theta(\cdot,\lambda) + \delta\varphi(\cdot,\lambda) \notin H$ for all $\gamma, \delta \in \mathbb{C}$, then $\lambda \in C\sigma(T_\alpha)$.

4. $\lambda \in PC\sigma(T_\alpha)$ if and only if there exists a real number $s < 0$ such that $\lim_{\nu \to 0} i\nu\, m_\alpha(\lambda + i\nu) = s$ and $m_\alpha(\lambda') - s(\lambda' - \lambda)^{-1}$ is not analytic at λ; in this case $\varphi(\cdot,\lambda) \in H$.

If (7.1.5) is regular at a and limit-circle at b, any self-adjoint extension of $T(\tau)$ has a discrete spectrum consisting only of isolated eigenvalues. In this case, the resolvent of T_α is

$$(R_\lambda(T_\alpha)f)(x) = \theta(x,\lambda) \int_a^x \varphi(t,\lambda) f(t) k(t) dt + \varphi(x,\lambda) \int_x^b \theta(t,\lambda) f(t) k(t) dt. \tag{7.1.35}$$

This is an integral operator with kernel in $H \times H$ since $\varphi, \theta \in H$. Consequently $R_\lambda(T_\alpha)$ is Hilbert-Schmidt, and hence compact. It follows that the spectrum of T_α is discrete, and if $\lambda_n, n \in \mathbb{N}$ are its non-zero eigenvalues, then $\sum_{n \in \mathbb{N}} \lambda_n^{-2} < \infty$. The outcome is similar if (7.1.5) is regular at b and limit-circle at a, or limit-circle at both a and b; see [190], Section 15.4.

7.2 KVB Theory for Positive Sturm–Liouville Operators

7.2.1 Semi-boundedness and Oscillation Theory

In [29] the KVB theory is applied to derive the positive self-adjoint extensions of the minimal Sturm–Liouville operator $T(\tau)$ of Section 7.1 under the minimal assumptions in (7.1.2) and (7.1.3) on the coefficients of the expression τ; we shall denote $\mathcal{D}(\tau)$ by $\mathcal{D}$ and $T(\tau)$ by T for simplicity throughout this section. A prominent part is played by a result of Kalf in [122] concerning the Friedrichs extension T_F of T under these assumptions. Kalf's theorem will be stated and proved in the next section, and this section is devoted to discussing preparatory results. We shall assume throughout that the equation (7.1.5) is regular at a; the case of b regular can be treated similarly, and indeed, so can that when both a and b are singular, as was the case in [122].

Some background material is needed. The equation $(\tau - \lambda)u = 0$ is said to be *oscillatory* at b if there exists a solution which has an infinite number of zeros in $[a, b)$ accumulating at b. Otherwise the equation is said to be *non-oscillatory*. Since we are assuming that $(\tau - \lambda)u = 0$ is regular at a, the equation is non-oscillatory at a. The following proposition is well-known; see [104].

Proposition 7.2.1 *Let $(\tau - \lambda)u = 0$ be non-oscillatory at b for some $\lambda \in \mathbb{R}$. Then there exists a real-valued fundamental system $\{f, g\}$ of $(\tau - \lambda)u = 0$ with the following properties:*

1. $f(x) \neq 0$, $g(x) \neq 0$ on $[c, b)$ for some $c \in [a, b)$;
2. $\lim_{c \to b-} \frac{f(x)}{g(x)} = 0$;
3. $\int_c^b \frac{1}{pf^2} dx = \infty$, $\int_c^b \frac{1}{pg^2} dx < \infty$.

The functions f and g are called the *principal* and *non-principal* solutions respectively, of $(\tau - \lambda)u = 0$ at b.

The next proposition was established by Rellich in [175].

Proposition 7.2.2

1. *If $(\tau - \lambda)u = 0$ is non-oscillatory at b for some $\lambda \in \mathbb{R}$ then T is bounded below.*
2. *If T is bounded below, with lower bound α, and $\lambda \leq \alpha$ then any non-trivial solution of $(\tau - \lambda)u = 0$ has at most one zero.*

Kalf observed in [122], Remark 2, that Rellich's result has the consequence

Proposition 7.2.3 *Suppose that $p > 0$ on (a, b). Then T is bounded below if and only if there exist $\mu \in \mathbb{R}$ and a function $h \in AC_{loc}[a, b)$ such that $h^{[1]} \in AC_{loc}[a, b)$, $h > 0$ near b, $\int_c^b (ph^2)^{-1} dx = \infty$ for some $c \in (a, b)$ and*

$$q \geq \mu k + \frac{(h^{[1]})'}{h} \quad \text{a.e. near } b. \tag{7.2.1}$$

7.2 KVB Theory for Positive Sturm–Liouville Operators

It is shown in [122], section 3 that Proposition 7.2.3 remains true if the condition $\int_c^b (ph^2)^{-1} dx = \infty$ is replaced by $\int_c^b (ph^2)^{-1} dx < \infty$. Thus h can behave like a principal or non-principal solution of $(\tau - \lambda)u = 0$ at b for some λ. Rellich characterises the Friedrichs extension of T in [175] as the restriction of T^* to the domain $\mathcal{D}(T_F)$ whose elements behave like a non-principal solution at b:

Proposition 7.2.4 *The Friedrichs extension of the lower semi-bounded operator T has domain*

$$\mathcal{D}(T_F) = \left\{ u : u \in \mathcal{D}, \lim_{x \to b-} \frac{u(x)}{g(x)} = 0 \right\}, \qquad (7.2.2)$$

where g is a non-principal solution of $(\tau - \lambda)u = 0$ at b for some $\lambda \in \mathbb{R}$. The characterisation is independent of the choice of g and λ.

The above results in Propositions 7.2.2, 7.2.3 and 7.2.4 were proved by Rellich in [175] under more restrictive conditions than those of (7.1.2) and (7.1.4), but it was shown in [178] that they continue to hold under our minimal conditions on the coefficients of τ; see also Section 11 in [53] where τ is replaced by the more general expression

$$\tau u := \frac{1}{k} \left(-(p[f' + sf])' + sp[f' + sf] + qf \right),$$

and, in addition to the conditions in (7.1.1), it is assumed that s is real, $s \in L_{1,loc}(a, b)$, and the quasi derivative (7.1.4) is replaced by

$$u^{[1]} := p[u' + su];$$

thus (7.1.1) is the special case $s = 0$.

We shall see in Kalf's theorem below, that $\mathcal{D}(T_F)$ may also be characterised in terms of the principal solution, and indeed, by an alternative to (7.2.2) involving a non-principal solution. The proof of Kalf's theorem rests on the following lemma, which is a one-dimensional special case of a Hardy-type inequality from [120] tailored for our needs.

Lemma 7.2.5 *Let $P > 0$, $1/P \in L_{1,loc}[a, b)$, and for $x \in (a, b)$,*

$$H(x) := \int_a^x \frac{dt}{P(t)}.$$

Let $v \in AC_{loc}(a, b)$ and $\sqrt{P} v' \in L_2(a, b)$.
(i) If $H(b) = \infty$, then, for $c \in (a, b)$,

$$\int_c^b \frac{|v|^2}{PH^2} dt < \infty \quad \text{and} \quad \lim_{x \to b-} \frac{|v(x)|^2}{H(x)} = 0. \qquad (7.2.3)$$

(ii) If $H(b) < \infty$ and $\liminf_{x \to b-} |v(x)| = 0$, then, with $c = a$ now possible,

$$\int_a^b \frac{|v|^2}{PH^2}dt < \infty \quad \text{and} \quad \lim_{x \to b-} \frac{|v(x)|^2}{H(x)} = 0. \tag{7.2.4}$$

Proof. We may suppose, without loss of generality, that v is real.
(i) Let $\psi = vH^{-1/2}$. Then

$$\int_c^x P|v'|^2 dt = \int_c^x \left\{ \sqrt{P}H^{1/2}\psi' + \frac{v}{2\sqrt{PH}} \right\}^2 dt$$

$$= \int_c^x PH|\psi'|^2 dt + \int_c^x \frac{|v|^2}{4PH^2}dt + \int_c^x \psi'\psi \, dt$$

$$= \int_c^x PH^2|\psi'|^2 dt + \frac{1}{4}\int_c^x \frac{|v|^2}{PH^2}dt + \frac{1}{2}\left\{\psi^2(x) - \psi^2(c)\right\}. \tag{7.2.5}$$

Hence

$$\int_c^x P|v'|^2 dt \geq \frac{1}{4}\int_c^x \frac{|v|^2}{PH^2}dt + \frac{1}{2}\left\{\psi^2(x) - \psi^2(c)\right\}. \tag{7.2.6}$$

Since $\sqrt{P}v' \in L_2(a,b)$ it follows that $\int_c^b \frac{|v|^2}{PH^2}dt < \infty$. Also,

$$\int_c^x PH|\psi'|^2 dt = \int_c^x \left| \sqrt{P}v' - \frac{v}{2\sqrt{PH}} \right|^2 dt$$

$$\leq 2\int_c^x P|v'|^2 dt + \frac{1}{2}\int_c^x \frac{v^2}{PH^2}dt$$

$$< \infty. \tag{7.2.7}$$

From (7.2.5) we therefore infer that $\lim_{x \to b-} \psi(x)$ exists. Moreover,

$$\int_c^x \frac{1}{PH}dt = \int_c^x \frac{H'}{H}dt \to \infty$$

as x tends to b, and as

$$\int_c^x \frac{v^2}{PH^2}dt = \int_c^x \frac{\psi^2}{PH}dt$$

we conclude that $\lim_{x \to b-} \psi^2(x) = \lim_{x \to b-} \frac{v^2(x)}{H(x)} = 0$.

(ii) We again set $\psi = vH^{-1/2}$ and deduce from (7.2.5) and (7.2.7), that $\int_c^b \frac{v^2}{PH^2}dt < \infty$ and $\lim_{x \to b} \psi^2(x)$ exists. Part (ii) follows, with now $c = a$ allowed, and the proof is complete. ∎

7.2.2 Kalf's Theorem

Armed with the results from the last section we can now proceed to state and prove Theorem 1 from [122].

Theorem 7.2.6 *Suppose that the conditions (7.1.2) and (7.1.3) are satisfied, but now with $p > 0$, and that (7.1.5) is regular at a. Then the minimal operator T in H is bounded below if and only if there exist $\mu \in \mathbb{R}$ and a function $h \in AC_{loc}[a, b)$ satisfying*

$$ph' \in AC_{loc}[a, b), \quad h > 0 \text{ in } [c, b), \quad \int_c^b \frac{1}{ph^2} dt = \infty, \quad (7.2.8)$$

for some $c \in (a, b)$, with respect to which,

$$q \geq \mu k + \frac{(ph')'}{h}, \quad x \in [c, b). \quad (7.2.9)$$

The Friedrichs extension T_F of T has domain

$$\mathcal{D}(T_F) := \left\{ u; u \in \mathcal{D}, \int_c^b ph^2 \left| \left(\frac{u}{h}\right)' \right|^2 dx < \infty \right\}, \quad (7.2.10)$$

and $T_F u = \tau u$ for $u \in \mathcal{D}(T_F)$. Moreover, for all $u \in \mathcal{D}(T_F)$

$$\int_c^b \left| q - \frac{(ph')'}{h} \right|^2 |u|^2 dx < \infty, \quad (7.2.11)$$

and if $\mu > 0$, the form of T is

$$t_F[u] = \int_a^b \left\{ ph^2 \left| \left(\frac{u}{h}\right)' \right|^2 + q_h |u|^2 \right\} dx, \quad u \in Q(T), \quad (7.2.12)$$

where

$$q_h = q - \frac{(ph')'}{h}.$$

Proof. Let $\mathcal{D}(S)$ denote the subspace (7.2.10) and S the operator defined by $Su := \tau u$ for $u \in \mathcal{D}(S)$. The functions in $\mathcal{D}$ which are compactly supported in $[a, b)$, constitute the domain of the operator T' of Section 7.1.2 which is densely defined in H and whose closure is T. It follows that $T \subset S$ and, in particular, S is densely defined. The Rellich Proposition 7.2.3 asserts the equivalence of T being bounded below and the conditions (7.2.8) and (7.2.9) for some function $h \in AC_{loc}[a, b)$. Hence it will be sufficient to prove that S is symmetric and $T_F \subset S$.

Use is made of the Jacobi factorisation identity

$$-(pu')' + \frac{(ph')'}{h}u = -\frac{1}{h}\left[ph^2\left(\frac{u}{h}\right)'\right]' \tag{7.2.13}$$

and so

$$\tau u = \frac{1}{k}\left\{-\frac{1}{h}\left[ph^2\left(\frac{u}{h}\right)'\right]' + q_h u\right\} \tag{7.2.14}$$

where $q_h := q - \frac{(ph')'}{h}$. This gives, for any $c \in (a,b)$,

$$\int_a^c (Su)\bar{u}k\,dx = \int_a^c \left\{-\left[ph^2\left(\frac{u}{h}\right)'\right]'\frac{\bar{u}}{h} + q_h|u|^2\right\}dx$$
$$= \left[-ph^2\left(\frac{u}{h}\right)'\frac{\bar{u}}{h}\right]_a^c + \int_a^c \left\{ph^2\left|\left(\frac{u}{h}\right)'\right|^2 + q_h|u|^2\right\}dx. \tag{7.2.15}$$

For $u \in \mathcal{D}(S)$, $v = u/h$ and $P = ph^2$ we have that $v \in AC_{loc}[a,b)$ and $\sqrt{P}v' \in L_2(a,b)$. It follows from Lemma 7.2.5(i) with $P := ph^2$ that

$$\int_a^b \frac{|v|^2}{PH^2}dx < \infty$$

and hence

$$\int_a^b \frac{P|\bar{v}v'|}{PH}dx \le \left(\int_a^b \frac{|v|^2}{PH^2}dx\right)^{1/2}\left(\int_a^b P|v'|^2 dx\right)^{1/2} < \infty.$$

Since $H(b) = \infty$ implies that $\int_c^b \frac{dx}{PH} = \int_c^b \frac{H'}{H}dx = \infty$, we have

$$\liminf_{x \to b-} P(x)|\overline{v(x)}v'(x)| = \liminf_{x \to b-}\left|ph^2\left(\frac{u}{h}\right)'\frac{\bar{u}}{h}\right| = 0. \tag{7.2.16}$$

Therefore there exists a sequence (b_n) tending to b such that in (7.2.15)

$$\lim_{n \to \infty}\left\{im \int_a^{b_n}(Su)\bar{u}k\,dx\right\} = 0$$

The limit

$$\lim_{t \to b}\left\{im \int_a^t (Su)\bar{u}k\,dx\right\}$$

clearly exists and we have therefore shown that (Su,u) is real and consequently that S is symmetric.

Next we observe that for all $u \in \mathcal{D}(S)$, (7.2.11) is a consequence of (7.2.15) and (7.2.16).

7.2 KVB Theory for Positive Sturm–Liouville Operators

The final step is to prove that $T_F \subset S$. Since $T_F \subset T^*$, we have that $\mathcal{D}(T_F) \subset \mathcal{D}$ and it is therefore sufficient to prove that

$$\int_c^b ph^2 \left|\left(\frac{u}{h}\right)'\right|^2 < \infty \quad (u \in \mathcal{D}(T_F)) \tag{7.2.17}$$

for $c \in [a, b)$ such that $h > 0$ in $[c, b)$. We shall use the fact established in (6.3.7) that

$$\mathcal{D}(T_F) = \mathcal{D}(T^*) \cap Q(T),$$

where $Q(T)$ is the form domain of T, i.e., the completion of $\mathcal{D}(T)$ with respect to the norm defined by

$$(Tu, u) - \alpha \|u\|^2 + \|u\|^2,$$

if $T \geq \alpha I$. Thus $u \in \mathcal{D}(T_F)$ if and only if $u \in \mathcal{D}(T^*)$ and there is a sequence $(u_n), u_n \in \mathcal{D}(T)$ which converges to u in H and

$$(T(u_n - u_m), u_n - u_m) \to 0 \text{ as } n, m \to \infty. \tag{7.2.18}$$

Let $u \in \mathcal{D}(T_F)$ and set $u_{jk} := u_j - u_k$, where (u_n) is the sequence in (7.2.18). For $w \in \mathcal{D}$ and $x \in [c, b)$

$$\int_c^x \left(p|w'|^2 + q|w|^2\right) dt = \left[p\frac{h'}{h}|w|^2\right]_x^c + \int_x^c \left\{ph^2\left|\left(\frac{w}{h}\right)'\right|^2 + q_h |w|^2\right\} dt; \tag{7.2.19}$$

this follows from the identity

$$h'\left(\frac{|w|^2}{h}\right)' + h^2 \left|\left(\frac{w'}{h}\right)'\right|^2 = |w'|^2$$

which is readily verified. For all $j, k \in \mathbb{N}$, we therefore have

$$(Tu_{jk}, u_{jk}) = \int_c^b \left\{ph^2\left|\left(\frac{u_{jk}}{h}\right)'\right|^2 + q_h |u_{jk}|^2\right\} dx$$
$$- \left(p\frac{h'}{h}|u_{jk}|^2\right)(c) + \int_a^c \left(p|u'_{jk}|^2 + q|u_{jk}|^2\right) dx,$$

since the u_j, being in $\mathcal{D}(T)$, vanish at a. It is proved in [175], p. 350, that regular Sturm–Liouville operators are bounded from below, and so there exists $\nu \in \mathbb{R}$ such that

$$-\left(p\frac{h'}{h}|u_{jk}|^2\right)(c) + \int_a^c \left(p|u'_{jk}|^2 + q|u_{jk}|^2\right) dx \geq \nu \int_a^c |u_{jk}|^2 k\, dx;$$

this is also proved in [53] (A34), under conditions which include those we assume. Hence, on putting $\alpha := |\mu| + |\nu|$, we derive

$$(Tu_{jk}, u_{jk}) + \alpha \|u_{jk}\|^2 \geq \int_c^b ph^2 \left|\left(\frac{u_{jk}}{h}\right)'\right|^2 dx.$$

Since the left-hand side tends to zero as $j, k \to \infty$ we conclude that there exists $\tilde{u}$ such that

$$\lim_{j \to \infty} \int_c^b ph^2 \left|\left(\frac{u_j}{h}\right)' - \tilde{u}\right|^2 dx = 0.$$

Therefore $\tilde{u} = (u/h)'$ a.e. on $[c, b)$ which establishes (7.2.17) and hence that $T_F \subset S$. Furthermore, (7.2.12) follows from (7.2.15) and this completes the proof. ∎

Kalf also gives the following alternative version of his theorem in which the principal solution h is assumed to be a non-principal solution rather than a principal one.

Theorem 7.2.7 *Suppose that the conditions (7.1.2) and (7.1.3) are satisfied, but now with $p > 0$, and that (7.1.5) is regular at a. Then the minimal operator T in H is bounded below if and only if there exist $\mu \in \mathbb{R}$ and a function $h \in AC_{loc}[a, b)$ satisfying*

$$ph' \in AC_{loc}[a, b), \quad h > 0 \text{ in } [c, b), \quad \int_c^b \frac{1}{ph^2} dt < \infty, \tag{7.2.20}$$

for some $c \in (a, b)$, with respect to which,

$$q \geq \mu k + \frac{(ph')'}{h}, \quad x \in [c, b). \tag{7.2.21}$$

The Friedrichs extension T_F of T has domain

$$\mathcal{D}(T_F) := \left\{ u; u \in \mathcal{D}, \int_c^b ph^2 \left|\left(\frac{u}{h}\right)'\right|^2 dx < \infty, \lim_{x \to b-} \frac{|u(x)|}{h(x)} = 0 \right\}, \tag{7.2.22}$$

and $T_F u = \tau u$ for $u \in \mathcal{D}(T_F)$. Moreover, for all $u \in \mathcal{D}(T_F)$,

$$\int_c^b \left| q - \frac{(ph')'}{h} \right|^2 |u|^2 dx < \infty \tag{7.2.23}$$

and (7.2.12) holds for $\mu > 0$.

Proof. We now apply part (ii) of Lemma 7.2.5. It is easily checked that the only property of (7.2.22) which requires proof is

7.2 KVB Theory for Positive Sturm–Liouville Operators

$$\lim_{x \to b-} \frac{|u(x)|}{h(x)} = 0, \quad \text{for all } u \in \mathcal{D}(T_F), \tag{7.2.24}$$

since the rest will follow as in the previous theorem. So, for $u \in \mathcal{D}(T_F)$, let (u_n) be a sequence in $\mathcal{D}(T)$ which converges to u in H and which satisfies (7.2.18), and let (u_{nm}) be a subsequence which converges pointwise a.e to u on (a, b). Then, for a.e $x \in (a, b)$ and all $m \in \mathbb{N}$,

$$\left|\frac{u_{nm}(x)}{h(x)}\right|^2 = \left|\int_x^b \sqrt{ph}\left(\frac{u_{nm}}{h}\right)' \frac{1}{\sqrt{ph}} dt\right|^2 \leq H(x) \int_x^b ph^2 \left|\left(\frac{u_{nm}}{h}\right)'\right|^2 dt.$$

This yields (7.2.24) and the theorem is proved. ∎

7.3 Application of the KVB Theory

We shall assume that $\mu > 0$ in (7.2.9) so that T is positive in Theorem 7.2.6, and similarly in Theorem 7.2.7. The results in this section are based on ones in [29].

Our principal tool is the result from the Krein–Vishik–Birman theory given by Theorem 6.4.6 and Theorem 6.4.7 (see Remark 6.4.8), that there is a one to one correspondence between the set of all positive self-adjoint extensions of T and the set of pairs $\{N_B, B\}$, where N_B is a subspace of $\mathcal{N} := \ker T^*$, the kernel of T^*, and B is a positive self-adjoint operator acting in $\mathcal{N}$. If $\tilde{T}$ is a positive self-adjoint extension of T, then $\tilde{T} = T_B$ for some B, where T_B is associated with a form $t_B[\cdot]$ which satisfies

$$t_B = t_F \dot{+} \mathfrak{b}, \quad Q(T_B) = Q(T_F) \dot{+} Q(B) \tag{7.3.1}$$

where $Q(\cdot)$ denotes the form domain and $\mathfrak{b}[\cdot]$ is the form of B. Thus any $v \in Q(T_B)$ can be written $v = u + \eta$, where $u \in Q(T_F)$ and $\eta \in Q(B) \subseteq \mathcal{N}$. Furthermore

$$t_B[v] = t_F[u] + (B\eta, \eta). \tag{7.3.2}$$

The Krein–von Neumann extension corresponds to $B = 0$; thus $Q(B) = \mathcal{N}$ and $t_B[v] = t_F[u]$. The Friedrichs extension corresponds to $B = \infty$; thus $Q(B) = \{0\}$ and $t_B = t_F$.

7.3.1 The Limit-Point Case at b

Since a is regular and $T \geq \mu > 0$, the deficiency index of $(T - \lambda I)$, and hence the dimension of $\ker(T^* - \lambda I)$, is constant for all $\lambda \in \mathbb{R} \setminus [\mu, \infty)$. In the limit-point case of $\tau u = \lambda u$ at b it therefore follows, in particular, that $\mathcal{N} := \ker(T^*)$ is of dimension

1, and so, any $\eta \in \mathcal{N}$ can be written $\eta = c\psi$, where $\psi \in H := L^2[a, b; k)$ is real, $c \in \mathbb{C}$ and $\tau\psi = 0$. Our main result in the limit-point case is

Theorem 7.3.1 *Let τ be in the limit point case at b. Then there is a one-one correspondence between the positive self-adjoint extensions of T and the one-parameter family of operators $\{T_l\}, 0 \leq l \leq \infty$, where T_l is the restriction of T^* to the domain*

$$\mathcal{D}(T_l) := \{v : v \in \mathcal{D}(T_0^*), (pv')(a) = [p(a)\psi'(a) + l\|\psi\|^2]v(a)\}. \quad (7.3.3)$$

The operator T_l with $l = 0$, is the Krein–von Neumann extension and T_∞ is the Friedrichs extension.

Proof. If $\psi(a) = 0$, then ψ is an eigenfunction at zero of the self-adjoint extension T_0 of T determined by the Dirichlet boundary condition $u(a) = 0$. Thus $\mathcal{N} = \ker T_0$ and $\mathcal{D}(T) \dotplus \mathcal{N} \subseteq \mathcal{D}(T_0)$. Since $\mathcal{D}(T) \dotplus \mathcal{N}$ is the domain of T_K, by (6.4.10), it follows that $T_0 = T_K$. However, by [220, Theorem 5], T_0 is the Friedrichs extension of T and therefore has no null space. Consequently $\psi(a) = 0$ is not possible.

Thus we may assume hereafter that $\psi(a) \neq 0$ and, without loss of generality, that

$$\psi(a) = 1. \quad (7.3.4)$$

Consequently, from (7.3.1), any $v = u + \eta \in Q(T_B)$ is uniquely expressible as

$$v(x) = v(x) - v(a)\psi(x) + v(a)\psi(x)$$

with

$$u(x) = v(x) - v(a)\psi(x) \in Q(T_F)$$

and

$$\eta(x) = v(a)\psi(x) \in Q(B).$$

Also, from (7.3.2)

$$t_B[v] = \int_a^b \{ph^2|\left(\frac{u}{h}\right)'|^2 + q_h|u|^2\}dx + l|v(a)|^2\|\psi\|^2,$$

where $q_h = q - \dfrac{(ph')'}{h}$ and $b[\eta] = (B\eta, \eta) = l\|\eta\|^2, 0 \leq l \leq \infty$. Moreover, for $\varphi = \theta + \xi = \varphi - \varphi(a)\psi + \varphi(a)\psi \in Q(T_B)$, the sesquilinear form identity for $t_B[v, \varphi]$ associated with (7.3.2) is

$$t_B[v, \varphi] = \int_a^b \left\{ph^2 \left(\frac{v - v(a)\psi}{h}\right)' \overline{\left(\frac{\varphi - \varphi(a)\psi}{h}\right)'}\right\} dx \quad (7.3.5)$$

$$+ \int_a^b \left\{q_h(v - v(a)\psi)\overline{(\varphi - \varphi(a)\psi)}\right\} dx + lv(a)\overline{\varphi(a)}\|\psi\|^2$$

7.3 Application of the KVB Theory

$$= \int_a^b \left\{ ph^2 \left(\frac{v}{h}\right)' \left(\frac{\overline{\varphi}}{h}\right)' + q_h v \overline{\varphi} \right\} dx$$

$$- \overline{\varphi(a)} \int_a^b \left\{ ph^2 \left(\frac{v}{h}\right)' \left(\frac{\overline{\varphi}}{h}\right)' + q_h v \overline{\varphi} \right\} dx$$

$$- v(a) \int_a^b \left\{ ph^2 \left(\frac{\psi}{h}\right)' \left(\overline{\frac{\varphi - \varphi(a)\psi}{h}}\right)' + q_h \psi \overline{(\varphi - \varphi(a)\psi)} \right\} dx$$

$$+ lv(a)\overline{\varphi(a)}\|\psi\|^2$$

$$=: I_1 + I_2 + I_3 + I_4 \qquad (7.3.6)$$

say. It follows on integration by parts, and since $\theta(a) = \varphi(a) - \varphi(a)\psi(a) = 0$, that

$$I_2 = -\left[ph^2 \left(\frac{v}{h}\right)' \left(\frac{\overline{\xi}}{h}\right) \right]_a^b, \quad \xi = \varphi(a)\psi,$$

$$I_3 = -v(a) \left[ph^2 \left(\frac{\psi}{h}\right)' \left(\overline{\frac{\varphi - \varphi(a)\psi}{h}}\right) \right]_a^b = -v(a) \left[ph^2 \left(\frac{\psi}{h}\right)' \left(\overline{\frac{\varphi - \varphi(a)\psi}{h}}\right) \right](b).$$

For $v \in \mathcal{D}(T_B) \subset \mathcal{D}(T^*)$ we have by (7.2.13)

$$t_B[v, \varphi] = (T_B v, \varphi) = \int_a^b \{(-(pv')' + qv)\overline{\varphi}\} dx$$

$$= \int_a^b \left\{ -\frac{1}{h}[ph^2 \left(\frac{v}{h}\right)']' + q_h v \right\} \overline{\varphi} dx$$

$$= -\left[ph^2 \left(\frac{v}{h}\right)' \left(\frac{\overline{\varphi}}{h}\right) \right]_a^b + I_1. \qquad (7.3.7)$$

Thus

$$-\left[ph^2 \left(\frac{v}{h}\right)' \left(\frac{\overline{\varphi}}{h}\right) \right]_a^b = -\overline{\varphi(a)} \left[ph^2 \left(\frac{v}{h}\right)' \left(\frac{\overline{\psi}}{h}\right) \right]_a^b$$

$$- v(a) \left[ph^2 \left(\frac{\psi}{h}\right)' \left(\overline{\frac{\varphi - \varphi(a)\psi}{h}}\right) \right](b) + lv(a)\overline{\varphi(a)}\|\psi\|^2$$

and

$$0 = ph^2 \left[\left(\frac{v}{h}\right)' \left(\frac{\overline{\varphi}}{h}\right) - \left(\frac{v}{h}\right) \left(\frac{\overline{\varphi(a)\psi}}{h}\right)' \right](b) - v(a) \left[ph^2 \left(\frac{\psi}{h}\right)' \left(\overline{\frac{\varphi - \overline{\varphi}(a)\psi}{h}}\right) \right](b)$$

$$- ph^2 \left[\left(\frac{v}{h}\right)' \left(\frac{\overline{\varphi}}{h}\right) - \left(\frac{v}{h}\right) \left(\frac{\overline{\varphi(a)\psi}}{h}\right)' \right](a) + lv(a)\overline{\varphi(a)}\|\psi\|^2. \qquad (7.3.8)$$

The value at b of the right-hand side is

$$ph^2\left[\left\{\left(\frac{v}{h}\right)'\left(\frac{\overline{\varphi}}{h}\right) - \left(\frac{v}{h}\right)\left(\frac{\overline{\varphi}}{h}\right)'\right\} + \left(\frac{v}{h}\right)\left(\frac{\overline{\theta}}{h}\right)' - \left(\frac{\eta}{h}\right)'\left(\frac{\overline{\theta}}{h}\right)\right]$$

$$= ph^2\left[\left\{\left(\frac{v}{h}\right)'\left(\frac{\overline{\varphi}}{h}\right) - \left(\frac{v}{h}\right)\left(\frac{\overline{\varphi}}{h}\right)'\right\} + \left\{\left(\frac{\eta}{h}\right)\left(\frac{\overline{\theta}}{h}\right)' - \left(\frac{\eta}{h}\right)'\left(\frac{\overline{\theta}}{h}\right)\right\} + \left(\frac{u}{h}\right)\left(\frac{\overline{\theta}}{h}\right)'\right]$$

$$= p[v'\overline{\varphi} - v\overline{\varphi}'](b) + p[\eta\overline{\theta}' - \eta'\overline{\theta}](b) + \left[ph^2\left(\frac{u}{h}\right)\left(\frac{\overline{\theta}}{h}\right)'\right](b).$$

Suppose now that $\varphi \in (T_B)$. Then the functions $v, u, \eta, \varphi, \theta, \xi$ are members of $\mathcal{D}(T^*)$ and thus
$$p[v'\overline{\varphi} - v\overline{\varphi}'](b) = p[\eta\overline{\theta}' - \eta'\overline{\theta}](b) = 0,$$
since we have assumed the limit-point condition at b; see [54], Theorem III.10.13 or [158], Section 18.3. Hence the value of the right-hand side of (7.3.8) at b is
$$\left[ph^2\left(\frac{\overline{\theta}}{h}\right)'\left(\frac{u}{h}\right)\right](b).$$

Since $u, \theta \in \mathcal{D}(T_F)$, we have
$$(T_F\theta, u) = \int_a^b \left\{ph^2\left(\frac{\theta}{h}\right)'\left(\frac{\overline{u}}{h}\right)' + q_h\theta\overline{u}\right\} dx =: I_5.$$

But as $u(a) = \theta(a) = 0$,
$$(T_F\theta, u) = \int_a^b \left\{-\frac{1}{h}\left[ph^2\left(\frac{\theta}{h}\right)'\right]' + q_h\theta\right\}\overline{u}\,dx = \left[ph^2\left(\frac{\theta}{h}\right)'\left(\frac{\overline{u}}{h}\right)\right](b) + I_5.$$

Therefore, it follows that
$$\left[ph^2\left(\frac{\theta}{h}\right)'\left(\frac{\overline{u}}{h}\right)\right](b) = 0. \tag{7.3.9}$$

Hence, we infer from (7.3.8) that if $v \in \mathcal{D}(T_B)$, then for all $\varphi \in \mathcal{D}(T_B)$,
$$\overline{\varphi}(a)\left\{(pv')(a) - v(a)p(a)\overline{\psi}'(a) - lv(a)\|\psi\|^2\right\} = 0.$$

If T_B is not the Friedrichs extension of T we have from [220] that there exists $\varphi \in \mathcal{D}(T_B)$ such that $\varphi(a) \neq 0$. Hence we have that any $v \in \mathcal{D}(T_B)$ satisfies the boundary condition
$$(pv')(a) = [p(a)\psi'(a) + l\|\psi\|^2]v(a). \tag{7.3.10}$$

7.3 Application of the KVB Theory

The real constants l parameterise the operators T_B: $l = 0$ corresponds to $B = 0$ and thus the Krein–von Neumann extension of T, while $l = \infty$ corresponds to $B = \infty$ and hence the Friedrichs extension. ∎

7.3.2 The Case of b Regular or Limit Circle, and $\tau u = 0$ Non-oscillatory at b

Let f, g be the principal and non-principal solutions respectively of $\tau u = 0$. From (7.2.24),

$$\frac{u}{g}(b) := \lim_{x \to b} \frac{u(x)}{g(x)} = 0, \quad \text{for all } u \in \mathcal{D}(T_F), \tag{7.3.11}$$

and from Corollary 1 in [177],

$$\frac{u}{f}(b) := \lim_{x \to b} \frac{u(x)}{f(x)} \text{ exists, for all } u \in \mathcal{D}(T_F). \tag{7.3.12}$$

We still have that $u(a) = 0$ for $u \in \mathcal{D}(T_F)$, and both

$$t_F[u] = \int_a^b \left\{ ph^2 \left| \left(\frac{u}{g}\right)' \right|^2 + q_g |u|^2 \right\} dx, \tag{7.3.13}$$

where $q_g = q - \dfrac{(pg')'}{g} > 0$, and

$$t_F[u] = \int_a^b \left\{ ph^2 \left| \left(\frac{u}{f}\right)' \right|^2 + q_f |u|^2 \right\} dx, \tag{7.3.14}$$

where $q_f = q - \dfrac{(pf')'}{f} > 0$, are valid.

We now have dim $\mathcal{N} = 2$, and $\{f, g\}$ is a fundamental system of solutions of $\tau u = 0$. The self-adjoint operators B act in subspaces N_B of $\mathcal{N}$ which therefore may be of dimension 1 or 2. In the case dim $N_B = 2$ of the next theorem, $\{\psi_1, \psi_2\}$ is a real orthonormal basis of N_B.

Theorem 7.3.2 *The positive self-adjoint extensions of T which correspond to operators B in the Krein–Vishik–Birman theory with dim $N_B = 1$ form a one-parameter family T_β of restrictions of T^* with domains*

$$\mathcal{D}(T_\beta) := \left\{ v \in \mathcal{D}(T^*) : \left[pg^2 \left\{ \left(\frac{v}{g}\right) \left(\frac{\psi}{g}\right)' - \left(\frac{v}{g}\right)' \left(\frac{\psi}{g}\right) \right\} \right]_a^b = \beta v(a) \|\psi\|^2 \right\}, \tag{7.3.15}$$

where ψ is a real basis of N_B with $\psi(a) = 1$.

The self-adjoint extensions corresponding to operators B with $\dim N_B = 2$ form a family T_β, where now β is a matrix $(b_{j,k})_{j,k=1,2}$ of parameters, and T_β is the restriction of T^* to the domain $\mathcal{D}(T_\beta)$ of functions $v \in \mathcal{D}(T^*)$ which satisfy the boundary conditions

$$\left[pg^2 \left\{ \left(\frac{v}{g}\right) \left(\frac{\psi_k}{g}\right)' - \left(\frac{v}{g}\right)' \left(\frac{\psi_k}{g}\right) \right\} \right]_a^b = \sum_{j=1}^{2} b_{k,j} c_j, \quad k = 1, 2, \qquad (7.3.16)$$

where c_1 and c_2 are determined in terms of the values of v at a and b by

$$\frac{v}{g}(a) = \sum_{j=1}^{2} c_j \frac{\psi_j}{g}(a),$$

$$\frac{v}{g}(b) = \sum_{j=1}^{2} c_j \frac{\psi_j}{g}(b). \qquad (7.3.17)$$

Proof.
The case $\dim N_B = 1$

Let the real function ψ be a basis of N_B, $\eta = c\psi \in N_B$, ($c \in \mathbb{C}$) and $B\eta = \beta c\psi$ for $\beta \in \mathbb{R}_+$. Then, $\psi = c_1 f + c_2 g$ for some $c_1, c_2 \in \mathbb{R}$, and $\lim_{x \to b} \frac{\psi(x)}{g(x)} = c_2$. Suppose that $\psi(a) = 1$.

Let $v, \varphi \in Q(T_B)$; then $v = u + \eta$, $\varphi = \theta + \xi$, where $u, \theta \in Q(T_F)$, $\eta, \xi \in Q(B)$ and as

$$v = v - v(a)\psi + v(a)\psi, \quad \varphi = \varphi - \varphi(a)\psi + \varphi(a)\psi,$$

we have that

$$u = v - v(a)\psi, \quad \eta = v(a)\psi, \quad \theta = \varphi - \varphi(a)\psi, \quad \xi = \varphi(a)\psi.$$

If $v, \varphi \in Q(T_B) \cap \mathcal{D}(T^*)$, then $u, \theta \in Q(T_F) \cap \mathcal{D}(T^*) = \mathcal{D}(T_F)$. Hence on using (7.3.13), we have from (7.3.2) that

$$t_B[v, \varphi] = \int_a^b \left\{ pg^2 \left(\frac{u}{g}\right)' \left(\frac{\overline{\theta}}{g}\right)' + q_g u \overline{\theta} \right\} dx + \mathfrak{b}(\eta, \xi). \qquad (7.3.18)$$

The argument following (7.3.6) can be repeated, with g replacing h and using the facts that now $\left(\frac{u}{g}\right)(b) = \left(\frac{\theta}{g}\right)(b) = 0$, as well as $u(a) = \theta(a) = 0$. The term corresponding to I_3 is now zero and the result is that

7.3 Application of the KVB Theory

$$t_B[v, \varphi] = \int_a^b \left\{ pg^2 \left(\frac{v}{g}\right)' \left(\frac{\overline{\varphi}}{g}\right)' + q_g v\overline{\varphi} \right\} dx$$
$$- \overline{\varphi(a)} \left[pg^2 \left(\frac{v}{g}\right)' \left(\frac{\psi}{g}\right)' \right]_a^b + \beta v(a)\overline{\varphi(a)} \|\psi\|^2. \qquad (7.3.19)$$

If $v \in \mathcal{D}(T_B)$, since $T_B \subset T^*$,

$$t_B[v, \varphi] = (T_B v, \varphi) = \int_a^b (\tau v)\overline{\varphi} dx$$
$$= \int_a^m \left\{ -\frac{1}{g}[pg^2 \left(\frac{v}{g}\right)']' + q_g v \right\} \overline{\varphi} dx$$
$$= -\left[pg^2 \left(\frac{v}{g}\right)' \left(\frac{\overline{\varphi}}{g}\right) \right]_a^b + \int_a^b \left\{ pg^2 \left(\frac{v}{g}\right)' \left(\frac{\overline{\varphi}}{g}\right)' + q_g v\overline{\varphi} \right\} dx$$
$$= -\left[pg^2 \left(\frac{v}{g}\right)' \left(\frac{\overline{\xi}}{g}\right) \right]_a^b + \int_a^b \left\{ pg^2 \left(\frac{v}{g}\right)' \left(\frac{\overline{\varphi}}{g}\right)' + q_g v\overline{\varphi} \right\} dx, \qquad (7.3.20)$$

since $\frac{\theta}{g}(a) = \frac{\theta}{g}(b) = 0$. We conclude from (7.3.19) and (7.3.20) that

$$\left[pg^2 \left\{ \left(\frac{v}{g}\right) \left(\overline{\varphi(a)}\frac{\psi}{g}\right)' - \left(\frac{v}{g}\right)' \left(\overline{\varphi(a)}\frac{\psi}{g}\right) \right\} \right]_a^b = \beta v(a)\overline{\varphi(a)} \|\psi\|^2.$$

Since $\varphi(a) = d\psi(a) = d$ for arbitrary $d \in \mathbb{C}$, it follows that

$$\left[pg^2 \left\{ \left(\frac{v}{g}\right) \left(\frac{\psi}{g}\right)' - \left(\frac{v}{g}\right)' \left(\frac{\psi}{g}\right) \right\} \right]_a^b = \beta v(a) \|\psi\|^2. \qquad (7.3.21)$$

The case dim $N_B = 2$

Let $\{\psi_1, \psi_2\}$ be a real orthonormal basis for $\mathcal{N}$. Then

$$B\psi_j = \sum_{k=1}^2 b_{k,j}\psi_k, \quad j = 1, 2,$$

where

$$b_{j,k} = \overline{b_{k,j}}, \quad (B\psi_j, \psi_k) = b_{k,j}.$$

If $\eta = \sum_{j=1}^2 c_j\psi_j$, $\xi = \sum_{j=1}^2 d_j\psi_j$, then

$$(B\eta, \xi) = \sum_{j,k=1}^2 b_{k,j}c_j\overline{d_k}, \quad (\eta, \xi) = \sum_{j=1}^2 c_j\overline{d_j}.$$

Furthermore, for some $\mu_{jk} \in \mathbb{R}$, $j,k = 1,2$,

$$\psi_j = \mu_{j,1} f + \mu_{j,2} g, \quad j = 1,2, \tag{7.3.22}$$

so that

$$\frac{\psi_j}{g}(b) = \mu_{j,2}, \quad j = 1,2. \tag{7.3.23}$$

From (7.3.19) and (7.3.20), we now have

$$\left[pg^2 \left\{ \left(\frac{v}{g}\right)\left(\frac{\bar{\xi}}{g}\right)' - \left(\frac{v}{g}\right)'\left(\frac{\bar{\xi}}{g}\right) \right\} \right]_a^b = (B\eta, \xi) = \sum_{j,k=1}^{2} b_{k,j} c_j \overline{d_k}$$

and so

$$\sum_{k=1}^{2} \overline{d_k} \left[pg^2 \left\{ \left(\frac{v}{g}\right)\left(\frac{\psi_k}{g}\right)' - \left(\frac{v}{g}\right)'\left(\frac{\psi_k}{g}\right) \right\} \right]_a^b$$

$$= (B\eta, \xi) = \sum_{j,k=1}^{2} b_{k,j} c_j \overline{d_k}. \tag{7.3.24}$$

Since d_1 and d_2 are arbitrary, we have

$$\left[pg^2 \left\{ \left(\frac{v}{g}\right)\left(\frac{\psi_k}{g}\right)' - \left(\frac{v}{g}\right)'\left(\frac{\psi_k}{g}\right) \right\} \right]_a^b = \sum_{j=1}^{2} b_{k,j} c_j, \quad k = 1,2. \tag{7.3.25}$$

In (7.3.25), c_1 and c_2 are determined by the values of v at a and b, for $v(a) = \eta(a)$ and $\left(\frac{v}{g}\right)(b) = \left(\frac{\eta}{g}\right)(b)$ by (7.3.11). To be specific,

$$\frac{v}{g}(a) = \sum_{j=1}^{2} c_j \frac{\psi_j}{g}(a),$$

$$\frac{v}{g}(b) = \sum_{j=1}^{2} c_j \frac{\psi_j}{g}(b). \tag{7.3.26}$$

∎

Remark 7.3.3

If in the case $\dim N_B = 1$ of the preceding theorem $\psi = c_1 f$, then $\frac{\psi}{g}(b) = 0$ and the boundary condition becomes

$$(pv')(a) = \left(\beta \|\psi\|^2\right) v(a),$$

7.3 Application of the KVB Theory

with no contribution from b as in the limit-point case. We could then repeat the above analysis with g replaced by f to get

$$\left[pf^2\left\{\left(\frac{v}{f}\right)\left(\frac{\overline{\varphi}}{f}\right)' - \left(\frac{v}{f}\right)'\left(\frac{\overline{\varphi}}{f}\right)\right\}\right]_a^b = \beta v(a)\overline{\varphi}(a)\|\psi\|^2, \tag{7.3.27}$$

where $\frac{\overline{\varphi}}{f}(b) = \frac{\overline{\theta}}{f}(b) + c_1$. The equation (7.3.27) has to be satisfied for all $\theta \in Q(T_F)$.

7.3.3 Limit-Point and Limit-Circle Criteria

Let

$$\tau = -\frac{d}{dx}\left(p\frac{d}{dx}\right) + q, \ p > 0, \ x \in (0, \infty),$$

and suppose that (7.1.2) and (7.1.3) are satisfied with $k = 1$. The determination of criteria for τ to be in the limit-point or limit-circle condition at the interval endpoints has been the subject of intensive research since its importance in the theory of self-adjoint extensions of the minimal operator T was realised. Results of great generality are known, but we select just a few of particular significance.

1. **Levinson [145]**; see proof in [158], Theorem 6 in Section 23.
   Let M be a positive, non-decreasing function such that

$$\int_c^\infty \frac{dx}{\sqrt{pM}} = \infty \tag{7.3.28}$$

and

$$\limsup_{x\to\infty} \frac{M'\sqrt{pM}}{M^2} < \infty. \tag{7.3.29}$$

Furthermore, suppose that, for all sufficiently large values of x,

$$q(x) > -KM(x), \tag{7.3.30}$$

where K is a positive constant. Then τ is in the limit point case at ∞.
A special case, obtained by taking $p = 1$ and $M(x) = x^2$, is that $\tau = -\frac{d^2}{dx^2} + q$ is in the limit-point case at ∞ if

$$q(x) > -Kx^2 \tag{7.3.31}$$

for sufficiently large values of x. The result (7.3.31) is best possible if we restrict to powers of x, for if $q(x) = -Kx^c$, $c < 2$, then all solutions of $\tau u = 0$ in $(1, \infty)$

can be shown to satisfy $u(x) = O(x^{-c/4})$ and are therefore in $L_2(1, \infty)$; see [54], Theorem III.10.28. This ensures that τ is in the limit-circle case at ∞.

2. **Kalf and Walter [120], Section 4, Corollary 3**

Suppose that for $x \in (0, l]$, where l is arbitrary,

$$\int_0^x p(t)^{-1/2} dt < \infty \tag{7.3.32}$$

and

$$\liminf_{x \to 0+} \left\{ \left(\int_0^x p(t)^{-1/2} dt \right) \left[q(x) + \frac{1}{4p(x)h^2(x)} \right] \right\} > 1, \tag{7.3.33}$$

where $h(x) = \int_x^1 p(t)^{-1} dt$. Then τ is in the limit-point case at 0. The special case

$$p(x) = x^{2\mu}, \quad \mu < 1/2,$$

yields the limit-point criterion (at 0)

$$\liminf_{x \to 0+} x^{2-2\mu} q(x) > \frac{3}{4} - \mu. \tag{7.3.34}$$

In particular, if $q(x) = c/x^2$, $c > 3/4$, then $\tau = -\frac{d^2}{dx^2} + q$ is in the limit-point case at 0. Moreover, the equation

$$\tau u(x) = -u''(x) + \frac{c}{x^2} u(x) = 0$$

has the linearly independent solutions $u_j = x^{\alpha_j}$ $(j = 1, 2)$, where

$$\alpha_1 = \frac{1}{2} \left[1 + \sqrt{(1+4c)} \right], \quad \alpha_2 = \frac{1}{2} \left[1 - \sqrt{(1+4c)} \right].$$

Near zero $u_1 \in L_2$, while $u_2 \in L_2$ if, and only if, $\alpha_2 > -1/2$, i.e., $c < 3/4$. Consequently τ is in the limit-point case at zero if and only if $c \geq 3/4$.

7.4 Coercive Sectorial Operators

Let

$$\tau = -\frac{d}{dx} \left(p \frac{d}{dx} \right) + q, \quad q = q_1 + i q_2, \tag{7.4.1}$$

and

$$\tau^+ = -\frac{d}{dx} \left(p \frac{d}{dx} \right) + \overline{q}, \tag{7.4.2}$$

7.4 Coercive Sectorial Operators

on the interval $[a, b)$, where the coefficients satisfy (7.1.2) and (7.1.3), except that q is now complex-valued. Suppose also that

$$q_{1,h} := q_1 - \frac{(ph')'}{h} \geq \mu k, \quad |q_2| \leq (\tan \theta) \left\{ q_1 - \frac{(ph')'}{h} \right\}, \tag{7.4.3}$$

for some $\mu > 0$ and $\theta \in (0, \pi/2)$. The minimal operators $T(\tau)$, $T(\tau^+)$ are then coercive and sectorial with their numerical ranges in the sector

$$\Sigma(\mu, \theta) := \{ z = x + iy \in \mathbb{C} : x \geq \mu > 0, \; |y| \leq \tan \theta (x - \mu) \}. \tag{7.4.4}$$

The maximal operators are the adjoints $T(\tau^+)^*$, $T(\tau)^*$ of the minimal operators, and we have

$$T(\tau) \subset T(\tau^+)^*, \quad T(\tau^+) \subset T(\tau)^* \tag{7.4.5}$$

so that $T(\tau)$, $T(\tau^+)$ form an adjoint pair. They are also J-symmetric with respect to the conjugation $J : u \to \bar{u}$, i.e.,

$$JT(\tau)J \subset T(\tau)^*, \quad JT(\tau^+)J \subset T(\tau^+)^*. \tag{7.4.6}$$

It follows from [54], Theorem III.10.7, that

$$2 \leq \operatorname{def} T(\tau) + \operatorname{def} T(\tau^+) \leq 4,$$

and since $\operatorname{def} T(\tau)$ and $\operatorname{def} T(\tau^+)$ are equal, being the dimensions of the kernels of $T(\tau)$ and $T(\tau^+)^*$ respectively since $\mu > 0$, we have that

$$1 \leq \dim(\ker T(\tau)^*) = \dim(\ker T(\tau^+)^*) \leq 2. \tag{7.4.7}$$

From Theorem 6.6.4, the Krein–von Neumann extension $T_K(\tau)$ satisfies the following:

$$Q(T_K(\tau)) = Q(T(\tau)) \dotplus \mathcal{N}, \quad \mathcal{N} := \ker(T(\tau)^*) \tag{7.4.8}$$

and

$$t_K(\tau)[u, v] = t(\tau)[Pu, Pv], \quad u, v \in Q(T_K(\tau)), \tag{7.4.9}$$

where $t_K(\tau)$, $t(\tau)$ are the forms of $T_k(\tau)$, $T(\tau)$, respectively, and P is the projection of $Q(T_K(\tau))$ onto $Q(T(\tau))$ in the decomposition (7.4.8). Furthermore, by (6.6.17),

$$\mathcal{D}(T_K(\tau)) = \mathcal{D}(T(\tau)) \dotplus \mathcal{N}, \quad T_K(\tau)(f + v) = T(\tau)f, \quad f \in \mathcal{D}(T(\tau)), v \in \mathcal{N}. \tag{7.4.10}$$

The identities (7.4.8)–(7.4.10) have exact analogues when τ is replaced by τ^+.

The case $\dim(\ker T(\tau)^*) = 1$.

The following theorem is Theorem 3.1 in [29]}.

Theorem 7.4.1 *Let $\psi \in \ker T(\tau)^*$ be such that $\psi(a) = 1$. Then $T_K(\tau)$ has domain*

$$\mathcal{D}(T_K(\tau)) := \{v : v - v(a)\psi \in \mathcal{D}(T_F(\tau)), \ v'(a) - v(a)\psi'(a) = 0\}, \quad (7.4.11)$$

and for all $v \in \mathcal{D}(T_K(\tau))$,

$$T_K(\tau)v = \tau(v - v(a)\psi). \quad (7.4.12)$$

Moreover, the form domain is

$$Q(T_K(\tau)) = \{v : u = v - v(a)\psi \in Q(T(\tau))\} \quad (7.4.13)$$

and

$$t_K(\tau)[v] = t_F(\tau)[u] = \int_a^b \left\{ ph^2 \left| \left(\frac{u}{h}\right)' \right|^2 + q_h |u|^2 \right\} dx, \ u = v - v(a)\psi. \quad (7.4.14)$$

Proof. Any $v \in Q(T_K(\tau))$ can be written as

$$v = v - v(a)\psi + v(a)\psi.$$

For all $u \in Q(T_F(\tau))$, $u(a) = 0$ and so, from (7.4.8), v has the unique representation $v = u + \xi$, where $u = v - v(a)\psi$ and $\xi = v(a)\psi$. Consequently, from (7.4.9)

$$t_K(\tau)[v, \varphi] = t_F(\tau)[u, \theta,] \quad (7.4.15)$$

for all $\varphi \in Q(T_K(\tau))$, with $P\varphi = \theta = \varphi - \varphi(a)\psi$. Therefore (7.4.13) and (7.4.14) are satisfied. From (7.4.10), and since $u(a) = (pu')(a) = 0$ for $u \in \mathcal{D}(T(\tau))$, we have that any $v \in \mathcal{D}(T_K(\tau))$ satisfies the boundary condition

$$v'(a) - v(a)\psi'(a) = 0.$$

Also,

$$T_K(\tau)v = T(\tau)(v - v(a)\psi) = \tau(v - v(a)\psi).$$

∎

In [29], Theorem 3.2, Arlinskii's analysis in [11], Section 3.1, is applied to characterise all the coercive m-sectorial extensions of $T(\tau)$. We give a brief sketch, but refer to Arlinskii's paper for the background abstract results and the details of their application to differential operators such as ours.

Let X_0 denote $Q(T(\tau))$, with norm

7.4 Coercive Sectorial Operators

$$\|u\|_X = \left\{ \int_a^b ph^2 \left|\left(\frac{u}{h}\right)'\right|^2 + q_{1,h}|u|^2 \right\}^{1/2} =: t^R[u]^{1/2}$$

and

$$X_1 = \{u : u \in AC_{loc}(a, b) : \|u\|_X < \infty\},$$

with the norm $\|\cdot\|_X$.

Arlynskii expresses $T(\tau)$ in the *divergence form*

$$T(\tau) = L_2^* R L_1,$$

where

1. L_1, L_2 are closed, densely defined operators with domains in $H := L^2(a, b; k)$ and ranges in $H_2 := H \oplus H$;
2. R is a bounded and coercive operator on H;
3. $\mathcal{D}(L_1) \cap \mathcal{D}(L_2^* R L_2)$ is dense in $\mathcal{D}(L_1)$.

In our application,

$$\mathcal{D}(L_1) = X_0; \quad L_1 u = \begin{pmatrix} u \\ hp^{1/2}\left(\frac{u}{h}\right)' \end{pmatrix};$$

$$\mathcal{D}(L_2) = X_1; \quad L_2 u = \begin{pmatrix} u \\ hp^{1/2}\left(\frac{u}{h}\right)' \end{pmatrix};$$

$$R = \frac{1}{k}\begin{pmatrix} q_h & 0 \\ 0 & 1 \end{pmatrix}$$

where $q_h = q - \frac{(ph')'}{h}$. The adjoint operators are given by

$$D(L_1^*) = H \oplus X_1, \quad D(L_2^*) = H \oplus X_0;$$

$$L_j^* \frac{1}{k}\begin{pmatrix} f_1 \\ f_2 \end{pmatrix} = f_1 - \frac{1}{h}\left(hp^{1/2} f_2\right)', \quad \text{for } j = 1, 2.$$

Then

$$L_1^* R^* L_2 u = \overline{q_h} u - \frac{1}{h}\left(ph^2 \left(\frac{u}{h}\right)'\right)' u = \tau^+ u = T^* u.$$

Arlinskii's approach yields the following result; ψ is the solution of $\tau^+ \psi = 0$ with $\psi(a) = 1$.

Theorem 7.4.2 *The formulae*

$$\mathcal{D}(\tilde{T}(\tau)) = \{v \in X_1 : v - (\psi - 2y)v(a) \in D(T_F(\tau));$$
$$\left[pv' - p(\psi' - 2y')v(a) - (\psi - 2y)v(a)\right](a) = wv(a)\},$$
$$\tilde{T}(\tau)v = \tau(v - (\psi - 2y)v(a)),$$

establish a one to one correspondence between all coercive m-sectorial extensions $\tilde{T}(\tau)$ of $T(\tau)$, excepting $T_F(\tau)$ and $T_K(\tau)$, and the set of all pairs $\langle w, y \rangle$, where w is a complex number with a positive real part, and $y \in X_0$ satisfies

$$\max\{re\left[t^R(\tau)[(2y - \varphi, \varphi)], \varphi \in X_0\}\right] < re\,[w].$$

The associated closed form is given by

$$\tilde{t}(\tau)[v] = t^R(\tau)[v - (\psi - 2y)v(a), v] + wv(a)\overline{v(a)}, \quad v \in X_1.$$

The Friedrichs and Krein–von Neumann extensions correspond to the pairs $\langle \infty, 0 \rangle$ and $\langle 0, 0 \rangle$ respectively.

7.4.1 The Case dim(ker T^*) = 2.

Let $\{\psi_1, \psi_2\}$ be a basis for ker $T(\tau)^*$ and let $\psi = \sum_{j=1}^{2} c_j \psi_j$ be such that $\psi(a) = 1$. Then Theorem 7.4.2 establishes a one-one correspondence between all coercive m-sectorial extensions of $T(\tau)$ and the set $< w, y, c_1, c_2 >$. This determines a vector in ker $T(\tau)^*$, and the complex number w in the open right half plane determines a one-dimensional coercive operator $W(\lambda \psi) = \lambda w \psi$. In general, the parameter w can be a $2\times$ coercive sectorial matrix which determines a linear operator, and in this case, y is a linear operator.

7.5 Realisations of Second-Order Elliptic Operators on Domains

In [96], Grubb applies her abstract theory for determining all the closed realisations of an adjoint pair of operators to elliptic differential operators generated by expressions A and their formal adjoint A' defined on a domain $\Omega \subset \mathbb{R}^n (n \geq 2)$ with smooth boundary. The operators A_0, A_1 of Section 6.5 are now defined to be the minimal and maximal operators associated with A, and A'_0, A'_1 are the analogous ones for A'. The outcome of the application is that any closed realisation of $\tilde{A}$ of A satisfying $A_0 \subset \tilde{A} \subset A_1$ (and analogously for $\tilde{A}'$ of A' satisfying $A'_0 \subset \tilde{A}' \subset A'_1$) is given by

7.5 Realisations of Second-Order Elliptic Operators on Domains

boundary conditions on the boundary $\partial\Omega$ expressed in terms of differential operators acting between function spaces defined on $\partial\Omega$. In this section we outline the main aspects of Grubb's results which are relevant to us, in the special case of second-order expressions A; our account is intended to give a flavour of Grubb's results and serve as an appetiser for the comprehensive treatment in [96] and in subsequent papers by Grubb and others.

Let Ω be a bounded domain in $\mathbb{R}^n$ whose boundary $\partial\Omega$ is a manifold of class C^∞, and let A, A' be the formally adjoint differential expressions

$$A := -\sum_{i,j=1}^{n} D_i(a_{ij}D_j) + q, \quad A' := -\sum_{i,j=1}^{n} D_i(\overline{a_{ji}}D_j) + \overline{q}, \qquad (7.5.1)$$

where $a_{ij}, q \in C_0^\infty(\overline{\Omega})$. We shall assume that A and A' satisfy the condition that for some $c > 0$,

$$re\left\{\sum_{i,j=1}^{n} a_{ij}(x)\xi_i\xi_j + q(x)\right\} \geq c|\xi|^2, \quad \text{for all } x \in \Omega, \text{ and } \xi \in \mathbb{R}^n; \qquad (7.5.2)$$

when $q = 0$, this means that A and A' are uniformly elliptic.

Some preliminary results and remarks concerning spaces defined on Ω and $\partial\Omega$ are required; these are established in [152], p. 30–39 and [204], pp. 317, 332. In the case $p = 2$, we shall follow standard practice and denote the Sobolev spaces $W_p^s(\Omega)$, $\overset{0}{W}_p^s(\Omega)$, for $s \in \mathbb{R}$, by $H^s(\Omega)$, $\overset{0}{H}^2(\Omega)$, respectively. The standard inner product and norm on $L_2(\Omega)$ will be denoted by $(\cdot, \cdot)$ and $\|\cdot\|$ respectively.

Remark 7.5.1

1. For $s \in \mathbb{R}$,

$$H^s(\mathbb{R}^n) := \{u : u \in \mathcal{S}'(\mathbb{R}^n), \ (1+|\xi|^2)^{s/2}\hat{u}(\xi) \in L_2(\mathbb{R}^n)\},$$

with Hilbert space norm

$$\|u\|_s = \|(1+|\xi|^2)^{s/2}\hat{u}(\xi)\|_{L_2(\mathbb{R}^n)}.$$

2.
$$H^s(\Omega) := \{u : \exists U \in H^s(\mathbb{R}^n), \text{ such that } u = U \text{ in } \Omega\},$$

$$\|u\|_{s,\Omega} := \inf_{U \in H^s(\mathbb{R}^n)} \{\|U\|_s : U = u \text{ in } \Omega\}.$$

For $s_1 < s_2$, $H^{s_2}(\Omega) \subset H^{s_1}(\Omega)$ algebraically and topologically (i.e., the embedding is continuous), and $H^{s_2}(\Omega)$ is dense in $H^{s_1}(\Omega)$. Furthermore, the embedding

$H^{s_2}(\Omega) \hookrightarrow H^{s_1}(\Omega)$ is compact. This implies that the spaces $H^s(\Omega)$ are different for all $s \in \mathbb{R}$. For if $H^{s_1}(\Omega) = H^{s_2}(\Omega)$ and $s_1 < s_2$ say, the identity map of $H^{s_1}(\Omega)$ to itself would be compact. As the closed unit ball in this space would therefore be compact, $H^{s_1}(\Omega)$ would be finite dimensional, hence a contradiction.

3.
$$\overset{0}{H^s}(\Omega) := \{u \in H^s(\mathbb{R}^n), \; \text{supp}\, u \subset \overline{\Omega}\};$$

$C^\infty(\overline{\Omega})$ is dense in $H^s(\Omega)$ and $C_0^\infty(\Omega)$ is dense in $\overset{0}{H^s}(\Omega)$. The map

$$f : \varphi \mapsto \int_\Omega \psi\overline{\varphi}\, dx$$

is a continuous, conjugate linear functional on $H^s(\Omega)$ and $H^{-s}(\Omega)$ is identified with the adjoint of $H^s(\Omega)$, i.e., the space of continuous, conjugate linear functionals on $H^s(\Omega)$. The duality between $H^s(\Omega)$ and $H^{-s}(\Omega)$ is given by

$$\langle u, v \rangle_{s,-s} := \int_\Omega u\overline{v}\, dx, \quad u \in H^s(\Omega), \; v \in H^{-s}(\Omega),$$

this being an extension of the $L_2(\Omega)$ inner product on $C_0^\infty(\Omega)$. For $s' > s > 0$, we have the continuous injections

$$H^{s'}(\Omega) \subset H^s(\Omega) \subset L_2(\Omega) \subset H^{-s}(\Omega) \subset H^{-s'}(\Omega).$$

4. $H^s(\partial\Omega)$ is defined by local co-ordinates using the definition of $H^s(\mathbb{R}^{n-1})$. It contains $C_0^\infty(\partial\Omega)$ as a dense subspace, and it follows from the boundedness of $\partial\Omega$ that a distribution $u \in \mathcal{D}'(\partial\Omega)$ lies in $H^s(\partial\Omega)$ for some s.
If $0 < s_1 < s_2 < \infty$, $H^{s_2}(\partial\Omega)$ is compactly embedded in $H^{s_1}(\partial\Omega)$; see [102], Proposition 4.22. It follows as in item 2 above that the spaces $H^s(\partial\Omega)$ are all distinct for $s > 0$.

5. $H^s(\partial\Omega)$ and $H^{-s}(\partial\Omega)$ are adjoint spaces with respect to an extension of

$$\int_{\partial\Omega} \varphi\overline{\psi}\, d\sigma, \quad \varphi, \psi \in C_0^\infty(\partial\Omega),$$

the duality being also denoted by $\langle \varphi, \psi \rangle_{s,-s}$; confusion with the notation for the spaces $H^s(\Omega), H^{-s}(\Omega)$, is unlikely as the meaning will be clear from the context.

6.
$$D_{s,A}(\Omega) := \{u : u \in H^s(\Omega), \; Au \in L_2(\Omega)\}$$

with norm

$$\|u\|_{D_{s,A}(\Omega)} := \left(\|u\|_s^2 + \|Au\|^2\right)^{1/2}.$$

7.5 Realisations of Second-Order Elliptic Operators on Domains

Hence, in particular, $D_{0,A}(\Omega) = \mathcal{D}(A_1)$, with the graph norm. For all $s \in \mathbb{R}$, $D_{s,A}(\Omega)$ is continuously embedded in $H^s(\Omega)$; when $s \geq 2$, $D_{s,A}(\Omega) = H^s(\Omega)$ since $Au \in H^{s-2}(\Omega) \subset L_2(\Omega)$ for $u \in H^s(\Omega)$.

7.
$$\mathcal{N}_{s,A}(\Omega) := \{u : u \in H^s(\Omega),\ Au = 0\}$$

This is a closed subspace of $D_{s,A}(\Omega)$. Note that $\mathcal{N}_{0,A}(\Omega) = \mathcal{N}(A_1)$, the null space of A_1.

8. *Trace theorem.* Let $\gamma_0 u$ denote the value of u on $\partial\Omega$ and $\gamma_1 u = \gamma_0(\partial u/\partial \nu)$, the value of the normal derivative of u on $\partial\Omega$ with respect to the outward unit normal $\nu = (\nu_1, \nu_2, \cdots, \nu_n)$. The trace maps γ_0 and γ_1, defined initially on $C_0^\infty(\overline{\Omega})$, have extensions which are continuous linear maps of $H^s(\Omega)$ onto $H^{s-1/2}(\partial\Omega)$ (for $s > 1/2$) and $H^{s-3/2}(\partial\Omega)$ (for $s > 3/2$), respectively; see [96], Theorem I.2.1. Consequently we have the continuous surjection

$$\{\gamma_0, \gamma_1\} : H^2(\Omega) \to H^{3/2}(\partial\Omega) \times H^{1/2}(\partial\Omega). \tag{7.5.3}$$

Note that for $u \in H^2(\Omega)$, $\{\gamma_0, \gamma_1\}u = \gamma_0 u \times \gamma_1 u$, and the surjectivity of the maps in (7.5.3) means, for instance, that if $\varphi \in H^{1/2}(\partial\Omega)$, and hence $\{0, \varphi\} \in H^{3/2}(\partial\Omega) \times H^{1/2}(\partial\Omega)$, then there exists $u \in H^2(\Omega)$ such that $\gamma_0 u = 0$ and $\gamma_1 u = \varphi$.

The minimal operators A_0, A_0' are the closures of $A \restriction_{C_0^\infty(\Omega)}$, $A' \restriction_{C_0^\infty(\Omega)}$ respectively; the domains of the minimal and maximal operators are as follows:

$$\mathcal{D}(A_0) = \mathcal{D}(A_0') = \overset{0}{H}{}^2(\Omega),$$
$$\mathcal{D}(A_1) = \{u : u \in L_2(\Omega),\ Au \in L_2(\Omega)\}, \tag{7.5.4}$$
$$\mathcal{D}(A_1') = \{u : u \in L_2(\Omega),\ A'u \in L_2(\Omega)\}, \tag{7.5.5}$$

where Au and $A'u$ are understood in the distributional sense, i.e.,

$$(Au, \varphi) := (u, A'\varphi)\ \text{for all}\ \varphi \in C_0^\infty(\Omega),$$

and similarly for $A'u$. It will be shown in Corollary 7.5.5 below that

$$\mathcal{D}(A_1) \not\subseteq H^2(\Omega),\ \mathcal{D}(A_1') \not\subseteq H^2(\Omega).$$

The form domain $Q(A_0)$ of A_0 is $\overset{0}{H}{}^1(\Omega)$ and the domains of the Friedrichs extension A_F and Krein–von Neumann extension A_K of A_0 are

$$\mathcal{D}(A_F) = \overset{0}{H}{}^1(\Omega) \cap \mathcal{D}(A_1),$$
$$\mathcal{D}(A_K) = \overset{0}{H}{}^2(\Omega) \dotplus \mathcal{N}(A_1),$$

where $\mathcal{N}(A_1) = \ker A_1$.

Lemma 7.5.2
$$(A_0')^* = A_1, \quad A_0^* = A_1'. \tag{7.5.6}$$

Therefore A, A' form an adjoint pair. Furthermore, the ellipticity assumption implies

$$\overset{0}{H^1}(\Omega) \cap \mathcal{D}(A_1) = \overset{0}{H^1}(\Omega) \cap \mathcal{D}(A_1'). \tag{7.5.7}$$

Proof. Let $u \in \mathcal{D}(A_1)$. Then for all $\varphi \in C_0^\infty(\Omega)$,

$$(Au, \varphi) = (u, A'\varphi)$$

and hence, since A_0' is the closure of $A' \upharpoonright_{C_0^\infty(\Omega)}$, it follows that $A_1 \subset (A_0')^*$. Next, let $v \in \mathcal{D}((A_0')^*)$. Then $v \in L_2(\Omega)$ and for all $\varphi \in \mathcal{D}(A_0')$,

$$(A_0'\varphi, v) = (\varphi, (A_0')^*v).$$

In particular, this holds for all $\varphi \in C_0^\infty(\Omega)$ and

$$(A_0'\varphi, v) = (\varphi, Av).$$

It follows that $(A_0')^*v = Av$. Thus $v \in \mathcal{D}(A_1)$ and $(A_0')^* \subset A_1$, which completes the proof of $(A_0')^* = A_1$.

The second identity in (7.5.6) is established in the same way. ∎

Following closely the treatment in [96], we shall describe in this section the set $\mathcal{M}$ of all the closed operators $\tilde{A}$ satisfying $A_0 \subset \tilde{A} \subset A_1$ given abstractly in Theorem 6.5.6, in terms of boundary operators acting between spaces on $\partial\Omega$. The special case of 2nd-order differential operators considered here avoids complications which have to be addressed in the problem for general $2m$th-order differential operators in [96], when certain admissibility conditions have to be imposed on the differential operators and compatible boundary differential operators to define well-posed problems.

For $u, v \in H^2(\Omega)$, Green's formula gives

$$\int_\Omega \left(Au\overline{v} - u(\overline{A'v})\right) dx = \int_{\partial\Omega} \left\{(\gamma_0 u)\,\overline{(\gamma_a' v)} - (\gamma_a u)\,\overline{(\gamma_0 v)}\right\} d\sigma, \tag{7.5.8}$$

where γ_0 is the trace operator which maps u into its value on $\partial\Omega$, $\nu = (\nu_1, \nu_2, \cdots, \nu_n)$ is the outward unit normal to $\partial\Omega$,

7.5 Realisations of Second-Order Elliptic Operators on Domains

$$\gamma_a u = \left(\sum_{i,j=1}^{n} \left(\gamma_0[\nu_i a_{ij} D_j u] \right) \right), \tag{7.5.9}$$

$$\gamma'_a v = \left(\sum_{i,j=1}^{n} \left(\gamma_0[\nu_i \overline{a_{ji}} D_j v] \right) \right), \tag{7.5.10}$$

and $d\sigma$ is the surface measure on $\partial\Omega$. Thus, when $A = -\Delta$, $\gamma_a u = \gamma_0 ([\nabla \cdot \nu]u) = \gamma_0 (\partial u/\partial \nu) =: \gamma_1 u$, the value of the normal derivative of u on $\partial\Omega$.

The following result of Lions and Magenes, [151], Theorem 5.4 (see also [96], Theorem I.3.3), gives a precise description of non-homogeneous boundary problems for A when the the coefficients $a_{i,j}$ of A and the boundary $\partial\Omega$ of Ω satisfy our smoothness assumptions.

Theorem 7.5.3

1. For $s \leq 2$, the maps $\{A, \gamma_a\} : u \mapsto \{Au, \gamma_a u\}$ and $\{A, \gamma_0\} : u \mapsto \{Au, \gamma_0 u\}$ are isomorphisms of $\mathcal{D}_{s,A}(\Omega)$ onto $L_2(\Omega) \times H^{s-3/2}(\partial\Omega)$ and $L_2(\Omega) \times H^{s-1/2}(\partial\Omega)$ respectively. Similarly, $\{A', \gamma'_a\}$ and $\{A', \gamma_0\}$ are isomorphisms of $\mathcal{D}_{s,A'}(\Omega)$ onto $L_2(\Omega) \times H^{s-3/2}(\partial\Omega)$ and $L_2(\Omega) \times H^{s-1/2}(\partial\Omega)$ respectively.
2. For $s \geq 2$, the maps $\{A, \gamma_a\}$ and $\{A, \gamma_0\}$ are isomorphisms of $H^s(\Omega)$ onto $H^{s-2}(\Omega) \times H^{s-3/2}(\partial\Omega)$ and $H^{s-2}(\Omega) \times H^{s-1/2}(\partial\Omega)$, respectively. Similarly, $\{A', \gamma'_a\}$ and $\{A', \gamma_0\}$ are isomorphisms of $H^s(\Omega)$ onto $H^{s-2}(\Omega) \times H^{s-3/2}(\partial\Omega)$ and $H^{s-2}(\Omega) \times H^{s-1/2}(\partial\Omega)$, respectively.

Note the difference between the condition on s imposed in Theorem 7.5.3(1) and the condition $s > 1/2$ for γ_0 and $s > 3/2$ for γ_a required in the trace theorem mentioned in Remark 7.5.1(8) above. The price paid for being able to allow s to go all the way down to 0 is that in Theorem 7.5.3(1) the trace operators are not defined on the whole of $H^s(\Omega)$ but merely on $\mathcal{D}_{s,A}(\Omega)$.

Theorem 7.5.4 *For all $s \in \mathbb{R}$, the maps $u \mapsto \gamma_a u$ and $u \mapsto \gamma_0 u$ are isomorphisms of $\mathcal{N}_{s,A}(\Omega)$ onto $H^{s-3/2}(\partial\Omega)$ and $H^{s-1/2}(\partial\Omega)$ respectively. Similarly, $u \mapsto \gamma'_a u$ is an isomorphism of $\mathcal{N}_{s,A'}(\Omega)$ onto $H^{s-3/2}(\partial\Omega)$.*

Proof.

In Theorem 7.5.3, when $s \leq 2$, the inverse of $\{A, \gamma_a\}$ maps $\{0\} \times H^{s-3/2}(\partial\Omega)$ onto a closed subspace of $\mathcal{D}_{s,A}(\Omega)$, and this is $\{u \in \mathcal{D}_{s,A}(\Omega) : Au = 0\}$ with the topology of $\mathcal{D}_{s,A}(\Omega)$; by the definitions in items 6 and 7 in Remark 7.5.1, this is $\mathcal{N}_{s,A}(\Omega)$. The other cases are similar.

When $s \geq 2$, we have from Theorem 7.5.3 that the inverse of $\{A, \gamma_a\}$ maps $\{0\} \times H^{s-3/2}(\partial\Omega)$ onto $\{u \in H^s\Omega) : Au = 0\}$ with the topology of $H^s(\Omega)$, and this is also $\mathcal{N}_{s,A}(\Omega)$. The same argument applies to the other cases. ∎

This allows us to give the following striking result.

Corollary 7.5.5 *For all $s > 0$, $\mathcal{N}(A_1) \not\subseteq H^s(\Omega)$. In particular, it follows that $\mathcal{N}(A_1) \not\subseteq H^2(\Omega)$ and hence $H^2(\Omega) \subset \mathcal{D}(A_1) \not\subseteq H^2(\Omega)$.*

Proof. Suppose that $\mathcal{N}(A_1) \subset H^s(\Omega)$ for some $s > 0$. Then

$$\mathcal{N}(A_1) \subset \mathcal{N}_{s,A}(\Omega)$$

and hence, by Theorem 7.5.4,

$$H^{-1/2}(\partial\Omega) = \gamma_0\left(\mathcal{N}(A_1)\right) \subset \gamma_0\left(\mathcal{N}_{s,A}(\Omega)\right) = H^{s-1/2}(\partial\Omega),$$

which contradicts Remark 7.5.1(4). ∎

On applying Theorem 7.5.4, (7.5.8) can be extended to

Corollary 7.5.6 *For all $s \leq 2$, $u \in \mathcal{D}_{s,A}(\Omega)$ and $v \in \mathcal{D}_{2-s,A'}(\Omega)$,*

$$(Au, v) - (u, A'v) = \langle \gamma_0 u, \gamma_a' v \rangle_{s-1/2,-s+1/2} - \langle \gamma_a u, \gamma_0 v \rangle_{s-3/2,-s+3/2}. \quad (7.5.11)$$

The extension (7.5.11) of (7.5.8) is the best possible. For suppose $u \in \mathcal{D}_{s,A}(\Omega)$ and $v \in \mathcal{D}_{s',A'}(\Omega)$. The formula only makes sense if each pair $\{\gamma_0 u, \gamma_a' v\}$ and $\{\gamma_a u, \gamma_0 v\}$, lie in dual spaces, which in view of Theorem 7.5.3 is true if and only if $s' - 3/2 = -s + 1/2$; hence $s' = 2 - s$. However, we seek an analogue of (7.5.11) which holds for all $u \in \mathcal{D}(A_1)$ and $v \in \mathcal{D}(A_1')$, to obtain an identity like that in Lemma 6.5.2 which had such an important role in the abstract theory. Recall that the latter identity was expressed in terms of reference operators A_β, A_β' which are such that $0 \in \rho(A_\beta) \cap \rho(A_\beta')$ and $A_0 \subset A_\beta \subset A_1$, $A_0' \subset A_\beta' \subset A_1'$. The appropriate operators for us are of the form

$$A_\beta = A_1 \upharpoonright_{\mathcal{D}(A_\beta)}: \mathcal{D}(A_\beta) = \{u : u \in \mathcal{D}(A_1), \ Bu = 0\}, \quad (7.5.12)$$
$$A_\beta' = A_1' \upharpoonright_{\mathcal{D}(A_\beta')}: \mathcal{D}(A_\beta') = \{u : u \in \mathcal{D}(A_1'), \ B'u = 0\}, \quad (7.5.13)$$

where either $B = B' = \gamma_0$ or $B = \gamma_a$, $B' = \gamma_a'$; the first choice determines the Dirichlet realisations of A and A' and the second the Neumann realisations of A and A'. For $u \in H^2(\Omega)$, $Bu = 0$ and $B'u = 0$ are defined by (7.5.3). For general $u \in \mathcal{D}(A_1)$, $Bu = 0$ is defined in the following weak sense:

$$Bu = 0 \text{ if and only if } u \in \mathcal{D}((A_\beta')^*), \quad (7.5.14)$$

and analogously for $B'u$ when $u \in \mathcal{D}(A_1')$. It is proved in [184] and [3] that $u \in \mathcal{D}(A_1)$ and $Bu = 0$ in this weak sense imply that $u \in H^2(\Omega)$ with $Bu = 0$ in the sense of the extended trace maps γ_0 and γ_a in Remark 7.5.1(8).

It follows from [96], Proposition III.5.2 (i), that with $s \leq 2$, and $r < s$,

$$\{u : u \in \mathcal{D}_{r,A}(\Omega), \ \gamma_0 u \in H^{s-1/2}(\partial\Omega)\} \subset \mathcal{D}_{s,A}(\Omega), \quad (7.5.15)$$
$$\{u : u \in \mathcal{D}_{r,A}(\Omega), \ \gamma_a u \in H^{s-3/2}(\partial\Omega)\} \subset \mathcal{D}_{s,A}(\Omega). \quad (7.5.16)$$

Thus if $r = 0$ and $s = 2$, we have from (7.5.15), since $\mathcal{D}_{2,A}(\Omega) \subset H^2(\Omega)$,

7.5 Realisations of Second-Order Elliptic Operators on Domains

$$\{u : u \in \mathcal{D}(A_1), \; \gamma_0 u \in H^{3/2}(\partial\Omega)\} \subset H^2(\Omega),$$

which in view of (7.5.3) yields

$$\{u : u \in \mathcal{D}(A_1), \; \gamma_0 u \in H^{3/2}(\partial\Omega)\} = H^2(\Omega); \tag{7.5.17}$$

the same identity with A replaced by A' is also clearly valid. Similarly, from (7.5.16) and (7.5.3),

$$\{u : u \in \mathcal{D}(A_1), \; \gamma_a u \in H^{1/2}(\partial\Omega)\} = H^2(\Omega), \tag{7.5.18}$$

and analogously

$$\{u : u \in \mathcal{D}(A'_1), \; \gamma'_a u \in H^{1/2}(\partial\Omega)\} = H^2(\Omega). \tag{7.5.19}$$

It therefore follows from (7.5.17) and (7.5.18) that

$$\mathcal{D}(A_\beta) \subset \{u : u \in H^2(\Omega), \; Bu = 0\} =: \mathcal{D}(\tilde{A}_\beta) \tag{7.5.20}$$

and similarly

$$\mathcal{D}(A'_\beta) \subset \{u : u \in H^2(\Omega), \; B'u = 0\} =: \mathcal{D}(\tilde{A}'_\beta) \tag{7.5.21}$$

Theorem 7.5.7 *When* $B = \gamma_0$ *or* γ_a, *and* $B' = \gamma_0$ *or* γ'_a, *we have*

$$\mathcal{D}(A_\beta) = \{u \in H^2(\Omega) : Bu = 0\}, \quad \mathcal{D}(A'_\beta) = \{u \in H^2(\Omega) : B'u = 0\}. \tag{7.5.22}$$

Also, $A_0 \subset A_\beta \subset A_1$, $A'_0 \subset A'_\beta \subset A'_1$, *and* A_β, A'_β *form an adjoint pair; in fact* $A^*_\beta = A'_\beta$. *Furthermore,* $0 \in \rho(A_\beta) \cap \rho(A'_\beta)$.

Proof. Let $\tilde{A}_\beta$, $\tilde{A}'_\beta$ be the restrictions of A, A' to $\mathcal{D}(\tilde{A}_\beta)$, $\mathcal{D}(\tilde{A}'_\beta)$, respectively. Then it is easily verified from (7.5.8) that $\tilde{A}_\beta$, $\tilde{A}'_\beta$ form an adjoint pair. Also

$$A_\beta \subset \tilde{A}_\beta, \quad A'_\beta \subset \tilde{A}'_\beta$$

imply that

$$A_\beta \subset \tilde{A}_\beta \subset \left(\tilde{A}'_\beta\right)^* \subset \left(A'_\beta\right)^*;$$

(7.5.13) and $A^*_\beta = A'_\beta$ are then consequences of (7.5.14). Since $\mathcal{D}(A_0) = \mathcal{D}(A'_0) = \overset{0}{H^2}(\Omega)$, it follows that $A_0 \subset A_\beta \subset A_1$, $A'_0 \subset A'_\beta \subset A'_1$. The final property is a simple consequence of (7.5.2). ∎

Note that $\mathcal{D}(A_\beta)$ for $B = \gamma_0$ and $B = \gamma_a$ are, respectively, the domains of the Dirichlet operator A_{DIR} and Neumann operator A_{NEU} realisations of A: thus

$$\mathcal{D}(A_{DIR}) = \{u \in \mathcal{D}(A_1), \gamma_0 u = 0\} = \{u \in H^2(\Omega) : \gamma_0 u = 0\} = H^2(\Omega) \cap \overset{0}{H^1}(\Omega). \tag{7.5.23}$$

and

$$\mathcal{D}(A_{NEU}) = \{u \in \mathcal{D}(A_1), \gamma_a u = 0\} = \{u \in H^2(\Omega) : \gamma_a u = 0\}. \tag{7.5.24}$$

A proof of (7.5.23) and (7.5.24) may also be found in [102], Theorem 5.31 when $A = -\Delta$. It remains valid if Ω is a Lipschitz domain which locally satisfies a uniform exterior ball condition or is of class $C^{1,r}$ for some $r > 1/2$; see [88], p. 57, and [205]. In [96], a general result is quoted in Theorem I.3.1, with reference made to [184] and [3].

The identity in Lemma 6.5.2 of central importance in the abstract theory is

$$(A_1 u, v) - (u, A_1' v) = (A_1 u, v_{\eta'}) - (u_\eta, A_1' v), \quad u \in \mathcal{D}(A_1), v \in \mathcal{D}(A_1'). \tag{7.5.25}$$

In (7.5.25), $u_\eta = P_\eta u$, where P_η is the projection of $\mathcal{D}(A_1)$ onto $\mathcal{N}(A_1)$ in the decomposition

$$\mathcal{D}(A_1) = \mathcal{D}(A_\beta) \dotplus \mathcal{N}(A_1),$$

and similarly for $v_{\eta'} = P_{\eta'} v$; see (6.5.2). To overcome the difficulty that (7.5.11) does not in general hold for all $u \in \mathcal{D}(A_1)$ and $v \in \mathcal{D}(A_1')$, new boundary operators M, M', related to γ_a, γ_a' and γ_0 are introduced in [96]. They are given in terms of the following operators P and P'. In the definition, 2 cases are considered, involving choices of boundary operators B, B', C, C' and $m \in \{0, 1\}$.

1. **Case 1: Neumann**

$$B = \gamma_a, \ C = C' = \gamma_0, \ B' = \gamma_a'; \ m = 1.$$

2. **Case 2: Dirichlet**

$$B = B' = \gamma_0, \ C = \gamma_a, \ C' = \gamma_a'; \ m = 0.$$

Note that by Theorem 7.5.4, for all $s \in \mathbb{R}$, B, C map $\mathcal{N}_{s,A}(\Omega)$ isomorphically onto $H^{s-m-1/2}(\partial\Omega)$, $H^{s+m-3/2}(\partial\Omega)$ respectively, and B', C' map $\mathcal{N}_{s,A'}(\Omega)$ isomorphically onto $H^{s-m-1/2}(\partial\Omega)$, $H^{s+m-3/2}(\partial\Omega)$ respectively.

Definition 7.5.8 *Let $s \in \mathbb{R}$. Then*

$$P : H^{s-m-1/2}(\partial\Omega) \to H^{s+m-3/2}(\partial\Omega), \ \varphi \mapsto Cu,$$

where $u \in \mathcal{N}_{s,A}(\Omega)$ is such that $Bu = \varphi$;

$$P' : H^{s-m-1/2}(\partial\Omega) \to H^{s+m-3/2}(\partial\Omega), \ \psi \mapsto C'v,$$

7.5 Realisations of Second-Order Elliptic Operators on Domains

where $v \in \mathcal{N}_{s,A'}(\Omega)$ is such that $B'v = \psi$.

For functions $u \in \mathcal{N}_{s,A}(\Omega)$, in Case 1, P is the Neumann to Dirichlet map $\gamma_a u \mapsto \gamma_0 u$, while in Case 2, P is the Dirichlet to Neumann map $\gamma_0 u \mapsto \gamma_a u$. It follows from Theorem 7.5.4 that the operator P in Case 1 is the inverse of that in Case 2.

The comment preceding the definition ensures that the definition is valid in both cases, and also that P and P' are continuous.

Definition 7.5.9 Let $u \in \mathcal{D}(A_1)$, $v \in \mathcal{D}(A'_1)$. Then

$$Mu := Cu - PBu, \quad M'v := C'v - P'B'v.$$

In view of Theorem 7.5.3, M maps $\mathcal{D}(A_1)$ continuously into $H^{m-3/2}(\partial\Omega)$. Similarly, M' maps $\mathcal{D}(A'_1)$ continuously into $H^{m-3/2}(\partial\Omega)$.

The following theorem is the special case of Theorem III.1.2 in [96] for the second-order differential expressions A and A' in (7.5.1); it establishes properties of M and M' which have a crucial role in connecting the abstract theory and the application being made in this section. We recall the notation of Section 2.5: P_β denotes the projection of $\mathcal{D}(A_1)$ onto $\mathcal{D}(A_\beta)$ in the direct sum (6.5.2), and P'_β has the analogous role with respect to (6.5.3).

Theorem 7.5.10 **A.** *The following definitions of M are equivalent:*

1. *For $u \in \mathcal{D}(A_1)$, $Mu = Cu - PBu$;*
2. *For $u \in \mathcal{D}(A_1)$, $Mu = Cu_\beta$, $u_\beta := P_\beta u$;*
3. *For $u \in \mathcal{D}(A_1)$, Mu is the unique element of $H^{m+1/2}(\partial\Omega)$ such that*

$$(Au, v) = \langle Mu, B'v \rangle_{m+1/2, -m-1/2} \text{ for all } v \in \mathcal{N}(A'_1).$$

B. *On $\mathcal{D}(A_\beta)$, the maps M and C coincide, and with $\mathcal{D}(A_1)$ and $\mathcal{D}(A_\beta)$ endowed with their graph topologies, $M : \mathcal{D}(A_1) \to H^{m+1/2}(\partial\Omega)$ is a continuous surjection.*
C. *The kernel of $M : \mathcal{D}(A_1) \to H^{m+1/2}(\partial\Omega)$ is $\mathcal{D}(A_0) + \mathcal{N}(A_1)$.*
D. *The properties of $M' = C' - P'B'$ are similar to those of M, and for all $u \in \mathcal{D}(A_1)$ and $v \in \mathcal{D}(A'_1)$*

$$(A_1 u, v) - (u, A'_1 v) = \langle Mu, B'v \rangle_{m+1/2, -m-1/2} - \langle Bu, M'v \rangle_{-m-1/2, m+1/2}. \tag{7.5.26}$$

E. *Define*

$$\mathcal{D}(A_M) := \{u : u \in \mathcal{D}(A_1), Mu = 0\}, \tag{7.5.27}$$

$$\mathcal{D}(A'_{M'}) := \{u : u \in \mathcal{D}(A'_1), M'u = 0\}. \tag{7.5.28}$$

Then $A_M \in \mathcal{M}$, $A'_{M'} \in \mathcal{M}'$; and $A_M^* = A'_{M'}$

Proof. **A.** By (6.5.2), any $u \in \mathcal{D}(A_1)$ can be written as $u = u_\beta + u_\eta$, where $u_\beta = P_\beta u \in \mathcal{D}(A_\beta)$ and $u_\eta = P_\eta u \in \mathcal{N}(A_1)$. Since $Bu_\beta = 0$ by (7.5.12), it follows that

$Cu - PBu = Cu_\beta + Cu_\eta - PBu_\eta = Cu_\beta$ by Definition 7.5.8. Consequently A(1) and A(2) are equivalent.

If $v \in \mathcal{N}(A')$, (7.5.11) with $s = 2$ yields

$$(A_1 u, v) = (Au_\beta, v) = (Au_\beta, v) - (u_\beta, A'v)$$
$$= \langle Cu_\beta, B'v \rangle_{m+1/2, -m-1/2} - \langle Bu_\beta, C'v \rangle_{3/2-m, -3/2+m}$$
$$= \langle Cu_\beta, B'v \rangle_{m+1/2, -m-1/2},$$

since $Bu_\beta = 0$. Thus if $\psi = Cu_\beta$ we have shown that

$$(A_1 u, v) = \langle Cu_\beta, B'v \rangle_{m+1/2, -m-1/2} \tag{7.5.29}$$

for all $v \in \mathcal{N}(A_1')$. Since B' maps $\mathcal{N}(A_1')$ isomorphically onto $H^{-m-1/2}(\partial\Omega)$, by Theorem 7.5.4 with $s = 0$, (7.5.29) determines $\psi \in H^{m+1/2}(\partial\Omega)$ uniquely for each $u \in \mathcal{D}(A_1)$. Thus A(2) and A(3) are equivalent definitions of M.

B. We know from Theorem 7.5.7 that $\mathcal{D}(A_\beta)$, endowed with its graph topology, is a closed subspace of $H^2(\Omega)$, and hence P_β maps $\mathcal{D}(A_1)$ continuously onto $\mathcal{D}(A_\beta)$. Also, by (7.5.3), C maps $H^2(\partial\Omega)$ continuously onto $H^{m+1/2}(\partial\Omega)$. Therefore $M = CP_\beta$ maps $\mathcal{D}(A_1)$ continuously into $H^{m+1/2}(\partial\Omega)$. By A(2), $Mu = Cu$ when $u \in \mathcal{D}(A_\beta)$ and so M maps $\mathcal{D}(A_\beta)$ continuously into $H^{m+1/2}(\partial\Omega)$; it is surjective, for given $\varphi \in H^{m+1/2}(\partial\Omega)$ there exists $u \in H^2(\Omega)$ such that $Cu = \varphi$ and $Bu = 0$ by (7.5.3), and so $Mu = Cu = \varphi$.

C. If $u \in \mathcal{D}(A_0)$, then $Bu = Cu = 0$ and hence $Mu = 0$ by A(1). If $u \in \mathcal{N}(A_1)$ then $Mu = 0$ by A(3). Hence $\mathcal{D}(A_0) + \mathcal{N}(A_1) \subset \ker M$. Conversely, let $u \in \ker M$. Then, by A(3), $(Au, v) = 0$ for all $v \in \mathcal{N}(A_1')$, i.e., $Au \in (\mathcal{N}(A_1'))^\perp = \mathcal{R}(A_0)$. Thus $u = u_0 + u_1$, where $u_0 \in \mathcal{D}(A_0)$ and $u_1 \in \mathcal{N}(A_1)$, whence $\mathcal{D}(A_0) + \mathcal{N}(A_1) = \ker M$.

D. $M' = C' - P'B'$ has analogous properties to M. For $u \in \mathcal{D}(A_1)$, $v \in \mathcal{D}(A_1')$, we obtain from (7.5.25) and A(3)

$$(A_1 u, v) - (u, A_1' v) = (A_1 u, v_{\eta'}) - (u_\eta, A_1' v)$$
$$= \langle Mu, B'v \rangle_{m+1/2, -m, -1/2} - \langle Bu, M'v \rangle_{-m-1/2, m+1/2}$$

since $B'v_\beta = 0$ and $Bu_\beta = 0$.

E. For $u \in \mathcal{D}(A_M)$, $v \in \mathcal{D}(A'_{M'})$, (7.5.26) yields

$$(Au, v) - (u, A'v) = 0.$$

Thus A_M, $A'_{M'}$ form an adjoint pair and $A_M \in \mathcal{M}$, $A'_{M'} \in \mathcal{M}'$. Let $u \in \mathcal{D}([A'_{M'}]^*)$. Then $u \in \mathcal{D}(A_1)$ and

$$(Au, v) - (u, A'v) = 0 \text{ for all } v \in \mathcal{D}(A'_{M'}).$$

Since $\mathcal{N}(A_1') \subset \mathcal{D}(A'_{M'})$, we have that

7.5 Realisations of Second-Order Elliptic Operators on Domains

$$(Au, v) = 0 \text{ for all } v \in \mathcal{N}(A'_1),$$

whence $Mu = 0$ by **A**(3). Consequently $u \in \mathcal{D}(A_M)$ and since $A_M \subset (A'_{M'})^*$, it follows that $A_M = (A'_{M'})^*$. The proof of $A'_{M'} = A_M^*$ is analogous. ∎

Remark 7.5.11

Since, by Corollary 7.5.5, $\mathcal{N}(A_1)$ is not contained in $H^s(\Omega)$ for any $s > 0$, it follows from Theorem 7.5.10(C) that the domain of the realisation A_M defined in (7.5.27) is not contained in $H^s(\Omega)$ for any $s > 0$, since it contains $\mathcal{N}(A_1)$.

Let V, W be closed subspaces of $\mathcal{N}(A_1)$, $\mathcal{N}(A'_1)$ respectively. Since B is an isomorphism of $\mathcal{N}(A_1)$ onto $H_{-m-1/2}(\partial\Omega)$, by Theorem 7.5.4 with $s = 0$, it follows that B maps V isomorphically onto a closed subspace, X say, of $H_{-m-1/2}(\partial\Omega)$. Since X is a a Hilbert space, it can be identified with its adjoint (conjugate dual) X'.

Since $X \subset H^{-m-1/2}(\partial\Omega)$, every element $\varphi \in H^{m+1/2}(\partial\Omega)$ defines a unique element $\varphi_1 \in X'$ by

$$\langle \varphi_1, \psi \rangle_{X',X} = \langle \varphi, \psi \rangle_{m+1/2,-m-1/2} \text{ for all } \psi \in X; \qquad (7.5.30)$$

φ_1 is the restriction of φ considered as a functional on $H^{-m-1/2}(\partial\Omega)$ to X. We shall write $\varphi = \varphi_1$ to mean the connection between φ and φ_1 in (7.5.30).

Moreover, we are able to infer the existence of an isomorphism E of V onto X' given by

$$\langle Ev_1, Bv_2 \rangle_{X',X} = (v_1, v_2)_V, \quad v_1, v_2 \in V;$$

$(\cdot, \cdot)_V$ denotes the inner-product on V. Similarly, setting $Y := B'W$, which is a closed subspace of $H^{-m-1/2}(\partial\Omega)$, there is an isomorphism F of W onto Y' given by

$$\langle Fw_1, B'w_2 \rangle_{Y',Y} = (w_1, w_2)_W, \quad w_1, w_2 \in W.$$

A linear operator $T : \mathcal{D}(T) \subset V \to \mathcal{R}(T) \subset W$ defines an operator $L : \mathcal{D}(L) \subset X \to Y'$ by

$$\mathcal{D}(L) = B\mathcal{D}(T), \quad LBv = FTv \text{ for } v \in \mathcal{D}(T).$$

Equivalently,

$$\langle LBv, B'w \rangle_{Y',Y} = (Tv, w)_W, \text{ for all } v \in \mathcal{D}(T), w \in W.$$

Similarly $T_1 : \mathcal{D}(T_1) \subset W \to \mathcal{R}(T_1) \subset V$ defines an operator $L_1; Y \to X'$ by

$$\mathcal{D}(L_1) = B'\mathcal{D}(T_1), \quad L_1 B'w = ET_1 w \text{ for } w \in \mathcal{D}(T_1);$$

thus

$$\langle L_1 B'w, Bv \rangle_{X',X} = (T_1 w, v)_W, \text{ for all } w \in \mathcal{D}(T_1), v \in V.$$

It is shown in [96], Section III.2 that the set of all such operators $T : \mathcal{D}(T) \subset V \to W$ is in one-one correspondence with the set of operators $L : \mathcal{D}(L) \subset X \to Y'$. Also if T is densely defined, then $T^* : W \to V$ corresponds to $L^* : Y \to X'$ and $L^* = L_1$, the operator defined above. This paves the way for the following application of Corollary 6.5.7 in which the realisations of the expressions A, A' in (7.5.1) are determined in terms of boundary conditions.

Theorem 7.5.12 *There is a one-one correspondence between all closed operators $\tilde{A} \in \mathcal{M}$ and all operators $L : X \to Y'$ satisfying*

1. *X and Y are closed subspaces of $H^{-m-1/2}(\partial\Omega)$;*
2. *L is densely defined and closed.*

The correspondence is given by

$$\mathcal{D}(\tilde{A}) = \{u : u \in \mathcal{D}(A_1), \ Bu \in \mathcal{D}(L), \ Mu = LBu \text{ on } Y\},$$

where $\mathcal{D}(L) = B\mathcal{D}(\tilde{A})$ and so $X = \overline{B\mathcal{D}(\tilde{A})}$. If $\tilde{A}$ corresponds to $L : X \to Y'$ in the above sense, then $\tilde{A}^*$ corresponds to $L^* : Y \to X'$ with

$$\mathcal{D}(\tilde{A}^*) = \{v \in \mathcal{D}(A'_1) : B'v \in \mathcal{D}(L^*), \ M'v = L^*B'v \text{ on } X\},$$

where $\mathcal{D}(L^*) = B'\mathcal{D}(\tilde{A}^*)$, and so $Y = \overline{B'\mathcal{D}(\tilde{A}^*)}$.

Suppose that A in (7.5.1) is formally self-adjoint, i.e., $A = A'$, and let $B' = B$, $C' = C$. Then the realisations $\tilde{A} \in \mathcal{M}$ of Theorem 7.5.12 are self-adjoint extensions of A_0. We now have

Theorem 7.5.13 *Let X be a closed subspace of $H^{-m-1/2}(\partial\Omega)$, and let $L : X \to X'$ be self-adjoint. Then the operator $\tilde{A} \in \mathcal{M}$ defined by*

$$\mathcal{D}(\tilde{A}) = \{u \in \mathcal{D}(A_1) : Bu \in \mathcal{D}(L), Mu = LBu \text{ on } X\} \quad (7.5.31)$$

is self-adjoint.

Conversely, any self-adjoint operator $\tilde{A} \in \mathcal{M}$ defines a self-adjoint operator $L : X \to X'$ by (7.5.31). In the correspondence, $\mathcal{D}(L) = B\mathcal{D}(\tilde{A})$, and $X = \overline{B\mathcal{D}(\tilde{A})}$.

In view of the assumption (7.5.4), A_0 is now positive, i.e., for some $c > 0$,

$$(A_0 u, u) \geq c\|u\|^2, \ \forall u \in \mathcal{D}(A_0). \quad (7.5.32)$$

Then

$$D_0 := \{u : u \in \mathcal{D}(A_1), \ \gamma_0 u = 0\} = \mathcal{D}(A_F), \quad (7.5.33)$$

the domain of the Friedrichs extension A_F of A_0. For by Theorem 7.5.7,

$$D_0 := \{u : u \in H^2(\Omega), \ \gamma_0 u = 0\}.$$

7.5 Realisations of Second-Order Elliptic Operators on Domains

Therefore
$$D_0 \subset \{u : u \in H^1(\Omega), \ \gamma_0 u = 0\} = \overset{0}{H}{}^1(\Omega)$$

and so $D_0 \subset \mathcal{D}(A_1) \cap \overset{0}{H}{}^1(\Omega) = \mathcal{D}(A_F)$. Since the restriction of A_1 to D_0 is self-adjoint by Theorem 7.5.7, (7.5.33) follows.

In Definition 7.5.8, take $m = 0$, $B = \gamma_0$ and $C = \gamma_a$, i.e., Case 2. The operator P is then the Dirichlet to Neumann map of $H^{s-1/2}(\partial\Omega)$ into $H^{s-3/2}(\partial\Omega)$ determined by $P\gamma_0 u = \gamma_1 u$ for $u \in \mathcal{N}_{s,A}(\Omega)$. With $M = \gamma_a - P\gamma_0$ in Definition 7.5.9, it follows from Theorem 7.5.10 (C) that

$$\mathcal{D}(A_M) = \mathcal{D}(A_0) \dotplus \mathcal{N}(A_1).$$

By (6.4.10), this is the domain of the Krein–von Neumann extension of A_0. Hence, A_K is the restriction of A_1 to the domain

$$\mathcal{D}(A_K) = \mathcal{D}(A_M) = \{u : u \in \mathcal{D}(A_1), \ P\gamma_0 = \gamma_1\}. \qquad (7.5.34)$$

7.6 Notes

1. In [1], the challenge of extending the results described in Section 7.5 to boundary-value problems for second-order elliptic operators with non-smooth coefficients in domains Ω with non-smooth boundaries is tackled. The operators are assumed to have coefficients in scales of Sobolev spaces and their generalisations to Besov and Bessel-potential spaces. The hypothesis on $\partial\Omega$ implies that it can be parameterised by functions in the Besov space $B_{p,2}^{3/2}$ for some $p > 2(n - 1)$; locally $B_{p,2}^{3/2}$ lies in between $C^{1+\tau}$ and $C^{3/2+\varepsilon}$ for $\tau = \frac{1}{2} - \frac{n-1}{p} > 0$ and any $\varepsilon > 0$.

Gesztesy and Mitrea in [88] had considered the same problem for the self-adjoint extensions of the Laplacian, in the case when Ω belongs to a sub-class of bounded Lipschitz domains, which they call *quasi-convex domains*. Such a domain is a hybrid of two types: it either behaves locally as a $C^{1,r}$ domain, with $r > 1$, or like a Lipschitz domain satisfying a uniform exterior ball condition. An analysis of the self-adjoint realisations of second-order problems on $C^{1,1}$ domains is made by Posilicano and Raimondi in [171], while Grubb treated non-selfadjoint realisations on such domains in [98] and allowed for Neumann-type boundary conditions of the form $\chi u = C\gamma_0 u$, where χ is related to a co-normal derivative and C is a differential operator of order 1. Behrndt and Micheler [23] show that the self-adoint extensions of the Laplacian on a Lipschitz domain can be obtained by use of a theory of *quasi*-boundary triplets introduced in [22].

2. In [14] a Weyl asymptotic formula is obtained for the non-zero eigenvalues $\lambda_{K,\Omega,j}$, $j \in \mathbb{N}$ of the perturbed Krein–von Neumann Laplacian $H_{K,\Omega} := -\Delta + V$ defined on $C_0^\infty(\Omega)$, where V is measurable, bounded and non-negative, and Ω is a

bounded domain in $\mathbb{R}^n$ belonging to a class of non-smooth domains containing all convex domains and those of class $C^{1,r}$, $r > 1/2$. It is proved that as $\lambda \to \infty$,

$$\{j \in \mathbb{N} : \lambda_{K,\Omega,j} \leq \lambda\} = (2\pi)^{-n} \omega_n |\Omega| \lambda^{n/2} + O\left(\lambda^{[n-(1/2)]/2}\right),$$

where $\omega_n = \frac{\pi^{n/2}}{\Gamma[(n/2)+1]}$ is the volume of the unit ball in $\mathbb{R}^n$, $|\Omega|$ is the volume of Ω and the eigenvalues $\lambda_{K,\Omega,j}$ are arranged in increasing order, counting multiplicities. This shows that the Weyl asymptotic formula remains valid for the non-smooth domains considered. It is also proved that if Ω is the exterior domain $\Omega = \mathbb{R}^n \setminus K$, where $K \subset \mathbb{R}^n$ is compact with zero Bessel capacity $B_{2,2}(K)$, then $-\Delta \restriction_{C_0^\infty(\Omega)}$ is essentially self-adjoint.

3. Let A_0 be the minimal operator defined by $A := \Delta + q$ in $L^2(\mathbb{R}^n \setminus \{0\})$, $n \geq 2$, where the coefficient q is real, in $L_{2,loc}(\mathbb{R}^n \setminus \{0\})$ and singular at the origin. An important problem in non-relativistic quantum mechanics is to determine conditions on q which imply that A_0 is essentially self-adjoint, i.e., has a unique self-adjoint extension. So-called *strongly singular potentials* q which behave like $|x|^{-2}$ at the origin are of special interest and have been studied by many authors. The following definitive result is proved by Simader in [191], an earlier version with a stronger condition on q having been proved by Kalf in [121].

Kalf–Simader Let $q \in L_{2,loc}(\mathbb{R}^n \setminus \{0\})$ and

$$q(x) \geq [1 - (\frac{1}{2}n - 1)^2]|x|^{-2} - c|x|^2 \quad (x \in \mathbb{R}^n \setminus \{0\}) \tag{7.6.1}$$

for some $c \in \mathbb{R}$. Then $A_0 := A \restriction C_0^\infty(\mathbb{R}^n \setminus \{0\})$ is essentially self-adjoint.

Chapter 8
The L_p Approach to the Laplace Operator

8.1 Preamble

In Chapter 5 we saw how a weak solution of the Dirichlet problem for the Poisson equation arises naturally as a minimiser of the functional E given by

$$E(u) = \frac{1}{2} \int_\Omega |\nabla u(x)|^2 \, dx - \int_\Omega f(x)u(x)dx, \ u \in \overset{0}{W^1_2}(\Omega),$$

where Ω is a bounded open subset of $\mathbb{R}^n$ and $f \in L_2(\Omega)$: all the functions appearing in this chapter are assumed to be real-valued. If the given function f does not belong to $L_2(\Omega)$ this procedure does not work. However, if $p \in (2, \infty)$ and $f \in L_{p'}(\Omega) \backslash L_2(\Omega)$, then it would be natural to look at the functional E_p on $\overset{0}{W^1_p}(\Omega)$ given by

$$E_p(u) = \frac{1}{p} \int_\Omega |\nabla u(x)|^p \, dx - \int_\Omega f(x)u(x)dx, \ u \in \overset{0}{W^1_p}(\Omega),$$

and seek to minimise it. Again the Friedrichs inequality (Theorem 1.3.7) can be deployed to show that for some constant c,

$$E_p(u) \geq \frac{1}{p} \|u\|^p_{1,p} - c \|u\|_{1,p}.$$

Just as in the case $p = 2$ it can now be shown that there is a mimimiser u of E_p, but this time u is a solution of the equation

$$\int_\Omega |\nabla u|^{p-2} \nabla u . \nabla v \, dx = \int_\Omega f v \, dx \text{ for all } v \in \overset{0}{W^1_p}(\Omega),$$

© Springer Nature Switzerland AG 2018
D. E. Edmunds and W. D. Evans, *Elliptic Differential Operators and Spectral Analysis*, Springer Monographs in Mathematics, https://doi.org/10.1007/978-3-030-02125-2_8

so that u is a weak solution not of the Poisson equation but of the equation

$$\Delta_p u := \sum_{i=1}^n D_i \left(|\nabla u|^{p-2} D_i u\right) = -f$$

involving the p-Laplacian Δ_p. Because of this difficulty with a variational approach, other techniques have to be used. A direct attack on the L_p version of the classical Dirichlet problem for uniformly elliptic operators of order $2m$ in a bounded open set Ω was given by Agmon in [2] using estimates obtained in [3]; these estimates were also used by Schechter in [185] and [186] to obtain the existence of weak solutions of various boundary-value problems for uniformly elliptic operators in an L_p setting. Before discussing this approach, however, we present in the next section a different approach used by Simader and Sohr in [192] to obtain existence of what we shall call a weak L_p solution of the Dirichlet problem for the Poisson equation, for simplicity restricting the discussion to the case of bounded Ω. This approach does not give optimal results, but its connection with the procedure used in section 7 for the L_2 case has some appeal, despite the inevitable technicalities.

8.2 Technical Results

Throughout we shall suppose that $p \in (1, \infty)$ and Ω is a bounded open subset of $\mathbb{R}^n$ with C^1 boundary; given any $R > 0$ we write $\Omega_R = \Omega \cap B_R$, where $B_R = B(0, R)$, the ball of radius R centred at the origin. For economy of expression the Sobolev space $\overset{0}{W}{}^1_p(\Omega)$ will be denoted by $X_p(\Omega)$, or even by X_p if the context is clear. We write

$$\langle u, v \rangle = \langle u, v \rangle_\Omega = \int_\Omega uv\, dx, \quad \langle \nabla u, \nabla v \rangle = \sum_{j=1}^n \langle D_j u, D_j v \rangle,$$

and for simplicity of appearance we shall sometimes write $\|\nabla u\|$ instead of the more accurate $\||\nabla u|\|$. Also $L_{p,loc}(\mathbb{R}^n)$ denotes the space of functions in $L_p(K)$ for every compact subset K of $\mathbb{R}^n$.

Lemma 8.2.1 *Let $p, r \in (1, \infty)$; suppose that $f \in L_{r,loc}(\mathbb{R}^n)$ is such that $|\nabla f| \in L_r(\mathbb{R}^n)$ and*

$$\sup \left\{ |\langle \nabla f, \nabla v \rangle| / \|\nabla v\|_{p', \mathbb{R}^n} : v \in C_0^\infty(\mathbb{R}^n) \setminus \{0\} \right\} < \infty. \tag{8.2.1}$$

Then $f \in L_{p,loc}(\mathbb{R}^n)$, $|\nabla f| \in L_p(\mathbb{R}^n)$ and there is a constant $C_1 = C_1(n, p)$ such that for all $g \in L_{p,loc}(\mathbb{R}^n)$ with $|\nabla g| \in L_p(\mathbb{R}^n)$,

$$\|\nabla g\|_{p, \mathbb{R}^n} \leq C_1 \sup \left\{ |\langle \nabla g, \nabla v \rangle| / \|\nabla v\|_{p', \mathbb{R}^n} : v \in C_0^\infty(\mathbb{R}^n) \setminus \{0\} \right\}. \tag{8.2.2}$$

Proof. We claim that $M := \{\Delta \phi : \phi \in C_0^\infty(\mathbb{R}^n)\}$ is dense in $L_{p'}(\mathbb{R}^n)$. For if not, by the Hahn-Banach theorem (Theorem 1.2.5) there would exist $H \in \left(L_{p'}(\mathbb{R}^n)\right)^*$, with

8.2 Technical Results

unit norm, such that $H \upharpoonright_M = 0$; corresponding to H there exists $h \in L_p(\mathbb{R}^n)$ such that $\|h\|_p = 1$ and $H(g) = \langle h, g \rangle$ for all $g \in L_{p'}(\mathbb{R}^n)$. But $0 = H(\Delta\phi) = \langle h, \Delta\phi \rangle$ for all $\phi \in C_0^\infty(\mathbb{R}^n)$: thus by Weyl's lemma (Theorem 2.2.10), h is equivalent to a function that is harmonic in $\mathbb{R}^n$; since $h \in L_p(\mathbb{R}^n)$ it must be the zero function (see the discussion following Corollary 1.4.8), contradicting the fact that $\|h\|_p = 1$.

For all $\phi \in C_0^\infty(\mathbb{R}^n)$, $\|D_i D_k \phi\|_{p'} \leq C \|\Delta\phi\|_{p'}$ (see Corollary 1.4.8). Define a functional L_i on M by $L_i(\Delta\phi) = \langle D_i f, \Delta\phi \rangle$. Since

$$\infty > \sup\left\{\frac{|\langle \nabla f, \nabla\phi \rangle|}{\|\nabla\phi\|_{p',\mathbb{R}^n}} : \phi \in C_0^\infty(\mathbb{R}^n) \setminus \{0\}\right\}$$

$$\geq \sup\left\{\frac{|\langle \nabla f, \nabla D_i w \rangle|}{\|\nabla D_i w\|_{p',\mathbb{R}^n}} : w \in C_0^\infty(\mathbb{R}^n) \setminus \{0\}\right\}$$

$$= \sup\left\{\frac{|\langle f, D_i \Delta w \rangle|}{\|\nabla D_i w\|_{p',\mathbb{R}^n}} : w \in C_0^\infty(\mathbb{R}^n) \setminus \{0\}\right\}$$

$$\geq C^{-1} \sup\left\{\frac{|\langle D_i f, \Delta w \rangle|}{\|\Delta w\|_{p',\mathbb{R}^n}} : w \in C_0^\infty(\mathbb{R}^n) \setminus \{0\}\right\}, \tag{8.2.3}$$

it follows that L_i is continuous on M when this subspace of $L_{p'}(\mathbb{R}^n)$ is equipped with the induced norm, and so it has a unique, norm-preserving extension to $L_{p'}(\mathbb{R}^n)$ that may be represented by $l_i \in L_p(\mathbb{R}^n)$. Hence $\langle D_i f - l_i, \Delta w \rangle = 0$ for all $w \in C_0^\infty(\mathbb{R}^n)$, so that, by Weyl's lemma again, $W := D_i f - l_i$ is harmonic on $\mathbb{R}^n$ and thus has the mean value property. Since, for each $x \in \mathbb{R}^n$ and each $R > 0$,

$$W(x) = \frac{1}{|B_R|} \int_{B_R} (D_i f(x) - l_i(x))\, dx,$$

use of Hölder's inequality gives

$$|W(x)| \leq |B_R|^{-1/r} \|D_i f\|_{r,\mathbb{R}^n} + |B_R|^{-1/p} \|l_i\|_{p,\mathbb{R}^n} \to 0 \text{ as } R \to \infty.$$

Hence $D_i f = l_i \in L_p(\mathbb{R}^n)$. Note that by continuity,

$$\sup\left\{\frac{|\langle D_i f, \Delta w \rangle|}{\|\Delta w\|_{p',\mathbb{R}^n}} : w \in C_0^\infty(\mathbb{R}^n) \setminus \{0\}\right\}$$

$$= \left\{\frac{|\langle D_i f, g \rangle|}{\|g\|_{p',\mathbb{R}^n}} : g \in L_{p'}(\mathbb{R}^n) \setminus \{0\}\right\} = \|D_i f\|_p.$$

Together with (8.2.3) this gives (8.2.2). Since $f \in L_{r,loc}(\mathbb{R}^n)$ it follows that $f \in L_{1,loc}(\mathbb{R}^n)$; for each $\varepsilon > 0$ put $f_\varepsilon(x) = \varepsilon^{-n} \int_{\mathbb{R}^n} \rho((x-y)/\varepsilon) f(y)\,dy$, where ρ is a mollifier. Let $R > 0$ and set $C_\varepsilon = |B_R|^{-1} \int_{B_R} f_\varepsilon(x)\,dx$. Then as $\varepsilon \to 0$, $C_\varepsilon \to C := |B_R|^{-1} \int_{B_R} f(x)\,dx$ and $\|\nabla f_\varepsilon - \nabla f\|_{p,B_R} \to 0$. By Poincaré's inequality (Theorem 1.3.9),

$$\|(f_\varepsilon - C_\varepsilon) - (f - C)\|_{p,B_R} \leq C(n,p,R) \|\nabla(f_\varepsilon - f)\|_{p,B_R}$$

and so $\|f - f_\varepsilon\|_{p,B_R} \to 0$. Thus there exists $\tilde{f} \in L_p(B_R)$ such that $\|\tilde{f} - f\|_{p,B_R} \to 0$; and since $\|f - f_\varepsilon\|_{1,B_R} \to 0$ it follows that $f = \tilde{f} \in L_p(B_R)$, that is, $f \in L_{p,loc}(\mathbb{R}^n)$. ∎

Lemma 8.2.2 *Let $x_0 \in \Omega$ and suppose that $R > 0$ is such that $\overline{B(x_0, R)} \subset \Omega$. Then given any $R' \in (0, R)$, there is a constant $C = C(R, R', p)$ such that*

$$\|\nabla(\eta u)\|_{p,\Omega} \leq C \sup\left\{ \frac{|\langle \nabla(\eta u), \nabla v \rangle|}{\|\nabla v\|_{p',B(x_0,R)}} : v \in C_0^\infty(B(x_0, R)) \setminus \{0\} \right\} \tag{8.2.4}$$

for all $u \in X_p$ and all $\eta \in C_0^\infty(B(x_0, R))$.

Proof. Let $\rho \in C_0^\infty(B(x_0, R))$ be such that $0 \leq \rho \leq 1$ and $\rho(x) = 1$ when $x \in B(x_0, R')$. Let $\phi \in C_0^\infty(\mathbb{R}^n)$ and put $c = |B(x_0, R)|^{-1} \int_{B(x_0, R)} \rho \, dx$, $v = \rho(\phi - c)$. By the Poincaré inequality (Theorem 1.3.9),

$$\|\phi - c\|_{p',B(x_0,R)} \leq C_R \|\nabla \phi\|_{p',B(x_0,R)}$$

for some constant C_R; hence

$$\|\nabla v\|_{p',B(x_0,R)} \leq (1 + C_R \|\nabla \rho\|_\infty) \|\nabla \phi\|_{p',\mathbb{R}^n}.$$

Plainly ∇v and $\nabla \phi$ coincide on $B(x_0, R')$. If $\phi, v \neq 0$, then

$$\frac{|\langle \nabla(\eta u), \nabla \phi \rangle|}{\|\nabla \phi\|_{p',\mathbb{R}^n}} \leq (1 + C_R \|\nabla \rho\|_\infty) \frac{|\langle \nabla(\eta u), \nabla v \rangle|}{\|\nabla v\|_{p',\mathbb{R}^n}},$$

and by Lemma 8.2.1,

$$\|\nabla(\eta u)\|_{p,\Omega} \leq C_1(p) \sup\left\{ \frac{|\langle \nabla(\eta u), \nabla \phi \rangle|}{\|\nabla \phi\|_{p',\mathbb{R}^n}} : \phi \in C_0^\infty(\mathbb{R}^n) \setminus \{0\} \right\}$$

$$\leq C_1(p)(1 + C_R \|\nabla \rho\|_\infty) \sup\left\{ \frac{|\langle \nabla(\eta u), \nabla v \rangle|}{\|\nabla v\|_{p',B(x_0,R)}} : v \in C_0^\infty(B(x_0, R)) \setminus \{0\} \right\},$$

as required. ∎

Lemma 8.2.3 *(i) Given any $x_0 \in \partial \Omega$, there are positive constants*

$R = R(p, x_0, \partial\Omega)$ *and* $C = C(R)$ *such that for all $u \in X_p$ and all $\eta \in C_0^\infty(B(x_0, R/2))$,*

$$\|\nabla(\eta u)\|_{p,\Omega} \leq C \sup\left\{ \frac{|\langle \nabla(\eta u), \nabla v \rangle|}{\|\nabla v\|_{p',\Omega_R(x_0)}} : v \in X_{p'}(\Omega_R(x_0)) \setminus \{0\} \right\}, \tag{8.2.5}$$

where $\Omega_R(x_0) = \Omega \cap B(x_0, R)$.

(ii) If $u \in X_p(\Omega)$ is such that $\langle \nabla u, \nabla \phi \rangle = 0$ for all $\phi \in C_0^\infty(\Omega)$, then given any $x_0 \in \partial\Omega$, there exists $R_0 = R_0(p, x_0, \partial\Omega) > 0$ such that $\eta u \in X_2(\Omega)$ whenever $\eta \in C_0^\infty(B(x_0, R_0))$.

The proof of this result is rather technical: full details are given in [192], Lemma 2.2.8. Note in particular the shift in (ii) from p to 2 in integration exponents.

8.3 Existence of a Weak L_p Solution

After the technical preparation of the last section we can now deal with the matter of existence of a certain type of weak solution of the Dirichlet problem for the Poisson equation. We begin with a uniqueness result. As before it is supposed that $p \in (1, \infty)$ and that Ω is a bounded open subset of $\mathbb{R}^n$ with C^1 boundary; $X_p = X_p(\Omega)$ will stand for the Sobolev space $\overset{0}{W}{}_p^1(\Omega)$. Given a Banach space X, the value of $x^* \in X^*$ at $x \in X$ will be denoted by $\langle x, x^* \rangle_X$ for simplicity, rather than $\langle x, x^* \rangle_{X,X^*}$ (not to be confused with $\langle \cdot, \cdot \rangle$ defined at the beginning of 8.2).

Theorem 8.3.1 *If $u \in X_p$ is such that*

$$\langle \nabla u, \nabla \phi \rangle = 0 \text{ for all } \phi \in X_{p'}, \tag{8.3.1}$$

then u is the zero function.

Proof. By Lemma 8.2.3 (ii), given any $x_0 \in \partial\Omega$, there exists $R_0 > 0$ such that $\eta u \in X_2(\Omega)$ whenever $\eta \in C_0^\infty(B(x_0, R_0))$. As $\partial\Omega$ is compact, it may be covered by a finite number of such balls, say $B_k = B(x_k, R_k)$ ($k = 1, ..., M$). Put $B_0 := \Omega \cap \left(\mathbb{R}^n \setminus \bigcup_{k=1}^M B_k\right)$ and let $\{\psi_k : k = 0, 1, ..., M\}$ be a partition of unity subordinate to the covering $\{B_0, B_1, ..., B_M\}$ of Ω, so that each $\psi_k \in C_0^\infty(B_k)$ and $\sum_{k=0}^M \psi_k(x) = 1$ when $x \in \overline{\Omega}$. By Lemma 8.2.3 (ii), $\psi_k u \in X_2(\Omega)$ when $1 \leq k \leq M$. Since $C_0^\infty(\Omega) \subset X_{p'}(\Omega)$, (8.3.1) and Weyl's lemma (Theorem 2.2.10) imply that u is equivalent to a function that is harmonic in Ω. Thus $\psi_0 u \in C_0^\infty(\Omega) \subset X_2(\Omega)$, and so $u = \sum_{k=0}^M \psi_k u \in X_2(\Omega)$. As (8.3.1) holds for all $\phi \in C_0^\infty(\Omega)$ and $u \in X_2(\Omega)$, it must also hold for all $\phi \in X_2(\Omega)$. The choice $\phi = u$ shows that $\|\nabla u\|_2 = 0$, from which it follows that $u = 0$ a.e. ∎

Now we can give a theorem of fundamental importance in this approach to the Dirichlet problem.

Theorem 8.3.2 *There is a constant C_p such that for all $u \in X_p(\Omega)$,*

$$\|\nabla u\|_p \leq C_p \sup \left\{ \frac{|\langle \nabla u, \nabla \phi \rangle|}{\|\nabla \phi\|_{p'}} : \phi \in X_{p'}(\Omega) \setminus \{0\} \right\}. \tag{8.3.2}$$

Proof. Suppose the result is false. Then there is a sequence $(u_k) \subset X_p(\Omega)$ such that $\|\nabla u_k\|_p = 1$ and

$$\varepsilon_k := \sup\left\{\frac{|\langle \nabla u_k, \nabla \phi \rangle|}{\|\nabla \phi\|_{p'}} : \phi \in X_{p'}(\Omega)\setminus\{0\}\right\} \to 0 \text{ as } k \to \infty.$$

Since $X_p(\Omega)$ is reflexive, there is a subsequence of (u_k), again denoted by (u_k), that converges weakly in $X_p(\Omega)$, to u, say. Thus for each $\phi \in X_{p'}(\Omega)$,

$$\langle \nabla u, \nabla \phi \rangle = \lim_{k \to \infty} \langle \nabla u_k, \nabla \phi \rangle = 0,$$

and so by Theorem 8.3.1, $u = 0$. By Theorem 1.3.5 (ii), $u_k \to 0$ in $L_p(\Omega)$.

By Lemma 8.2.3 (i) we see that given any $x_0 \in \partial\Omega$, there exists $R_0 > 0$ such that (8.2.5) holds for $u \in X_p(\Omega)$ and $\eta \in C_0^\infty(B(x_0, R_0/2))$. As $\partial\Omega$ is compact, there are points $x_i \in \partial\Omega$ and positive numbers R_i ($i = 1, ..., M$) such that $\partial\Omega$ is covered by the balls $B_i := B(x_i, R_i/4)$; and $\Omega_1 := \Omega \cap \cap_{i=1}^M (\mathbb{R}^n \setminus B_i)$ is compact. By Lemma 8.2.2, given any $x_0 \in \Omega$ there exists $R_0 > 0$ such that $B(x_0, R_0) \subset \Omega$ and (8.2.4) holds for $u \in X_p(\Omega)$ and $\eta \in C_0^\infty(B(x_0, R_0/2))$. The compactness of Ω_1 leads to the existence of points $x_i \in \Omega_1$ and positive numbers R_i ($i = M+1, ..., P$) such that Ω_1 is covered by the balls $B_i := B(x_i, R_i/4)$ ($i = M+1, ..., P$). For $i \in \{1, ..., P\}$ choose $\psi_i \in C_0^\infty(B_i')$, where $B_i' = B(x_i, R_i/2)$, so that $0 \le \psi_i \le 1$ and ψ_i has the value 1 on B_i. For $i \in \{1, ..., P\}$ set $\Omega_i = \Omega \cap B(x_i, R_i)$. By Lemmas 8.2.2 and 8.2.3 we have, for $i \in \{1, ..., P\}$,

$$\|\nabla u_k\|_{p, B_i} \le \|\nabla(\psi_i u_k)\|_p \le C_p^{(i)} \sup\left\{\frac{\langle \nabla(\psi_i u_k), \nabla v \rangle}{\|\nabla v\|_{p', \Omega_i}} : v \in X_{p'}(\Omega_i)\setminus\{0\}\right\}$$

$$:= d_k^{(i)}. \tag{8.3.3}$$

Fix $i \in \{1, ..., P\}$. For each $k \in \mathbb{N}$ there exists $v_k \in X_{p'}(\Omega)$ with $\|\nabla v_k\|_{p', \Omega_i} = 1$ and $0 \le d_k^{(i)} - \langle \nabla(\psi_i u_k), \nabla v_k \rangle \le 1/k$. Then

$$d_k^{(i)} \le 1/k + \langle \nabla u_k, \nabla(\psi_i v_k) \rangle - \langle \nabla u_k, v_k \nabla \psi_i \rangle + \langle u_k \nabla \psi_i, \nabla v_k \rangle$$

$$\le 1/k + \varepsilon_k \|\nabla(\psi_i v_k)\|_{p'} + |\langle \nabla u_k, v_k \nabla \psi_i \rangle| + |\langle u_k \nabla \psi_i, \nabla v_k \rangle|. \tag{8.3.4}$$

There is a subsequence of (v_k), again denoted by (v_k), and an element v of $X_{p'}(\Omega_i)$ such that $v_k \rightharpoonup v$ in $X_{p'}(\Omega_i)$ and hence $v_k \to v$ in $L_{p'}(\Omega_i)$. Note that

$$|\langle \nabla u_k, v_k \nabla \psi_i \rangle| \le |\langle \nabla u_k, (v_k - v)\nabla \psi_i \rangle| + |\langle \nabla u_k, v\nabla \psi_i \rangle|$$

$$\le \|\nabla u_k\|_{p, \Omega} \|(v_k - v)\nabla \psi_i\|_{p', \Omega} + |\langle \nabla u_k, v\nabla \psi_i \rangle|. \tag{8.3.5}$$

For each $u \in X_p(\Omega)$ set $F^*(u) = \langle v\nabla\psi_i, \nabla u \rangle$: plainly F^* is continuous and hence $F^*(u_k) \to 0$. Moreover, $\|\nabla u_k\|_{p, \Omega} = 1$ and for each $i \in \{1, ..., P\}$, $\|(v_k - v)\nabla \psi_i\|_{p', \Omega} \to 0$ as $k \to \infty$. Similarly,

8.3 Existence of a Weak L_p Solution

$$|\langle u_k \nabla \psi_i, \nabla v_k\rangle| \leq \|u_k \nabla \psi_i\|_{p,\Omega_i} \leq \|\nabla \psi_i\|_\infty \|u_k\|_{p,\Omega} \to 0 \text{ as } k \to \infty.$$

To estimate $\|\nabla(\psi_i v_k)\|_{p'}$, note that use of the Poincaré inequality (Theorem 1.3.9) shows that

$$\|\nabla(\psi_i v_k)\|_{p'} \leq C \|\nabla \psi_i\|_\infty + 1.$$

Putting all these estimates together we see that for each $i \in \{1, ..., P\}$, $d_k^{(i)} \to 0$ as $k \to \infty$. Thus by (8.3.3),

$$1 = \|\nabla u_k\|_{p,\Omega} \leq \sum_{i=1}^{P} \|\nabla u_k\|_{p,\Omega_i} \leq \sum_{i=1}^{P} d_k^{(i)} \to 0 \text{ as } k \to \infty.$$

This contradiction establishes the theorem. ∎

To show more concretely what this implies we need the extension of the Lax-Milgram lemma given in Theorem 1.2.8.

Theorem 8.3.3 *Let Ω be a bounded open subset of $\mathbb{R}^n$ with C^1 boundary and let $p \in (1, \infty)$; for simplicity, write X_p for $\overset{0}{W}{}^1_p(\Omega)$. Then to every $F^* \in (X_{p'})^*$ there corresponds a unique $f \in X_p$ such that*

$$\langle \phi, F^* \rangle_{X_{p'}} = \langle \nabla f, \nabla \phi \rangle \text{ for all } \phi \in X_p.$$

Proof. Immediate from Theorems 8.3.2 and 1.2.8. ∎

Note that

$$C_p^{-1} \|\nabla f\|_p \leq \|F^*\| \leq \|\nabla f\|_p.$$

Observe also that if $f \in L_p(\Omega)$, then use of Hölder's inequality and the Friedrichs inequality shows that the functional F^* defined by $\langle \phi, F^* \rangle_{X_{p'}} = \langle f, \phi \rangle$ ($\phi \in X_{p'}$) belongs to $(X_{p'})^*$. Theorem 8.3.3 thus implies that there is a unique $u \in X_p$ such that

$$\langle \nabla u, \nabla \phi \rangle = \langle f, \phi \rangle \quad (\phi \in X_{p'});$$

that is,

$$\int_\Omega \nabla u(x) . \nabla \phi(x) dx = \int_\Omega f(x)\phi(x) dx \text{ for all } \phi \in X_{p'}.$$

We shall refer to such a function u as a **weak L_p solution of the Dirichlet problem for the Poisson equation.** We summarise this result in the following theorem.

Theorem 8.3.4 *Let $f \in L_p(\Omega)$. Then there is a unique function $u \in \overset{0}{W}{}^1_p(\Omega)$ such that*

$$\int_\Omega \nabla u(x) . \nabla \phi(x) dx = \int_\Omega f(x)\phi(x) dx \text{ for all } \phi \in \overset{0}{W}{}^1_{p'}(\Omega).$$

A consequence of Theorem II.9.1 of [192] is that if, in addition to the hypotheses of Theorem 8.3.4, we suppose that $\partial\Omega \in C^2$, then given any $f \in L_p(\Omega)$ $(1 < p < \infty))$, there is a unique function $u \in \overset{0}{W}{}^1_p(\Omega) \cap W^2_p(\Omega)$ such that $-\Delta u = f$. When $p = 2$, the space $\overset{0}{W}{}^1_p(\Omega) \cap W^2_p(\Omega)$ is precisely the domain of the Dirichlet Laplacian A_{DIR} discussed in 7.5, and so there is a unique solution of the equation $A_{DIR} u = f$.

8.4 Other Procedures

As indicated at the beginning of this chapter, a direct attack on the Dirichlet problem for the Poisson equation on a smooth domain $\Omega \subset \mathbb{R}^n$, and indeed on much more general elliptic equations, has been made in [2] and [3], where the techniques involved include results for pseudodifferential operators and Calderón-Zygmund theory. A great deal of work has been done on the situation when Ω has a Lipschitz boundary. For example, suppose that $\varepsilon \in (0, 1]$ and consider the following conditions on parameters s and p:

$$2/(1+\varepsilon) < p < 2/(1-\varepsilon) \text{ and } 0 < s < 1; \tag{a}$$

$$1 < p \leq 2/(1+\varepsilon) \text{ and } (2/p) - 1 - \varepsilon < s < 1; \tag{b}$$

$$2/(1-\varepsilon) \leq p < \infty \text{ and } 0 < s < (2/p) + \varepsilon. \tag{c}$$

In [113] it is shown that given any bounded Lipschitz domain Ω, there exists $\varepsilon \in (0, 1]$ such that the Poisson equation

$$\Delta u = f \text{ in } \Omega,$$

with zero Dirichlet boundary conditions, has a unique solution $u \in \overset{0}{W}{}^{s+1/p}_p(\Omega)$ for any $f \in W^{s+1/p-2}_p(\Omega)$ if the pair (s, p) satisfies any one of conditions (a), (b) and (c). The techniques used rely essentially on estimates for the harmonic measure in Ω. The paper [75] develops a different approach to this problem, in which the use of harmonic measures is replaced by functional-analytic properties of boundary-layer potentials on Sobolev-Besov spaces: this has the advantage that both the Neumann and the Dirichlet problem can be handled in a unified way, as can systems of equations. As the technical background needed for all this is beyond the scope of this book, we refer the reader to the references given above and to papers mentioned in them.

8.5 Notes

1. If $p \geq 2$, then since $L_p(\Omega) \hookrightarrow L_2(\Omega)$, there is a unique weak solution (in the sense of Chapter 5) of the Poisson equation $-\Delta v = f$ ($f \in L_p(\Omega)$); that is, there exists a unique $v \in \overset{0}{W}{}^1_2(\Omega)$ such that

$$\int_\Omega \nabla v(x).\nabla \phi(x) dx = \int_\Omega f(x)\phi(x) dx \text{ for all } \phi \in \overset{0}{W}{}^1_2(\Omega).$$

The function $u \in \overset{0}{W}{}^1_p(\Omega)$ whose existence is guaranteed by Theorem 8.3.4 coincides with this weak solution v, but more can be said about it. For example, if $p > n \geq 2$, then by the Sobolev embedding theorem, $u \in C^{1-n/p}(\overline{\Omega})$; and as $\partial\Omega \in C^1$, the fact that $u \in C^{1-n/p}(\overline{\Omega}) \cap \overset{0}{W}{}^1_p(\Omega)$ means that u is zero on $\partial\Omega$ (see Theorem 1.3.16 (ii)). For further details of how u can even be shown to be a classical solution of the Poisson equation, given enough smoothness of the boundary $\partial\Omega$, we refer to [192]. When $1 < p < 2$, the material of Chapter 6 cannot be used: even though Theorem 8.3.4 ensures the existence of a function $u \in \overset{0}{W}{}^1_p(\Omega)$ such that

$$\int_\Omega \nabla v(x).\nabla \phi(x) dx = \int_\Omega f(x)\phi(x) dx \text{ for all } \phi \in C_0^\infty(\Omega),$$

the fact that f need not belong to $L_2(\Omega)$ prevents the use of the Hilbert space methods of Chapter 5. Again we refer to [192] for information about regularity questions.

2. For details of how to proceed when Ω is an exterior domain, see [192]. The technical complications are considerable.

Chapter 9
The p−Laplacian

9.1 Preamble

Given $p \in (1, \infty)$, the p−Laplacian Δ_p has already been briefly mentioned in the previous chapter as the map $u \longmapsto \Delta_p u := \sum_{i=1}^{n} D_i \left(|\nabla u|^{p-2} D_i u \right)$, where u belongs to some suitable space of real-valued functions on a bounded, open, connected subset Ω of $\mathbb{R}^n$. Just as the Laplace operator $\Delta = \Delta_2$ plays a prominent role in various branches of mathematics, so the p−Laplacian is important with regard to nonlinear phenomena. The literature on Δ_p is enormous: here we merely establish a few basic results and indicate some of the similarities to and differences from the Laplacian in the hope that this will encourage the reader to look further into specific points.

9.2 Preliminaries

The fact that the p−Laplacian is nonlinear forces us to consider ideas and methods that are well suited to this environment. Prominent among these is the notion of a convex function: see [176] for a detailed account of convex analysis. A function $f : \mathbb{R}^n \to \mathbb{R}$ is said to be **convex** if, for all $x, y \in \mathbb{R}^n$ and all $\lambda \in (0, 1)$,

$$f(\lambda x + (1 - \lambda)y) \leq \lambda f(x) + (1 - \lambda) f(y); \tag{9.2.1}$$

if this inequality is strict whenever $x \neq y$ and $\lambda \in (0, 1)$, the function is called **strictly convex**. For example, if $p \in [1, \infty)$, then since the map $x \longmapsto |x|^p$ of $\mathbb{R}^n$ to $\mathbb{R}$ is the composition of the strictly convex maps $x \longmapsto |x|$ and $t \longmapsto t^p$ it is strictly convex. If f is convex and differentiable at $x \in \mathbb{R}^n$, then

$$f(y) \geq f(x) + (y - x) \cdot \nabla f(x) \; (y \in \mathbb{R}^n), \tag{9.2.2}$$

where the dot indicates the usual scalar product of elements of $\mathbb{R}^n$. To establish this, note that if $\lambda \in (0, 1]$, then (9.2.1) can be rewritten as

$$f(y) \geq \frac{1}{\lambda} f\left((1-\lambda)x + \lambda y\right) - \left(\frac{1-\lambda}{\lambda}\right) f(x)$$
$$= \left(\frac{1-\lambda}{\lambda}\right) \{f(x + \lambda(y-x)) - f(x)\} + f((1-\lambda)x + \lambda y).$$

Now let $\lambda \to 0$:

$$f(y) \geq f'_x(y-x) + f(x) = (y-x) \cdot \nabla f(x) + f(x),$$

as claimed.

Application of (9.2.2) to the function $f : \mathbb{R}^n \to \mathbb{R}$ defined by $f(x) = |x|^p$ ($p \in [2, \infty)$) leads to the inequality

$$|y|^p > |x|^p + p|x|^{p-2} x \cdot (y-x) \qquad (9.2.3)$$

for all $x, y \in \mathbb{R}^n$ with $x \neq y$. The next Lemma sharpens this.

Lemma 9.2.1 *Let* $x, y \in \mathbb{R}^n$, $x \neq 0$. *Then if* $p \in [2, \infty)$,

$$|y|^p \geq |x|^p + p|x|^{p-2} x \cdot (y-x) + \frac{|x-y|^p}{2^{p-1} - 1}, \qquad (9.2.4)$$

while if $p \in (1, 2)$,

$$|y|^p \geq |x|^p + p|x|^{p-2} x \cdot (y-x) + \frac{3p(p-1)|x-y|^2}{16(|x|+|y|)^{2-p}}. \qquad (9.2.5)$$

Proof. First suppose that $p \in [2, \infty)$. By Clarkson's inequality ([108], pp. 225-226),

$$|x|^p + |y|^p \geq 2\left|\frac{x+y}{2}\right|^p + 2\left|\frac{x-y}{2}\right|^p. \qquad (9.2.6)$$

From (9.2.3),

$$\left|\frac{x+y}{2}\right|^p \geq |x|^p + \frac{1}{2} p |x|^{p-2} x \cdot (y-x),$$

and hence

$$|y|^p \geq |x|^p + p|x|^{p-2} x \cdot (y-x) + 2\left|\frac{x-y}{2}\right|^p. \qquad (9.2.7)$$

Now repeat this argument, starting with (9.2.6) but using (9.2.7) instead of (9.2.3): the constant 2^{1-p} in the final term in (9.2.7) is now improved to $2^{1-p} + 4^{1-p}$. Iteration now gives the constant to be

9.2 Preliminaries

$$2^{1-p} + 4^{1-p} + \ldots = \frac{1}{2^{p-1} - 1},$$

and (9.2.4) follows.

When $p \in (1, 2)$, fix $x, y \in \mathbb{R}^n$ and define $f : [0, 1] \to \mathbb{R}$ by $f(t) = |x + t(y - x)|^p$. By Taylor's theorem with integral remainder,

$$f(1) = f(0) + f'(0) + \int_0^1 (1 - t) f''(t) dt,$$

we have

$$|y|^p = |x|^p + p |x|^{p-2} x \cdot (y - x) + \int_0^1 (1 - t) f''(t) dt, \qquad (9.2.8)$$

provided that for all $t \in [0, 1]$, $|x + t(y - x)| \neq 0$. Since the result is easy to verify when $x + t(y - x) = 0$ for some $t \in [0, 1]$ we may suppose that this condition holds. Direct calculation plus the use of Schwarz's inequality gives

$$\begin{aligned} f''(t) = & p(p - 2) |x + t(y - x)|^{p-4} \{(x + t(y - x)) \cdot (y - x)\}^2 \\ & + p |x + t(y - x)|^{p-2} |x - y|^2 \\ \geq & p(p - 1) |x + t(y - x)|^{p-2} |x - y|^2. \end{aligned} \qquad (9.2.9)$$

Since

$$\int_0^1 (1 - t) f''(t) dt \geq \frac{3}{4} \int_0^{1/4} f''(t) dt$$

and $|x + t(y - x)| \leq |x| + |y|$, the desired result is obtained from (9.2.8) and (9.2.9). ∎

Corollary 9.2.2 *If $p \in [2, \infty)$, then there is a positive constant C, depending only on p, such that for all $x, y \in \mathbb{R}^n$,*

$$C |y - x|^p \leq \{|y|^{p-2} y - |x|^{p-2} x\} \cdot (y - x). \qquad (9.2.10)$$

Proof. Simply combine (9.2.4) with the corresponding inequality with x and y interchanged. ∎

9.3 The Dirichlet Problem

The focus is now on the p–Laplacian; throughout we assume that Ω is a bounded domain in $\mathbb{R}^n$, $p \in (1, \infty)$ and all the functions that appear are real-valued. We begin our brief account of this operator with the notion of a weakly p–harmonic function, by which we mean a function $u \in W_p^1(\Omega)$ such that

$$\int_\Omega |\nabla u|^{p-2} \nabla u . \nabla \phi \, dx = 0 \text{ for all } \phi \in C_0^\infty(\Omega).$$

Naturally, if $u \in C^2(\Omega)$ and $\Delta_p u = 0$ in Ω, we say that u is p–harmonic (in Ω).

Theorem 9.3.1 *Let $u \in W_p^1(\Omega)$. Then u is weakly p–harmonic in Ω if and only if it is minimising in the sense that*

$$\int_\Omega |\nabla u|^p \, dx \leq \int_\Omega |\nabla v|^p \, dx$$

for all v such that $v - u \in \overset{0}{W}{}_p^1(\Omega)$.

Proof. Suppose u is minimising, put $v = u + \varepsilon\phi$, where $\phi \in C_0^\infty(\Omega)$, and set

$$I(\varepsilon) = \int_\Omega |\nabla(u + \varepsilon\phi)|^p \, dx.$$

Since I attains its minimum when $\varepsilon = 0$, it follows that $I'(0) = 0$, and so u is weakly p–harmonic.

Conversely, suppose that u is weakly p–harmonic. By (9.2.3), if $v - u \in \overset{0}{W}{}_p^1(\Omega)$, then

$$\int_\Omega |\nabla v|^p \, dx \geq \int_\Omega |\nabla u|^p \, dx + p \int_\Omega |\nabla u|^{p-2} \nabla u . \nabla (v - u) dx,$$

from which, on taking $\phi = v - u$, we see that the final integral is zero and hence u is minimising. ∎

As for the existence of weakly p–harmonic functions with specified boundary values, we choose to establish this by the so-called direct method in the calculus of variations to illustrate the technique.

Theorem 9.3.2 *Let $g \in W_p^1(\Omega)$. Then there exists a unique function u that is weakly p–harmonic in Ω and satisfies $u - g \in \overset{0}{W}{}_p^1(\Omega)$.*

Proof. To establish uniqueness, suppose that u_1, u_2 are two functions satisfying the given conditions and with $\nabla u_1 \neq \nabla u_2$ on a set S of positive measure. Note that since the composition of convex functions is convex,

$$\left|\frac{\nabla u_1 + \nabla u_2}{2}\right|^p \leq \frac{|\nabla u_1|^p + |\nabla u_2|^p}{2};$$

the inequality is strict on S. Hence in view of Theorem 9.3.1,

9.3 The Dirichlet Problem

$$\int_\Omega |\nabla u_2|^p \, dx \leq \int_\Omega \left|\frac{\nabla u_1 + \nabla u_2}{2}\right|^p dx$$
$$< \frac{1}{2} \int_\Omega |\nabla u_1|^p \, dx + \frac{1}{2} \int_\Omega |\nabla u_2|^p \, dx = \int_\Omega |\nabla u_2|^p \, dx;$$

contradiction. It follows that $\nabla u_1 = \nabla u_2$ a.e. in Ω, so that $\|\nabla(u_1 - u_2)\|_{p,\Omega} = 0$. Since $u_1 - u_2 \in \overset{0}{W}{}^1_p(\Omega)$, use of the Friedrichs inequality (Theorem 1.3.7) now shows that $u_1 = u_2$.

For existence, put

$$I = \inf \int_\Omega |\nabla v|^p \, dx,$$

where the infimum is taken over all v such that $v - g \in \overset{0}{W}{}^1_p(\Omega)$. As g is such a function, $I \leq \int_\Omega |\nabla g|^p \, dx < \infty$. Let (v_j) be a minimising sequence such that

$$\int_\Omega |\nabla v_j|^p \, dx < I + j^{-1}$$

for all $j \in \mathbb{N}$. By the Friedrichs inequality (Theorem 1.3.7) there is a constant $C > 0$ such that

$$\|w\|_{p,\Omega} \leq C \|\nabla w\|_{p,\Omega}$$

for all $w \in \overset{0}{W}{}^1_p(\Omega)$ and hence for each function $v_j - g$. Thus

$$\|v_j - g\|_{p,\Omega} \leq C \left\{ \|\nabla v_j\|_{p,\Omega} + \|\nabla g\|_{p,\Omega} \right\}$$
$$\leq C \left\{ (I+1)^{1/p} + \|\nabla g\|_{p,\Omega} \right\}.$$

It follows that the sequence $\left(\|v_j\|_{p,\Omega} \right)$ is bounded above, and thus so is $\left(\|v_j\|_{1,p,\Omega} \right)$. Because $W^1_p(\Omega)$ and $L_p(\Omega)$ are reflexive, there are a subsequence of (v_j), still denoted by (v_j) for convenience, and a function $u \in W^1_p(\Omega)$, such that

$$v_j \rightharpoonup u \text{ and } \nabla v_j \rightharpoonup \nabla u \text{ in } L_p(\Omega).$$

Since the reflexive space $\overset{0}{W}{}^1_p(\Omega)$ is weakly complete, $u - g \in \overset{0}{W}{}^1_p(\Omega)$. As

$$\int_\Omega |\nabla u|^p \, dx \leq \lim_{j \to \infty} \int_\Omega |\nabla v_j|^p \, dx = I,$$

u is a minimiser. Thus

$$\int_\Omega |\nabla u|^p \, dx = \inf_{v-g \in \overset{0}{W}{}^1_p(\Omega)} \int_\Omega |\nabla v|^p \, dx = \inf_{v-u \in \overset{0}{W}{}^1_p(\Omega)} \int_\Omega |\nabla v|^p \, dx.$$

In view of Theorem 9.3.1 the proof is complete. ∎

As in the classical case when $p = 2$, it is desirable to know how smooth is a weakly p−harmonic function. If $p > n$ and $\partial\Omega$ is of Lipschitz class, Hölder continuity follows immediately from the embedding theorem $W^1_p(\Omega) \hookrightarrow C^{1-n/p}(\overline{\Omega})$ (see Theorem 1.3.4 (ii)); if u is the function of Theorem 9.3.2 corresponding to $g = 0$, the assumption about the boundary is unnecessary. The situation when p is not greater than n requires much more work. A useful result in this connection is the following inequality of Harnack type, for a proof of which we refer to [208], [147] and [78].

Theorem 9.3.3 *Let u be weakly p−harmonic in Ω, let $x_0 \in \Omega$, suppose that $B^{(2r)} := B(x_0, 2r) \subset \Omega$ and assume that $u \geq 0$ in $B^{(2r)}$. Then there is a constant C, depending only on n and p, such that*

$$\underset{B^{(r)}}{\mathrm{ess\,sup}}\, u \leq C \underset{B^{(r)}}{\mathrm{ess\,inf}}\, u.$$

With the aid of this result, Hölder continuity can be established (see [150]).

Theorem 9.3.4 *Let u be weakly p−harmonic in Ω, let $x_0 \in \Omega$ and suppose that $B^{(2r)} := B(x_0, 2r) \subset \Omega$. Then there are positive constants C and λ such that*

$$|u(x) - u(y)| \leq C |x - y|^\lambda$$

for a.e. $x, y \in B^{(r)}$.

As an immediate consequence there is the next result, of maximum principle type.

Corollary 9.3.5 *Let u be weakly p−harmonic and continuous in Ω. If it attains its supremum on Ω at a point $x_0 \in \Omega$, it is constant on Ω.*

Proof. The function $x \mapsto u(x_0) - u(x)$ is weakly p−harmonic and non-negative on Ω. By Theorem 9.3.3, $u(x) = u(x_0)$ whenever $|x - x_0| < 2 \,\mathrm{dist}\,(x_0, \partial\Omega)$. Now iterate this result. ∎

To conclude this brief discussion of the Dirichlet problem for the p−Laplacian we remark that there is a p−analogue of the Perron approach (discussed in Chapter 4) to the Dirichlet problem for the Laplace operator. For details we refer to [150].

9.4 An Eigenvalue Problem

Here we consider the problem

$$\Delta_p u = -\lambda |u|^{p-2} u \text{ in } \Omega, \ u|_{\partial\Omega} = 0.$$

9.4 An Eigenvalue Problem

By an eigenfunction of this we shall mean a function $u \in \overset{0}{W}{}^1_p(\Omega)\setminus\{0\}$ such that

$$\int_\Omega |\nabla u|^{p-2}\nabla u.\nabla\phi\, dx = \lambda \int_\Omega |u|^{p-2} u\phi\, dx \tag{9.4.1}$$

for all $\phi \in C_0^\infty(\Omega)$. The corresponding real number λ is called an eigenvalue. The choice of u as test function is permissible as $C_0^\infty(\Omega)$ is dense in $\overset{0}{W}{}^1_p(\Omega)$; it immediately shows that every eigenvalue λ is positive and that

$$\lambda = \frac{\int_\Omega |\nabla u|^p\, dx}{\int_\Omega |u|^p\, dx};$$

that is, λ is given by the Rayleigh quotient involving u.

Proposition 9.4.1 *Let λ be an eigenvalue of (9.4.1). Then*

$$\lambda \geq (\omega_n/|\Omega|)^{p/n},$$

where ω_n is the volume of the unit ball in $\mathbb{R}^n$.

Proof. Let u be an eigenfunction corresponding to λ. Then

$$\lambda = \frac{\int_\Omega |\nabla u|^p\, dx}{\int_\Omega |u|^p\, dx}.$$

By the Friedrichs inequality (Theorem 1.3.7),

$$\|u\|_{p,\Omega} \leq (|\Omega|/\omega_n)^{1/n} \|\nabla u\|_{p,\Omega}.$$

The result follows. ∎

Any eigenfunction u is continuous in Ω; if $\partial\Omega \in C^{1,\gamma}$ for some $\gamma \in (0,1]$, then $u \in C(\overline{\Omega})$ (in fact, Hölder-continuous for some exponent) and u vanishes on $\partial\Omega$. For these results, see [143], 5.1 and [147].

Theorem 9.4.2 *There is a least eigenvalue of (9.4.1).*

Proof. Put

$$\lambda_1 = \lambda_1(\Omega) = \inf \int_\Omega |\nabla v|^p\, dx,$$

where the infimum is taken over all $v \in X := \overset{0}{W}{}^1_p(\Omega)$ with $\|v\|_{p,\Omega} = 1$; plainly $\lambda_1 \geq 0$. Let (v_m) be a minimising sequence, so that

$$\|v_m\|_{p,\Omega} = 1, \quad \int_\Omega |\nabla v_m|^p\, dx = \lambda_1 + \varepsilon_m, \text{ where } \varepsilon_m \downarrow 0 \text{ as } m \to \infty.$$

We suppose that X is endowed with the norm

$$\|\cdot\|_X := \left(\int_\Omega |\nabla \cdot|^p \, dx\right)^{1/p}.$$

Then (v_m) is bounded in X, and so there is a subsequence of (v_m), again denoted by (v_m) for convenience, such that $v_m \rightharpoonup u \in X$. Since X is compactly embedded in $L_p(\Omega)$ (see Theorem 1.3.5 (ii)), $v_m \to u$ in $L_p(\Omega)$, so that $\|u\|_{p,\Omega} = 1$ and $u \neq 0$. Moreover, as $v_m \rightharpoonup u$ in X,

$$\|u\|_X^p \leq \liminf_{m\to\infty} \|v_m\|_X^p = \lambda_1^p.$$

Hence $\int_\Omega |\nabla u|^p \, dx = \lambda_1^p$; and as $u \in X\setminus\{0\}$ it cannot be constant. Thus $\lambda_1 > 0$ and the positive infimum is attained at u.

Note that if $v \in X$ is an eigenfunction with corresponding eigenvalue λ, we may put $\phi = v$ in (9.4.1) and obtain

$$\lambda \int_\Omega |v|^p \, dx = \int_\Omega |\nabla v|^p \, dx.$$

If $w := v/\|v\|_{p,\Omega}$, then $\|w\|_{p,\Omega} = 1$ and $\lambda = \int_\Omega |\nabla w|^p \, dx$; hence $\lambda \geq \lambda_1$. Thus no eigenvalue can be less than λ_1. It remains to show that λ_1 is an eigenvalue.

To do this, we observe that λ_1 arises from the minimisation of

$$E(v) := \int_\Omega |\nabla v|^p \, dx \ (v \in X)$$

subject to the constraint

$$G(v) := \int_\Omega |v|^p \, dx - 1 = 0.$$

We use the infinite-dimensional version of the Lagrange multiplier theorem (see [46], Theorem 26.1) to conclude that there is a real number μ such that for all $h \in X$,

$$0 = \langle h, E'(u) - \mu G'(u) \rangle_X$$
$$= p \int_\Omega \left\{|\nabla u|^{p-2} \nabla u . \nabla h - \mu |u|^{p-2} uh\right\} dx.$$

Hence u is an eigenfunction of (9.4.1) with corresponding eigenvalue μ. The choice $h = u$ shows that

$$\mu = \mu \int_\Omega |u|^p \, dx = \int_\Omega |\nabla u|^p \, dx = \lambda_1.$$

■

9.4 An Eigenvalue Problem

Any eigenfunction corresponding to the least eigenvalue λ_1 (often called the first eigenvalue) and obtained by the above minimisation process will be called a **minimiser**. In view of Theorem 1.3.11 we immediately have

Corollary 9.4.3 *If u is a minimiser, so is $|u|$.*

Further information can be obtained by use of the following Harnack inequality, similar to Theorem 9.3.3 (see [149], p. 37).

Theorem 9.4.4 *Let u be a non-negative eigenfunction. Then there is a constant $C = C(n, p)$ such that*
$$\max_{B^{(r)}} u \leq C \min_{B^{(r)}} u$$
whenever $B^{(2r)} \subset \Omega$. Here $B^{(r)}$ and $B^{(2r)}$ are concentric balls with radii r and $2r$.

Corollary 9.4.5 *Let u be a minimiser. Then either $u > 0$ or $u < 0$ in Ω.*

We now give further properties of eigenvalues and corresponding eigenvectors. In particular we show that positive eigenfunctions of the p−Laplacian must be associated with the first eigenvalue λ_1, and that they are unique modulo scaling. The ingenious arguments given below to establish this are due to Kawohl and Lindqvist [123].

Theorem 9.4.6 *Let u and v be minimisers correponding to the least eigenvalue λ_1. Then u is a constant multiple of v: the eigenvalue λ_1 is simple.*

Proof. In view of Corollaries 9.4.3 and 9.4.5 we may suppose that $u > 0$ and $v > 0$. Put $w = (u^p + v^p)^{1/p}$: we show below that $w \in \overset{0}{W}{}^1_p(\Omega)$. From this it follows that

$$\lambda_1 = \frac{\int_\Omega |\nabla u|^p\,dx}{\int_\Omega |u|^p\,dx} = \frac{\int_\Omega |\nabla v|^p\,dx}{\int_\Omega |v|^p\,dx} \leq \frac{\int_\Omega |\nabla w|^p\,dx}{\int_\Omega |w|^p\,dx}.$$

Since
$$\nabla w = w\left(\frac{u^p \nabla \log u + v^p \nabla \log v}{u^p + v^p}\right),$$

the convex combination of $\nabla \log u$ and $v^p \nabla \log v$ that appears means that we can apply Jensen's inequality (see [195], (4.45)) to obtain

$$|\nabla w|^p \leq w^p \left(\frac{u^p |\nabla \log u|^p}{u^p + v^p} + \frac{v^p |\nabla \log v|^p}{u^p + v^p}\right) = |\nabla u|^p + |\nabla v|^p,$$

the inequality being strict where $\nabla \log u \neq \nabla \log v$. If $\nabla \log u \neq \nabla \log v$ on a set of positive measure, then

$$\lambda_1 < \frac{\int_\Omega |\nabla u|^p\,dx + \int_\Omega |\nabla v|^p\,dx}{\int_\Omega |u|^p\,dx + \int_\Omega |v|^p\,dx} = \lambda_1.$$

Hence $\nabla \log u = \nabla \log v$ a.e. in Ω, and thus u is a positive multiple of v.

It remains to prove that $w \in \overset{0}{W}{}^1_p(\Omega)$. Let (u_m), (v_m) be sequences in $C_0^\infty(\Omega)$ that converge in $\overset{0}{W}{}^1_p(\Omega)$ to u, v respectively; by Corollary 1.3.13 we may suppose that each u_m and v_m is non-negative. Put $w_m = (u_m^p + v_m^p)^{1/p}$ and observe that

$$\int_\Omega w_m^p \, dx = \int_\Omega (u_m^p + v_m^p) \, dx \to \int_\Omega (u^p + v^p) \, dx,$$

while since

$$|\nabla w_m| = \left| \left(\frac{u_m}{w_m}\right)^{p-1} \nabla u_m + \left(\frac{v_m}{w_m}\right)^{p-1} \nabla v_m \right| \leq |\nabla u_m| + |\nabla v_m|$$

we see that

$$\int_\Omega |\nabla w_m|^p \, dx \leq 2^p \max\left\{ \int_\Omega |\nabla u_m|^p \, dx, \int_\Omega |\nabla u_m|^p \, dx \right\}.$$

Thus $(\|w_m\|_{1,p})$ is bounded. For each m, supp w_m is a compact subset of Ω and so, by Lemma 1.3.15, $w_m \in \overset{0}{W}{}^1_p(\Omega)$. By reflexivity we may assume, by passage to a subsequence if necessary, that $w_m \rightharpoonup W \in \overset{0}{W}{}^1_p(\Omega)$; hence by Theorem 1.3.5(ii), $w_m \to W$ in $L_p(\Omega)$. Since $u_m \to u$ and $v_m \to v$ in $L_p(\Omega)$, we may suppose that $w_m \to w$ a.e. Hence $w = W$ a.e. and our claim is justified. ∎

Proposition 9.4.7 Let $(\Omega_j)_{j \in \mathbb{N}}$ be an increasing sequence of open subsets of Ω with union Ω, and for each j let $\lambda_{1,j}$ be the least eigenvalue for the Dirichlet problem on Ω_j. Then

$$\lim_{j \to \infty} \lambda_{1,j} = \lambda_1.$$

Proof. Plainly $\lambda_{1,j}$ decreases as j increases: thus $\lim_{j \to \infty} \lambda_{1,j}$ exists and is $\geq \lambda_1$. Given $\varepsilon > 0$, there exists $\phi \in C_0^\infty(\Omega)$ such that

$$\frac{\int_\Omega |\nabla \phi|^p \, dx}{\int_\Omega |\phi|^p \, dx} < \lambda_1 + \varepsilon.$$

Let j be so large that supp $(\phi) \subset \Omega_j$. Then

$$\lambda_{1,j} \leq \frac{\int_{\Omega_j} |\nabla \phi|^p \, dx}{\int_{\Omega_j} |\phi|^p \, dx} = \frac{\int_\Omega |\nabla \phi|^p \, dx}{\int_\Omega |\phi|^p \, dx}.$$

9.4 An Eigenvalue Problem

Hence $\lambda_1 \leq \lambda_{1,j} < \lambda_1 + \varepsilon$. The result follows. ∎

Theorem 9.4.8 *If $\lambda > \lambda_1$, there is no positive eigenfunction with eigenvalue λ.*

Proof. Suppose that u is a positive eigenfunction with corresponding eigenvalue $\lambda > \lambda_1$. By Proposition 9.4.7 and Theorem V.4.20 of [54], there are a smoothly bounded domain Ω', with $\overline{\Omega'} \subset \Omega$ and a corresponding smallest eigenvalue $\lambda_1' = \lambda_1\left(\Omega'\right) < \lambda$; let $u_1' \geq 0$ be the corresponding minimiser. Since $\partial\Omega'$ is smooth, $u_1' \in C\left(\overline{\Omega'}\right)$ and $u_1' = 0$ on $\partial\Omega'$ (see the comment following Proposition 9.4.1). As u has a positive minimum on $\overline{\Omega'}$, by multiplication of u_1' by a small constant, if necessary, we have

$$u_1' \leq u \text{ in } \overline{\Omega'}.$$

Now define $\kappa = (\lambda_1'/\lambda)^{1/(p-1)}$: evidently $\kappa \in (0, 1)$. Let ϕ be a non-negative test function with support in Ω'. Then

$$\int_{\Omega'} \left|\nabla u_1'\right|^{p-2} \nabla u_1' \cdot \nabla \phi \, dx = \lambda_1' \int_{\Omega'} \left(u_1'\right)^{p-1} \phi \, dx$$
$$\leq \lambda_1' \int_{\Omega'} u^{p-1} \phi \, dx = \lambda \int_{\Omega^*} (\kappa u)^{p-1} \phi \, dx$$
$$= \int_{\Omega'} |\nabla(\kappa u)|^{p-2} \nabla(\kappa u) \cdot \nabla \phi \, dx.$$

The choice $\phi = \left(u_1' - \kappa u\right)_+$ is permissible and leads to

$$\int_{u_1' \geq \kappa u} \left\{\left|\nabla u_1'\right|^{p-2} \nabla u_1' - |\nabla(\kappa u)|^{p-2} \nabla(\kappa u)\right\} \cdot \nabla \left(u_1' - \kappa u\right) dx \leq 0.$$

Use of Corollary 9.2.2 now shows that $u_1' \leq \kappa u$. Repetition of this procedure with κu in place of u gives $u_1' \leq \kappa^2 u$. By iteration we have

$$0 \leq u_1' \leq \kappa^j u \to 0 \text{ as } j \to \infty.$$

Hence $u_1' = 0$ and we have a contradiction. ∎

Corollary 9.4.9 *Any positive eigenfunction is a minimiser of the Rayleigh quotient.*

Theorem 9.4.10 *The least eigenvalue is isolated.*

Proof. Suppose there is a sequence (μ_k) of eigenvalues with limit the least eigenvalue λ_1; let (u_k) be the sequence of corresponding normalised eigenvectors. Thus for each k,

$$\int_\Omega |\nabla u_k|^p \, dx = \mu_k, \int_\Omega |u_k|^p \, dx = 1.$$

Reflexivity of $\overset{0}{W}{}^1_p(\Omega)$ and the compactness of the embedding $\overset{0}{W}{}^1_p(\Omega) \hookrightarrow L_p(\Omega)$ (see Theorem 1.3.5 (ii)) show that there are a subsequence of (u_k), still denoted by (u_k), and a function $u \in \overset{0}{W}{}^1_p(\Omega)$, such that $\nabla u_k \rightharpoonup \nabla u$ in $L_p(\Omega)$, $u_k \to u$ in $L_p(\Omega)$ and $u(x) = \lim_{k\to\infty} u_k(x)$ a.e. in Ω. Hence

$$\frac{\int_\Omega |\nabla u|^p \, dx}{\int_\Omega |u|^p \, dx} \leq \lim_{k\to\infty} \mu_k = \lambda_1,$$

so that u is an eigenvector corresponding to the least eigenvalue λ_1. We may suppose that $u > 0$ in Ω.

If $\mu_k \neq \lambda_1$, then by Theorem 9.4.8, u_k must change sign in Ω. Hence both $\Omega_k^+ := \{x \in \Omega : u_k(x) > 0\}$ and $\Omega_k^- := \{x \in \Omega : u_k(x) < 0\}$ are non-empty. Let

$$\phi(x) := \begin{cases} u_k(x), & x \in \Omega_k^+, \\ 0, & x \in \Omega \setminus \Omega_k^+. \end{cases}$$

Then $\phi \in \overset{0}{W}{}^1_p(\Omega)$ (see Theorem 1.3.11). The choice of this ϕ as test function in (9.4.1) shows that

$$\mu_k = \frac{\int_{\Omega_k^+} |\nabla u_k|^p \, dx}{\int_{\Omega_k^+} |u_k|^p \, dx}.$$

By use of the Friedrichs inequality as in the proof of Proposition 9.4.1,

$$\|u_k\|_{p,\Omega_k^+} \leq \left(|\Omega_k^+|/\omega_n\right)^{1/n} \|\nabla u_k\|_{p,\Omega_k^+}.$$

Hence

$$|\Omega_k^+| \geq \omega_n \mu_k^{-n/p}.$$

In the same way it follows that $|\Omega_k^-| \geq \omega_n \mu_k^{-n/p}$. Now write

$$\Omega^+ = \limsup_k \Omega_k^+, \quad \Omega^- = \limsup_k \Omega_k^-.$$

Observe that

$$|\Omega^+| \geq \limsup_k |\Omega_k^+| > 0 \text{ and } |\Omega^-| \geq \limsup_k |\Omega_k^-| > 0$$

(see, for example, [52], p. 129). Since $u \geq 0$ a.e. in Ω^+ and $u \leq 0$ in Ω^-, this contradicts the fact that $u > 0$ in Ω. ∎

Next we discuss *nodal domains* of eigenfunctions. A famous theorem of Courant (see [42], VI.6) asserts that if u_k is an eigenfunction associated to the kth eigenvalue λ_k of $-\Delta$ (with Dirichlet boundary conditions), then u_k has at most k nodal domains:

9.4 An Eigenvalue Problem

for completeness we prove this result below. For the p-Laplacian with $p \neq 2$ a result of this sharpness is not known, but the following theorem (see [149]) gives an upper bound for the number of nodal domains of an arbitrary eigenvector.

Theorem 9.4.11 *Let λ be an eigenvalue of (9.4.1), let u be a corresponding eigenvector, and let $N(u)$ be the number of nodal domains N_j of u. Then*

$$N(u) \leq \lambda^{n/p} |\Omega| / \omega_n.$$

Proof. As in the proof of the preceding theorem,

$$|N_j| \geq \omega_n \lambda^{-n/p}.$$

Hence

$$|\Omega| \geq \sum_j |N_j| \geq \omega_n \lambda^{-n/p} \sum_j 1.$$

The result follows. ∎

For further information about nodal domains see [51] and [41]. Now we prove the Courant theorem.

Theorem 9.4.12 *Let Ω be a bounded, open connected subset of $\mathbb{R}^n$ and let $(\lambda_j)_{j \in \mathbb{N}}$ be the sequence of all eigenvalues (arranged in increasing order and repeated according to multiplicity) of $-\Delta_\Omega$, the positive self-adjoint operator in $L_2(\Omega)$ generated by $-\Delta$ with zero Dirichlet boundary conditions, as explained in Sect. 5.5; let $(\phi_j)_{j \in \mathbb{N}}$ be the sequence of corresponding eigenvectors. Let $k \in \mathbb{N}$ and suppose that ϕ_k has nodal domains $\Omega_1, ..., \Omega_l$. Then $l \leq k$.*

Proof. Note that by Theorem 9.4.11 the number of nodal domains of ϕ_k is finite. Suppose that $l > k$. For $j \in \{1, ..., k\}$, let $\phi_k^j := \chi_{\Omega_j} \phi_k$. By Theorem 1.3.17, $\phi_k^j \in \overset{0}{W}{}_2^1(\Omega_j)$. Choose real numbers $a_1, ..., a_k$, not all zero, such that

$$\phi := \sum_{j=1}^k a_j \phi_k^j \perp \text{sp}\{\phi_1, ..., \phi_{k-1}\}.$$

Then by Theorem 5.5.2,

$$\lambda_k \leq \frac{\int_\Omega |\nabla \phi|^2 \, dx}{\int_\Omega |\phi|^2 \, dx} = \frac{\sum_{j=1}^k a_j^2 \int_{\Omega_j} |\nabla \phi_k|^2 \, dx}{\sum_{j=1}^k a_j^2 \int_{\Omega_j} |\phi_k|^2 \, dx}.$$

But ϕ_k is the kth eigenfunction of $-\Delta_\Omega$. By Theorem 5.5.1, λ_k is an eigenvalue of $-\Delta_{\Omega_j}$; the corresponding eigenvector is ϕ_k^j. Hence

$$\int_{\Omega_j} |\nabla \phi_k|^2 \, dx = \int_{\Omega_j} \left|\nabla \phi_k^j\right|^2 dx = -\int_{\Omega_j} \phi_k^j \Delta \phi_k^j dx = \lambda_k \int_{\Omega_j} \left|\phi_k^j\right|^2 dx$$

$$= \lambda_k \int_{\Omega_j} |\phi_k|^2 \, dx,$$

which, together with the above upper bound for λ_k derived from Theorem 5.5.1, shows that

$$\lambda_k = \frac{\int_\Omega |\nabla \phi|^2 \, dx}{\int_\Omega |\phi|^2 \, dx},$$

so that ϕ is an eigenvector of $-\Delta_\Omega$. By Theorem 5.5.5, ϕ is analytic in Ω. However, ϕ is identically zero on Ω_{k+1} and must therefore be zero throughout Ω. This contradiction proves the theorem. ∎

After this brief digression we return to the p–Laplacian.

Theorem 9.4.13 *The set S of all eigenvalues of (9.4.1) is closed.*

Proof. Let $\mu \in \overline{S}$ and suppose that $(\mu_k)_{k \in \mathbb{N}}$ is a sequence in S with limit μ; for each k let u_k be an eigenvector, with unit norm, corresponding to μ_k. Thus for all $\phi \in C_0^\infty(\Omega)$,

$$\int_\Omega |\nabla u_k|^{p-2} \nabla u_k \cdot \nabla \phi \, dx = \mu_k \int_\Omega |u_k|^{p-2} u_k \phi \, dx, \quad \mu_k = \int_\Omega |\nabla u_k|^p \, dx. \quad (9.4.2)$$

Since $\overset{0}{W}{}^1_p(\Omega)$ is reflexive and compactly embedded in $L_p(\Omega)$, there are a subsequence of (u_k), again denoted by (u_k), and a function $u \in \overset{0}{W}{}^1_p(\Omega)$, such that $u_k \to u$ in $L_p(\Omega)$ and $\nabla u_k \rightharpoonup \nabla u$ in $L_p(\Omega)$. For each k,

$$\int_\Omega \left\{ |\nabla u_k|^{p-2} \nabla u_k - |\nabla u|^{p-2} \nabla u \right\} \cdot \nabla(u_k - u) \, dx$$

$$= \lambda_k \int_\Omega |u_k|^{p-2} u_k (u_k - u) \, dx - \int_\Omega |\nabla u|^{p-2} \nabla u \cdot (\nabla u_k - \nabla u) \, dx.$$

Hence

$$\lim_{k \to \infty} \int_\Omega \left\{ |\nabla u_k|^{p-2} \nabla u_k - |\nabla u|^{p-2} \nabla u \right\} \cdot \nabla(u_k - u) \, dx = 0.$$

If $p \in [2, \infty)$, then by Corollary 9.2.2 it follows that

$$\lim_{k \to \infty} \int_\Omega |\nabla(u_k - u)|^p \, dx = 0,$$

and so, from (9.4.2),

9.4 An Eigenvalue Problem

$$\int_\Omega |\nabla u|^{p-2} \nabla u \cdot \nabla \phi \, dx = \mu \int_\Omega |u|^{p-2} u \phi \, dx,$$

which shows that $\mu \in S$ and u is a corresponding eigenvector.

The case when $1 < p < 2$ can be handled in a similar way, making use of the appropriate part of Lemma 9.2.1. ∎

We have seen in 5.5 that the Courant min-max principle provides, by means of the Rayleigh quotient, a variational characterisation of the eigenvalues and eigenvectors of the Laplacian, and that all such eigenvalues and eigenvectors are obtained by this means. Something similar can be obtained for the p-Laplacian by means of the Liusternik-Schnirelmann procedure, which uses the notion of the **genus** of a set. Let X be a real Banach space and let A be a closed, symmetric (that is, $A = -A$) subset of X. The genus of A is defined to be

$$\gamma(A) = \inf \{m \in \mathbb{N} : \text{there is a continuous odd map } h : A \to \mathbb{R}^m \setminus \{0\}\},$$

with the understanding that $\gamma(A) = \infty$ if no such $m \in \mathbb{N}$ exists, and $\gamma(\emptyset) = 0$. Note that if $0 \in A$, then $\gamma(A) = 0$. To put some flesh on the bones of the definition and give an impression of what it means we list some of the basic properties of the genus.

(i) Let A, B be closed, symmetric subsets of X. Then $\gamma(A) \geq 0$, with equality if and only if $A = \emptyset$; $\gamma(A) \leq \gamma(B)$ if $A \subset B$; $\gamma(A \cup B) \leq \gamma(A) + \gamma(B)$; if $\gamma(B) < \infty$, then $\overline{A \setminus B}$ is symmetric and $\gamma\left(\overline{A \setminus B}\right) \geq \gamma(A) - \gamma(B)$; if $h : X \to X$ is continuous and odd, then $\gamma(A) \leq \gamma\left(\overline{h(A)}\right)$.

(ii) Let $X = \mathbb{R}^n$ and suppose that B is a bounded symmetric neighbourhood of the origin in $\mathbb{R}^n$. Then $\gamma(\partial B) = n$.

(iii) Let X be a Hilbert space and suppose that A is a compact symmetric subset of X with $\gamma(A) = m \in \mathbb{N}$. Then A contains at least m mutually orthogonal vectors.

Proofs of these assertions are given, for example, in [196], II.5. The genus may be thought of as a generalisation of the notion of dimension of a linear space.

For the application of this notion that we have in mind we take X to be $\overset{0}{W}{}^1_p(\Omega)$ and consider symmetric closed subsets A of

$$S := \left\{ u \in \overset{0}{W}{}^1_p(\Omega) : \|u\|_{p,\Omega} = 1 \right\}.$$

We also need the Rayleigh quotient given by

$$I(u) := \frac{\int_\Omega |\nabla u|^p \, dx}{\int_\Omega |u|^p \, dx} \quad \left(u \in \overset{0}{W}{}^1_p(\Omega) \right).$$

The eigenvalues and eigenvectors (in the weak sense) of $-\Delta_p$ correspond to the critical values and critical points of the restriction of I to S. For each $k \in \mathbb{N}$ let F_k be the set of all closed, symmetric subsets A of S with $\gamma(A) \geq k$, and let

$$\mu_k := \inf_{A \in F_k} \sup_{u \in A} I(u).$$

Then it turns out (see [40] and [197]) that $(\mu_k)_{k \in \mathbb{N}}$ is an increasing sequence of eigenvalues of $-\Delta_p$ tending to infinity, and that μ_1 coincides with the least eigenvalue λ_1 described earlier. It is not known whether or not all eigenvalues of $-\Delta_p$ are contained in this sequence, the members of which are called Liusternik-Schnirelmann eigenvalues.

9.5 More About the First Eigenvalue

As in the last section, we study the Dirichlet eigenvalues of the $p-$ Laplacian in a bounded open subset Ω of $\mathbb{R}^n$, namely those real λ for which there is a function u such that
$$-\Delta_p u = |\lambda|^{p-2} \lambda u \text{ in } \Omega, u|_{\partial \Omega} = 0.$$

More precisely, we require u to belong to $\overset{0}{W}{}^1_p(\Omega) \setminus \{0\}$ and the equation to be satisfied in the sense that for all $\phi \in C_0^\infty(\Omega)$,

$$\int_\Omega |\nabla u|^{p-2} \nabla u . \nabla \phi \, dx = \lambda \int_\Omega |u|^{p-2} u \phi \, dx.$$

The eigenvalue λ is given by the Rayleigh quotient

$$\lambda = \frac{\int_\Omega |\nabla u|^p \, dx}{\int_\Omega |u|^p \, dx}.$$

Our concern this time is with the first eigenvalue, that is the smallest such λ. From 9.3 we know that such a λ exists: here we shall denote it by $\lambda_1(\Omega, p)$ and obtain bounds for it.

When $n = 1$ and $\Omega = (-R, R)$ the position is absolutely clear:

$$\lambda_1((-R, R), p) = \frac{p-1}{R^p} \left(\frac{\pi}{p \sin(\pi/p)} \right)^p.$$

For this see, for example, [63], p. 60. Thus

$$\lim_{p \to \infty} \lambda_1((-R, R), p) = \begin{cases} \infty, & \text{if } 0 < R \leq 1, \\ 0, & \text{if } R > 1. \end{cases}$$

When $n > 1$, there is no exact formula for $\lambda_1(\Omega, p)$, but its behaviour as $p \to \infty$ can be determined. Thus in [87] and [116] it is shown that if Ω is bounded and

9.5 More About the First Eigenvalue

$$d(\Omega) := \max\{\text{dist}(x, \partial\Omega) : x \in \Omega\}$$

is the radius of the largest ball contained in Ω, then

$$\lim_{p\to\infty} \lambda_1(\Omega, p)^{1/p} = 1/d(\Omega).$$

Thus if $d(\Omega) < 1$, $\lim_{p\to\infty} \lambda_1(\Omega, p) = \infty$; while if $d(\Omega) > 1$, then $\lim_{p\to\infty} \lambda_1(\Omega, p) = 0$. It turns out that the *Picone identity* is extremely useful in the derivation of results similar to those just described, and to illustrate this technique we give an account here of results established by Benedikt and Drábek [20] that rely on this identity. The next two lemmas are based on [4].

Lemma 9.5.1 *(Picone's identity) Let Ω be an open subset of R^n, let $p \in (1, \infty)$, let $u, v \in C^1(\Omega)$ and suppose that $u \geq 0$ and $v > 0$ in Ω; put*

$$L(u, v) = |\nabla u|^p + (p-1)(u/v)^p |\nabla v|^p - p(u/v)^{p-1} |\nabla v|^{p-2} \nabla u \cdot \nabla v$$

and

$$R(u, v) = |\nabla u|^p - |\nabla v|^{p-2} \nabla\left(\frac{u^p}{v^{p-1}}\right) \cdot \nabla v.$$

Then $L(u, v) = R(u, v) \geq 0$ and $L(u, v) = 0$ a.e. in Ω if and only if $\nabla(u/v) = 0$ a.e. in Ω, which is true if and only if u is a constant multiple of v in each component of Ω.

Proof. That $L(u, v) = R(u, v)$ follows immediately on calculating $\nabla\left(u^p/v^{p-1}\right)$. Since

$$L(u, v) = |\nabla u|^p - p(u/v)^{p-1} |\nabla v|^{p-2} |\nabla u \cdot \nabla v| + (p-1)(u/v)^p |\nabla v|^p \tag{9.5.1}$$

$$+ p(u/v)^{p-1} |\nabla v|^{p-2} \{|\nabla u \cdot \nabla v| - \nabla u \cdot \nabla v\},$$

application of Young's inequality

$$ab \leq \frac{a^p}{p} + \frac{b^{p'}}{p'} \tag{9.5.2}$$

to the second term on the right-hand side of (9.4.1) shows that $L(u, v) \geq 0$. Suppose there exists $x_0 \in \Omega$ such that $L(u, v)(x_0) = 0$. If $u(x_0) \neq 0$, then from (9.4.1) we see that $|\nabla u(x_0) \cdot \nabla v(x_0)| = \nabla u(x_0) \cdot \nabla v(x_0)$ and, from the case of equality in (9.4.2), $|\nabla u(x_0)| = (u(x_0)/v(x_0)) |\nabla v(x_0)|$. Thus $\nabla(u(x_0)/v(x_0)) = 0$. However, if Z is the zero set of the non-negative function u, then $\nabla u = 0$ a.e. in Z (see [54], p. 220) and so $\nabla(u/v) = 0$ a.e. in Z. Hence $\nabla(u/v) = 0$ a.e. in Ω, and the Lemma follows. ∎

Lemma 9.5.2 *Let Ω be an open subset of R^n, let $p \in (1, \infty)$, let $v \in C^1(\Omega)$ be such that $v > 0$ and $\Delta_p v \in C(\Omega)$, and suppose there exists $\lambda > 0$ such that*

$$-\Delta_p v \geq \lambda v^{p-1} \text{ in } \Omega.$$

Then $\lambda \leq \lambda_1(\Omega, p)$.

Proof. Let $u \in C_0^\infty(\Omega)$, $u \geq 0$ and put $w = u^p/v^{p-1}$; observe that

$$|\nabla v|^{p-2} \nabla v \cdot \nabla w = \text{div}\left\{w |\nabla v|^{p-2} \nabla v\right\} - w \text{ div}\left\{|\nabla v|^{p-2} \nabla v\right\}.$$

Then, with the notation of Lemma 9.4.1, we see that

$$0 \leq \int_\Omega R(u,v)dx = \int_\Omega |\nabla u|^p \, dx - \int_\Omega |\nabla v|^{p-2} \nabla\left(\frac{u^p}{v^{p-1}}\right) \cdot \nabla v \, dx$$

$$= \int_\Omega |\nabla u|^p \, dx + \int_\Omega \frac{u^p}{v^{p-1}} \Delta_p v \, dx$$

$$\leq \int_\Omega |\nabla u|^p \, dx - \lambda \int_\Omega u^p dx.$$

The lemma now follows from the variational characterisation of λ_1. ∎

Theorem 9.5.3 *Let $p \in (1, \infty)$, $n \in N$, $k \in \{1, ..., n\}$, $R > 0$ and suppose that Ω is a bounded open subset of $B_k(0, R) \times R^{n-k}$, where $B_k(0, R)$ is the open ball in R^k with centre 0 and radius R. Then*

$$\lambda_1(\Omega, p) \geq kp/R^p.$$

Proof. Given any $x = (x_1, ..., x_n) \in \mathbb{R}^n$, let $|x|_k := \left(\sum_{j=1}^k x_i^2\right)^{1/2}$. Define $v : \Omega \to \mathbb{R}$ by $v(x) = \int_{|x|_k}^R (R^p - (R-t)^p)^{1/(p-1)} \, dt$ $(x \in \Omega)$ and let $\lambda = kp/R^p$. We claim that v and λ have the properties required by Theorem 9.4.2. Evidently $\lambda > 0$ and $v \in C^1(\Omega)$, $\Delta_p v \in C(\Omega)$, $v > 0$. Since

$$\lambda v^{p-1}(x) = \frac{kp}{R^p}\left(\int_{|x|_k}^R (R^p - (R-t)^p)^{1/(p-1)} \, dt\right)^{p-1}$$

$$\leq \frac{kp}{R^p}\left(\int_{|x|_k}^R R^{p/(p-1)} dt\right)^{p-1} = kp(R - |x|_k)^{p-1} \quad (x \in \Omega),$$

it is enough to show that

$$-\Delta_p v(x) \geq kp(R - |x|_k)^{p-1} \quad (9.5.3)$$

in Ω. Plainly

9.5 More About the First Eigenvalue

$$\nabla v(x) = \left(\frac{d}{ds}\int_s^R \left(R^p - (R-t)^p\right)^{1/(p-1)} dt\right)\bigg|_{s=|x|_k} \cdot \left(\frac{x_1}{|x|_k}, \ldots, \frac{x_k}{|x|_k}, 0, \ldots, 0\right)$$

$$= -\left(R^p - (R-|x|_k)^p\right)^{1/(p-1)} \left(\frac{x_1}{|x|_k}, \ldots, \frac{x_k}{|x|_k}, 0, \ldots, 0\right)$$

provided that $|x|_k \neq 0$; if $|x|_k = 0$ then $\nabla v(x) = 0$. Suppose that $|x|_k \neq 0$, in which case

$$|\nabla v(x)|^{p-2} \nabla v(x) = -\left(R^p - (R-|x|_k)^p\right) \left(\frac{x_1}{|x|_k}, \ldots, \frac{x_k}{|x|_k}, 0, \ldots, 0\right) \quad (9.5.4)$$

Then if $i \in \{1, \ldots, k\}$,

$$D_i\left(|\nabla v|^{p-2} D_i v\right)(x) =$$
$$- p\left(R - |x|_k\right)^{p-1} \frac{x_i^2}{|x|_k^2} - \left(R^p - (R-|x|_k)^p\right) \frac{1}{|x|_k}\left(1 - \frac{x_i^2}{|x|_k^2}\right),$$

while

$$D_i\left(|\nabla v|^{p-2} D_i v\right)(x) = 0 \text{ if } i \in \{k+1, \ldots, n\}.$$

Thus

$$-\Delta_p v(x) = p\left(R - |x|_k\right)^{p-1} + \left(R^p - (R-|x|_k)^p\right) \frac{(k-1)}{|x|_k}.$$

Since

$$R^p - (R-|x|_k)^p = \int_{R-|x|_k}^R pt^{p-1} dt \geq p\left(R-|x|_k\right)^{p-1} |x|_k,$$

it follows that

$$-\Delta_p v(x) \geq kp\left(R - |x|_k\right)^{p-1},$$

which gives (9.5.3) when $|x|_k \neq 0$.

Suppose now that $|x|_k = 0$. Then $|\nabla v(x)|^{p-2} \nabla v(x)$ is zero. Let h be the element of $\mathbb{R}^n$ whose only non-zero element is h_i (in the ith place). From the definition of D_i, together with (9.5.4) and use of l'Hôpital's rule, we see that if $i \leq k$,

$$D_i\left(|\nabla v|^{p-2} D_i v\right)(x) = \lim_{h_i \to 0} \frac{|\nabla v(x+h)|^{p-2} D_i v(x+h)}{h_i}$$
$$= \lim_{h_i \to 0} \frac{-\left(R^p - (R-|h|_k)^p\right) \frac{h_i}{|h_i|}}{h_i} = \lim_{h_i \to 0} \frac{-\left(R^p - (R-|h_i|)^p\right)}{|h_i|}$$
$$= \lim_{h_i \to 0} -p(R-|h_i|)^{p-1} = -pR^{p-1}.$$

If $i > k$, then $D_i \left(|\nabla v|^{p-2} D_i v\right)(x) = 0$. Thus (9.5.3) holds when $|x|_k = 0$, and the proof is complete. ∎

As an immediate consequence of this result we have

Corollary 9.5.4 *In Theorem 9.4.3 suppose that $0 < R \leq 1$. Then*

$$\lim_{p \to \infty} \lambda_1(\Omega, p) = \infty.$$

To deal with the case in which $R > 1$ we use

Theorem 9.5.5 *Let $R > 0$ and $p \in (1, \infty)$. Then*

$$\lambda_1(B(0, R), p) \leq \frac{(p+1)(p+2)\ldots(p+n)}{n! R^p}.$$

Proof. Define $v : B(0, R) \to \mathbb{R}$ by $v(x) = R - |x|$. The variational characterisation of $\lambda_1(B(0, R), p)$ shows that

$$\lambda_1(B(0, R), p) \leq \frac{\int_{B(0,R)} |\nabla v|^p \, dx}{\int_{B(0,R)} v^p \, dx} = \frac{|B(0, R)|_n}{\int_0^R (R-t)^p |\partial B(0,t)|_{n-1} \, dt}$$

$$= \frac{R^n |B(0, 1)|_n}{|\partial B(0, 1)|_{n-1} \int_0^R (R-t)^p t^{n-1} dt}$$

$$= \frac{R^n |B(0, 1)|_n}{|\partial B(0, 1)|_{n-1} R^{p+n} B(p+1, n)},$$

where $B(p+1, n)$ denote the beta function. Hence

$$\lambda_1(B(0, R), p) \leq \frac{\Gamma(p+1+n)}{n \Gamma(n) \Gamma(p+1) R^p},$$

and the theorem follows. ∎

Corollary 9.5.6 *Suppose that $p \in (1, \infty)$ and that for some $R > 1$, $B(0, R) \subset \Omega$. Then*

$$\lim_{p \to \infty} \lambda_1(\Omega, p) = 0.$$

For further results in this direction see [118].

9.6 Notes

1. For an excellent account of work in this area, highlighting significant developments, remarks on the history of the results reported here and problems requiring further study, the papers of Lindqvist (see, for example, [148–150]) should be consulted.

2. The study of the Dirichlet problem for the p-Laplacian is greatly simplified when $n = 1$. It is then known that for all $p \in (1, \infty)$, the sequence $(\mu_k)_{k \in \mathbb{N}}$ of Liusternik-Schnirelmann eigenvalues coincides with the sequence $(\lambda_k)_{k \in \mathbb{N}}$ of all eigenvalues obtained by the Courant min-max principle: see, for example, [40], where it is also shown that when $n > 1$, these two sequences also coincide if $p = 2$. The asymptotic behaviour of the μ_k is determined in [84]. Further information and details of what happens with different boundary conditions is given in [143]. For the more general problem

$$-\Delta_p u = \lambda \|u\|_{q,\Omega}^{p-q} |u|^{q-2} u,$$

where $p \in (1, \infty)$ and $1 < q < p^*$, with $p^* = np/(n-p)$ if $1 < p < n$ and $p^* = \infty$ if $p \geq n$, see [82].

3. The literature on nodal domains is now very extensive. We mention in particular the old result of Pleijel [170] that when Ω is a bounded open, connected subset of $\mathbb{R}^2$, then the number N of nodal domains of the kth eigenvector of the Dirichlet Laplacian on Ω satisfies

$$\lim_{k \to \infty} \sup \frac{N}{k} \leq \left(\frac{2}{j}\right)^2 = 0.691...,$$

where j is the smallest positive zero of the Bessel function J_0. This shows that equality in Courant's theorem can be attained at a finite number only of values of k. An improvement of this is given in [26].

Chapter 10
The Rellich Inequality

10.1 Preamble

The classical inequality due to Rellich [175] asserts that if $u \in C_0^\infty (\mathbb{R}^n \setminus \{0\})$ and $n \in \mathbb{N} \setminus \{2\}$, then

$$\frac{n^2(n-4)^2}{16} \int_{\mathbb{R}^n} \frac{|u(x)|^2}{|x|^4} dx \leq \int_{\mathbb{R}^n} |\Delta u(x)|^2 dx. \qquad (10.1.1)$$

The constant $n^2(n-4)^2/16$ is sharp. This inequality also holds when $n = 2$, but only for those $u \in C_0^\infty (\mathbb{R}^2 \setminus \{0\})$ that satisfy, in terms of polar coordinates (r, θ), the condition

$$\int_0^{2\pi} u(r,\theta) \cos\theta d\theta = \int_0^{2\pi} u(r,\theta) \sin\theta d\theta = 0 :$$

see [16], 6.4.2 for this result and also for an account of how (10.1.1) can be rescued from triviality when $n = 4$. If Ω is a proper, non-empty open subset of $\mathbb{R}^n$ ($n \geq 2$) and δ is the distance function defined by $\delta(x) = \text{dist}(x, \partial\Omega)$, then it is known (see Corollary 6.2.7 in [16]) that for all $u \in C_0^\infty(\Omega)$,

$$\frac{9}{16} \int_\Omega \frac{|u(x)|^2}{\delta(x)^4} dx \leq \int_\Omega |\Delta u(x)|^2 dx \qquad (10.1.2)$$

provided that $\Delta\delta \leq 0$ in the distributional sense; this condition is satisfied when Ω is convex. Refinements of (10.1.2) can be found in [16], [19] and [74].

All these results hold in an L_2 setting. Versions of (10.1.1) and (10.1.2) in L_p have been obtained by Davies and Hinz [45]. They rely on the existence of a positive function V such that $\Delta V < 0$ and $\Delta (V^a) \leq 0$ for some $a > 1$, and show that if $p \in (1, \infty)$, then for all $u \in C_0^\infty(\Omega)$,

$$\int_\Omega |\Delta V(x)| |u(x)|^p \, dx \leq \left(\frac{p^2}{(p-1)a+1}\right)^p \int_\Omega \frac{V^p(x)}{|\Delta V(x)|^{p-1}} |\Delta u(x)|^p \, dx.$$

When $\Omega = \mathbb{R}^n \setminus \{0\}$ and $n > 2$, such a function V is that given by $V(x) = |x|^{-(\alpha-2)}$, where $2 < \alpha < n$, and $a = (n-2)/(\alpha-2)$: this choice leads to the result that for all $u \in C_0^\infty(\mathbb{R}^n \setminus \{0\})$,

$$\int_{\mathbb{R}^n} \frac{|u(x)|^p}{|x|^\alpha} \, dx \leq \left(\frac{p^2}{(n-\alpha)\{(p-1)n+\alpha-2p\}}\right)^p \int_{\mathbb{R}^n} \frac{|\Delta u(x)|^p}{|x|^{\alpha-2p}} \, dx.$$

In particular, if $n > 2p$, then for all $u \in C_0^\infty(\mathbb{R}^n \setminus \{0\})$,

$$\int_{\mathbb{R}^n} \frac{|u(x)|^p}{|x|^{2p}} \, dx \leq \left(\frac{p^2}{n(p-1)(n-2p)}\right)^p \int_{\mathbb{R}^n} |\Delta u(x)|^p \, dx; \tag{10.1.3}$$

and the multiplicative constant appearing on the right-hand side is sharp. For general open sets it seems harder to construct a function V with the required properties.

Here we give an account of the work of [57] in which further Rellich-type L_p inequalities are obtained. These use a mean distance function $\delta_{M,2p}$, similar to that introduced by Davies [43] when $p = 2$ and Tidblom [200] for general p, to obtain an L_p analogue of (10.1.2). This function is obtained by averaging, in a certain sense, the distance to the boundary in all possible directions: when Ω is convex, it is bounded above by the usual distance function. Its usefulness in connection with the Hardy inequality is well known (see [16], for example); here it enables us to obtain a Rellich-type inequality of the form

$$\int_\Omega \frac{|u(x)|^p}{\delta_{M,2p}(x)^{2p}} \, dx \leq C \int_\Omega |\Delta u(x)|^p \, dx$$

for all $u \in C_0^\infty(\Omega)$, where C is an explicit function of n and p, independent of u. The main ingredients in the proof are properties of $\delta_{M,2p}$ and sharp estimates of the norms of Riesz transforms in L_p. When $n = 1$ we also give inequalities of a somewhat similar form with the Laplacian replaced by the p–Laplacian. The case $n > 1$ presents difficulties for the p–Laplacian that are only partially resolved here.

10.2 The Mean Distance Function

As usual, by S^{n-1} we denote the unit sphere of $\mathbb{R}^n$; its surface area is $n\omega_n$, where ω_n is the volume of the unit ball in $\mathbb{R}^n$. Throughout we suppose that $p \in (1, \infty)$. Given $\nu = (\nu_i) \in S^{n-1}$ and $x \in \Omega$, put

$$\delta_\nu(x) = \min\{\tau_\nu(x), \tau_{-\nu}(x)\}, \tag{10.2.1}$$

10.2 The Mean Distance Function

where

$$\tau_\nu(x) = \min\{s > 0 : x + s\nu \notin \Omega\}.$$

The usual distance function δ, where $\delta(x) = \text{dist}(x, \partial\Omega)$, is related to the δ_ν by

$$\delta(x) = \inf\{\delta_\nu(x) : \nu \in S^{n-1}\}. \tag{10.2.2}$$

We shall need an averaged version of δ_ν, namely the mean distance function $\delta_{M,p}$ defined by

$$\delta_{M,p}(x)^{-p} = B(n,p) \int_{S^{n-1}} \delta_\nu(x)^{-p} d\sigma(\nu) \quad (x \in \Omega), \tag{10.2.3}$$

where σ is the surface measure on S^{n-1} normalised so that $\int_{S^{n-1}} d\sigma(\nu) = 1$, and

$$B(n,p) = \frac{\sqrt{\pi}\,\Gamma\left(\frac{n+p}{2}\right)}{\Gamma\left(\frac{p+1}{2}\right)\Gamma\left(\frac{n}{2}\right)}. \tag{10.2.4}$$

Proposition 10.2.1 *For all $x \in \Omega$,*

$$\delta_{M,p}(x) \geq B(n,p)^{1/p} \delta(x); \tag{10.2.5}$$

and if Ω is convex, then

$$\delta_{M,p}(x) \leq \delta(x). \tag{10.2.6}$$

Proof. The first inequality is obvious from the definition of the mean distance function. As for the second, let $x \in \Omega$ and let $\mu \in S^{n-1}$ be such that $\delta_\mu(x) = \delta(x)$. The convexity of Ω implies that for all $\nu \in S^{n-1}$,

$$\delta_\nu(x) \cos(\mu \cdot \nu) \leq \delta(x).$$

Thus

$$\int_{S^{n-1}} \delta_\nu(x)^{-p} d\sigma(\nu) \geq \int_{S^{n-1}} |\cos(\mu \cdot \nu)|^p \delta(x)^{-p} d\sigma(\nu) = B(n,p) \delta(x)^{-p},$$

since a straightforward calculation shows that

$$\int_{S^{n-1}} |\cos(\mu \cdot \nu)|^p d\sigma(\nu) = B(n,p).$$

∎

We give two examples to illustrate the determination of the mean distance function.

Example 10.2.2

Here for technical convenience we replace p by $2p$. Let $n = 2$, $h > 0$, take $\Omega = [0, h] \times [0, 1]$ and let $(x, y) \in \Omega$. Set

$$m(x, h) = \min\{h - x, x\}, \quad q(y) = \min\{1 - y, y\},$$

and denote by θ ($\in [0, 2\pi)$) the angle made with the x-axis by the unit vector ν through (x, y). The lines through (x, y) parallel to the coordinate axes divide Ω into four regions. The region determined by $0 < \theta < \pi/2$ is divided into two subregions:

- $0 < \theta < \tan^{-1}\left(\frac{1-y}{h-x}\right)$, in which $\delta_\nu(x, y)\cos\theta = m(x, h)$;
- $\tan^{-1}\left(\frac{1-y}{h-x}\right) < \theta < \pi/2$, in which $\delta_\nu(x, y)\sin\theta = q(y)$.

The contribution of the first region to $\int_{S^1} \delta_\nu(x)^{-2p} d\sigma(\nu)$ is therefore

$$I_1 := \int_0^{\tan^{-1}\left(\frac{1-y}{h-x}\right)} \left(\frac{\cos\theta}{m(x, h)}\right)^{2p} d\theta + \int_{\tan^{-1}\left(\frac{1-y}{h-x}\right)}^{\pi/2} \left(\frac{\sin\theta}{q(y)}\right)^{2p} d\theta.$$

The other regions are handled similarly and give the following contributions:

$$I_2 := \int_{\pi - \tan^{-1}\left(\frac{1-y}{x}\right)}^{\pi} \left(\frac{\cos\theta}{m(x, h)}\right)^{2p} d\theta + \int_{\pi/2}^{\pi - \tan^{-1}\left(\frac{1-y}{x}\right)} \left(\frac{\sin\theta}{q(y)}\right)^{2p} d\theta,$$

$$I_3 := \int_\pi^{\pi + \tan^{-1}\left(\frac{y}{x}\right)} \left(\frac{\cos\theta}{m(x, h)}\right)^{2p} d\theta + \int_{\pi + \tan^{-1}\left(\frac{y}{x}\right)}^{3\pi/2} \left(\frac{\sin\theta}{q(y)}\right)^{2p} d\theta$$

and

$$I_4 := \int_{3\pi/2}^{3\pi/2 + \tan^{-1}\left(\frac{h-x}{y}\right)} \left(\frac{\sin\theta}{q(y)}\right)^{2p} d\theta + \int_{3\pi/2 + \tan^{-1}\left(\frac{h-x}{y}\right)}^{2\pi} \left(\frac{\cos\theta}{m(x, h)}\right)^{2p} d\theta.$$

Hence

$$\delta_{M,p}(x)^{-2p} = B(2, 2p) \sum_{j=1}^{4} I_j,$$

which as $h \to \infty$ converges to

$$\frac{B(2, 2p)}{q(y)^{2p}} \left\{ \int_0^{2\pi} \sin^{2p}\theta d\theta - \int_{\pi - \tan^{-1}\left(\frac{1-y}{x}\right)}^{\pi + \tan^{-1}\left(\frac{y}{x}\right)} \sin^{2p}\theta d\theta \right\}$$

$$+ \frac{B(2, 2p)}{x^{2p}} \int_{\pi - \tan^{-1}\left(\frac{1-y}{x}\right)}^{\pi + \tan^{-1}\left(\frac{y}{x}\right)} \cos^{2p}\theta d\theta.$$

Example 10.2.3

Let $n = 2$, $\Omega = B(0, R)$, where $R > 0$. Let $x \in B(0, R)$ and denote by θ the angle made with the horizontal by the ray ν with origin x. By the cosine rule, if $\theta \in (0, \pi/2)$,

$$\delta_\nu(x) = -r \cos\theta + \sqrt{R^2 - r^2 \sin^2\theta},$$

where $r = |x|$. This gives

$$\int_0^{\pi/2} \delta_\nu(x)^{-p} d\theta = r^{-p}(b^2 - 1)^{-p} \int_0^1 \frac{\left(t + \sqrt{b^2 - 1 + t^2}\right)^p}{\sqrt{1 - t^2}} dt, \quad (10.2.7)$$

where $b = R/r$. The integrals of δ_ν^{-p} over the intervals $(\pi/2, \pi)$, $(\pi, 3\pi/2)$ and $(3\pi/2, 2\pi)$ may be determined in a similar fashion, and an expression for the mean distance function follows, giving

$$\left(\frac{\delta(x)}{\delta_{M,,p}(x)}\right)^p = \frac{4B(2, p)}{2\pi(b+1)^p} \int_0^1 \frac{\left(t + \sqrt{b^2 - 1 + t^2}\right)^p}{\sqrt{1 - t^2}} dt.$$

For general values of p integrals such as that in (10.2.7) cannot be expressed in closed form, but when $p = 4$, representations in terms of hypergeometric functions are available. Note that for fixed x, as $b \to \infty$,

$$\left(\frac{\delta(x)}{\delta_{M,,p}(x)}\right)^p = B(2, p)(1 + o(1)) \to \frac{\sqrt{\pi}\Gamma(p/2 + 1)}{\Gamma\left(\frac{p+1}{2}\right)} > 1.$$

To make full use of the mean distance function some integral identities will be useful. Let $\nu = (\nu_i) \in S^{n-1}$. Then each component ν_i may be written as

$$\nu_i = \prod_{k=1}^{i-1} \sin\theta_k \cos\theta_i$$

for some $\theta_j \in [0, \pi]$ if $j = 1, , , , .n - 2$ and $\theta_{n-1} \in [0, 2\pi]$; the convention that $\prod_{k=r}^s = 1$ when $s < r$ is made here. Note that the element of surface area is given by

$$d\sigma(\nu) = \frac{(n-2)!!}{\gamma_n} \prod_{k=1}^{n-2} (\sin\theta_k)^{n-1-k} d\theta_k d\theta_{n-1},$$

where $n!! = n(n-2)(n-4)...2$ if n is even, $n!! = n(n-2)(n-4)...1$ if n is odd, and

$$\gamma_n = \begin{cases} 2(2\pi)^{(n-1)/2}, & n \text{ odd}, \\ (2\pi)^{n/2}, & n \text{ even}. \end{cases}$$

Routine calculations show that

$$\int_{S^{n-1}} |\nu_j \nu_k|^2 d\sigma(\nu) = \begin{cases} 1/n(n+2), & 1 \leq j < k \leq n, \\ 3/n(n+2), & j = k = 1, ..., n. \end{cases} \quad (10.2.8)$$

As there are n terms with $j = k$ and $n^2 - n$ terms with $j \neq k$,

$$\sum_{j,k=1}^{n} \int_{S^{n-1}} |\nu_j \nu_k|^2 d\sigma(\nu) = 1. \quad (10.2.9)$$

Proposition 10.2.4 *For all $u \in C_0^2(\Omega)$,*

$$\int_{\Omega} \int_{S^{n-1}} \left| \sum_{j,k=1}^{n} \nu_j \nu_k D_j D_k u(x) \right|^2 d\sigma(\nu) dx = \frac{3}{n+2} \int_{\Omega} |\Delta u(x)|^2 dx. \quad (10.2.10)$$

Proof. First note that

$$\sum_{j,k=1}^{n} \nu_j \nu_k D_j D_k u(x) = \sum_{j=1}^{n} \nu_j^2 D_j^2 u(x) + 2 \sum_{1 \leq j < k \leq n} \nu_j \nu_k D_j D_k u(x).$$

Insertion of this in (10.2.10) and use of the easily checked identities

$$\int_{S^{n-1}} (\nu_m)^2 \nu_j \nu_k d\sigma(\nu) = 0 \ (m = 1, ..., n; \ 1 \leq j < k \leq n)$$

and

$$\int_{S^{n-1}} \nu_r \nu_s \nu_j \nu_k d\sigma(\nu) \neq 0 \ (1 \leq r < s \leq n; \ 1 \leq j < k \leq n)$$

only if $j = r$ and $k = s$, leads to

$$\int_{S^{n-1}} \left| \sum_{j,k=1}^{n} \nu_j \nu_k D_j D_k u(x) \right|^2 d\sigma(\nu)$$
$$= \sum_{l,m=1}^{n} D_l^2 u(x) \overline{D_m^2 u(x)} \int_{S^{n-1}} \nu_l^2 \nu_m^2 d\sigma(\nu)$$
$$+ 4 \sum_{1 \leq r < s \leq n} |D_r^2 u(x)|^2 \int_{S^{n-1}} \nu_r^2 \nu_s^2 d\sigma(\nu).$$

Together with (10.2.8) this gives

$$\int_{S^{n-1}} \left| \sum_{j,k=1}^{n} \nu_j \nu_k D_j D_k u(x) \right|^2 d\sigma(\nu)$$
$$= \frac{1}{n(n+2)} \left\{ |\Delta u(x)|^2 + 2 \sum_{i,j=1}^{n} |D_i D_j u(x)|^2 \right\}. \quad (10.2.11)$$

10.2 The Mean Distance Function

Integration of this over Ω and use of the standard identity (obtainable by means of the Fourier transform)

$$\sum_{i,j=1}^{n} \int_{\Omega} |D_i D_j u(x)|^2 \, dx = \int_{\Omega} |\Delta u(x)|^2 \, dx,$$

concludes the proof. ∎

10.3 Results Involving the Laplace Operator

We begin with the one-dimensional situation in which $\Omega = (a, b)$ and the distance function δ is given by

$$\delta(x) = \min\{x - a, b - x\} \quad (a < x < b).$$

Proposition 10.3.1 *Let* $-\infty < a < b \leq \infty$ *and* $p \in (1, \infty)$. *Then for all* $u \in C_0^2(\Omega)$,

$$\int_a^b \frac{|u(t)|^p}{\delta(t)^{2p}} \, dt \leq c_p \int_a^b |u''(t)|^p \, dt, \tag{10.3.1}$$

where

$$c_p = \left(\frac{p}{2p-1}\right)^p \left(\frac{p}{p-1}\right)^p. \tag{10.3.2}$$

Proof. Put $c = \frac{1}{2}(a + b)$ when $b < \infty$, $c = \infty$ otherwise; let $\alpha > 1$. Using integration by parts we see that

$$\int_a^b \frac{|u(t)|^p}{(t-a)^\alpha} \, dt$$

equals

$$(\alpha - 1)^{-1} \int_a^c \left(p|u(t)|^{p-2} \Re\left[\overline{u}(t) u'(t)\right]\right) \left\{(t-a)^{1-\alpha} - (c-a)^{1-\alpha}\right\} dt$$

$$\leq \frac{p}{\alpha - 1} \int_a^c \frac{|u(t)|^{p-1} |u'(t)|}{(t-a)^{\alpha - 1}} \, dt$$

$$\leq \frac{p}{\alpha - 1} \left(\int_a^c \frac{|u(t)|^p}{(t-a)^\alpha} \, dt\right)^{1/p'} \left(\int_a^c \frac{|u'(t)|^p}{(t-a)^{\alpha - p}} \, dt\right)^{1/p}.$$

With α replaced by $\alpha + p$ this gives

$$\int_a^b \frac{|u(t)|^p}{(t-a)^{\alpha+p}} dt \le \left(\frac{p}{\alpha+p-1}\right)^p \int_a^c \frac{|u'(t)|^p}{(t-a)^\alpha} dt.$$

A similar inequality holds on (c, b) and so

$$\int_a^b \frac{|u(t)|^p}{\delta(t)^{\alpha+p}} dt \le \left(\frac{p}{\alpha+p-1}\right)^p \int_a^c \frac{|u'(t)|^p}{\delta(t)^\alpha} dt.$$

Now put $\alpha = p$ and use the Hardy inequality (see, for example, (3.3.4) in [16]) to obtain the result. ∎

This result is a one-dimensional form of Rellich's inequality. Note that $c_2 = 16/9$. To deal with the higher-dimensional situation we use the mean distance function $\delta_{M,2p}$ of the last section.

Theorem 10.3.2 *Let Ω be a non-empty, proper, open subset of $\mathbb{R}^n$, let $p \in (1, \infty)$ and suppose that $u \in C_0^2(\Omega)$. Then*

$$\int_\Omega \frac{|u(x)|^p}{\delta_{M,2p}(x)^{2p}} dx \le K(p,n) \int_\Omega |\Delta u(x)|^p \, dx, \tag{10.3.3}$$

where

$$K(p,n) = c_p B(n, 2p) n^d \cot^{2p}\left(\frac{\pi}{2p^*}\right). \tag{10.3.4}$$

Here

$$d = 2 \text{ if } 1 < p < 2, \ d = 2p/p' \text{ if } 2 \le p < \infty,$$

$p^* = \max\{p, p'\}$, *and* c_p, $B(n, 2p)$ *are given by (10.3.2), (10.2.4) respectively. If $p = 2$, then*

$$\int_\Omega \frac{|u(x)|^2}{\delta_{M,4}(x)^4} dx \le \frac{16}{9} \int_\Omega |\Delta u(x)|^2 \, dx. \tag{10.3.5}$$

Proof. From Proposition 10.3.1 we see that for any $\nu \in S^{n-1}$,

$$\int_\Omega \frac{|u(x)|^p}{\delta_\nu(x)^{2p}} dx \le c_p \int_\Omega \left|(\nabla.\nu)^2 u(x)\right|^p dx.$$

Integration of this with respect to ν over S^{n-1} gives

$$\int_\Omega \frac{|u(x)|^p}{\delta_{M,2p}(x)^{2p}} dx \le c'_p \int_\Omega \int_{S^{n-1}} \left|\sum_{j,k=1}^n \nu_j \nu_k D_j D_k u(x)\right|^p d\sigma(\nu) dx, \tag{10.3.6}$$

where $c'_p = c_p B(n, 2p)$. When $p = 2$, (10.3.6) together with (10.2.10) lead to (10.3.5). If Ω is convex, then by Proposition 10.2.1, $\delta_{M,4} \le \delta$ and we recover (10.1.2).

10.3 Results Involving the Laplace Operator

If $p \neq 2$, more effort is required. First suppose that $p \in (2, \infty)$. From (10.3.6) and Hölder's inequality we have

$$\int_\Omega \frac{|u(x)|^p}{\delta_{M,2p}(x)^{2p}} dx \leq c_p' \int_\Omega \int_{S^{n-1}} \left(\sum_{j,k} |\nu_j \nu_k|^p\right) \left(\sum_{j,k} |D_j D_k u|^{p'}\right)^{p/p'} d\sigma(\nu) dx$$

$$\leq c_p' \int_\Omega \int_{S^{n-1}} \left(\sum_{j,k} |\nu_j \nu_k|^2\right) \left(\sum_{j,k} |D_j D_k u|^{p'}\right)^{p/p'} d\sigma(\nu) dx$$

$$= c_p' \int_\Omega \left(\sum_{j,k} |D_j D_k u|^{p'}\right)^{p/p'} dx,$$

the final inequality following from (10.2.9). By Corollary 1.4.8,

$$\int_\Omega \left(\sum_{j,k} |D_j D_k u|^{p'}\right)^{p/p'} dx = \left\|\sum_{j,k} |D_j D_k u|^{p'}\right\|_{p/p'}^{p/p'}$$

$$\leq \left(\sum_{j,k} \left\||D_j D_k u|^{p'}\right\|_{p/p'}\right)^{p/p'}$$

$$= \left(\sum_{j,k} \|D_j D_k u\|_p^{p'}\right)^{p/p'} \leq C(p,n) \|\Delta u\|_p^p,$$

where

$$C(p,n) = n^{2p/p'} \cot^{2p}\left(\frac{\pi}{2p^*}\right).$$

Thus

$$\int_\Omega \frac{|u(x)|^p}{\delta_{M,2p}(x)^{2p}} dx \leq C(p,n) \|\Delta u\|_p^p.$$

When $1 < p < 2$ we use (10.3.6) and obtain

$$\int_\Omega \frac{|u(x)|^p}{\delta_{M,2p}(x)^{2p}} dx \leq c_p' \int_\Omega \int_{S^{n-1}} \left|\sum_{j,k=1}^n \nu_j \nu_k D_j D_k u(x)\right|^p d\sigma(\nu) dx$$

$$\leq c_p' \int_\Omega \int_{S^{n-1}} \left(\sum_{j,k} |\nu_j \nu_k|^{p'}\right)^{p/p'} \left(\sum_{j,k} |D_j D_k u(x)|^p\right) d\sigma(\nu) dx$$

$$\leq c_p' \int_\Omega \int_{S^{n-1}} \left(\sum_{j,k} |\nu_j \nu_k|^2\right)^{p/p'} \left(\sum_{j,k} |D_j D_k u(x)|^p\right) d\sigma(\nu) dx.$$

Since

$$\int_{S^{n-1}} \left(\sum_{j,k} |\nu_j \nu_k|^2\right)^{p/p'} d\sigma(\nu) \leq \left(\int_{S^{n-1}} \sum_{j,k} |\nu_j \nu_k|^2 d\sigma(\nu)\right)^{p/p'}$$

$$\times \left(\int_{S^{n-1}} d\sigma(\nu)\right)^{1-p/p'}$$

$$= 1,$$

we have

$$\int_\Omega \frac{|u(x)|^p}{\delta_{M,2p}(x)^{2p}} dx \le c'_p \int_\Omega \sum_{j,k} |D_j D_k u(x)|^p dx$$
$$\le c'_p n^2 \cot^{2p}\left(\frac{\pi}{2p^*}\right) \int_\Omega |\Delta u(x)|^p dx.$$

∎

Remark 10.3.3

If $1 < p \le 2$ and the measure $|\Omega|$ of Ω is finite, we can also proceed as follows from (10.3.6):

$$\int_\Omega \frac{|u(x)|^p}{\delta_{M,2p}(x)^{2p}} dx \le c'_p \int_\Omega \left(\int_{S^{n-1}} \left|\sum_{j,k=1}^n \nu_j \nu_k D_j D_k u(x)\right|^2 d\sigma(\nu)\right)^{p/2} dx$$
$$\le c'_p \left\{\int_\Omega \int_{S^{n-1}} \left|\sum_{j,k=1}^n \nu_j \nu_k D_j D_k u(x)\right|^2 d\sigma(\nu) dx\right\}^{p/2} |\Omega|^{1-p/2}$$
$$= c'_p |\Omega|^{1-p/2} \left\{\frac{3}{n(n+2)} \int_\Omega |\Delta u(x)|^2 dx\right\}^{p/2}.$$

Hence

$$\left(\int_\Omega \frac{|u(x)|^p}{\delta_{M,2p}(x)^{2p}} dx\right)^{1/p} \le \left(\frac{3}{n(n+2)}\right)^{1/2} (c'_p)^{1/p} |\Omega|^{\frac{1}{p}-\frac{1}{2}} \|\Delta u\|_2. \quad (10.3.7)$$

When $p = 2$ and Ω is bounded and convex, estimates sharper than (10.1.2) are known. These are of the form

$$\int_\Omega \frac{|u(x)|^2}{\delta(x)^4} dx + A(n) |\Omega|^{-4/n} \int_\Omega |u(x)|^2 dx \le \frac{16}{9} \int_\Omega |\Delta u(x)|^2 dx, \quad (10.3.8)$$

where $A(n)$ is an explicit positive constant. For these see [19] and (with a better constant for $n \ge 4$) [16] and [74].

An obvious application of Theorem 10.3.2 is to the Dirichlet eigenvalue problem for the p-biharmonic operator, namely

$$\Delta\left(|\Delta u|^{p-2} \Delta u\right) = \lambda |u|^{p-2} u \text{ in } \Omega, u = \frac{\partial u}{\partial \nu} = 0 \text{ on } \partial\Omega, \quad (10.3.9)$$

where Ω is a bounded open subset of $\mathbb{R}^n$ with smooth boundary, $\frac{\partial}{\partial \nu}$ denotes the derivative with respect to the outer normal of $\partial\Omega$, and $p \in (1, \infty)$. This problem is interpreted in the weak sense, so that by a solution of (10.3.9) is meant a function $u \in \overset{0}{W}{}^2_p(\Omega)$ such that for all $\phi \in C_0^\infty(\Omega)$,

10.3 Results Involving the Laplace Operator

$$\int_\Omega |\Delta u|^{p-2} \Delta u . \Delta \phi dx = \int_\Omega \lambda |u|^{p-2} u\phi dx.$$

As in the case of the Dirichlet problem for the $p-$Laplacian (see Theorem 9.3.2) it can be proved that there is a least eigenvalue $\lambda_1(\Omega, p)$ of (10.3.9) and that it is given by

$$\lambda_1(\Omega, p) = \min \frac{\int_\Omega |\Delta u|^p \, dx}{\int_\Omega |u|^p \, dx},$$

where the minimum is taken over all $u \in \overset{0}{W}{}^2_p(\Omega)\setminus\{0\}$. Thus from Theorem 10.3.2 we see that if $\Omega = B(0, R)$, so that $\delta(x) \leq R$ for all $x \in B(0, R)$, then

$$\lambda_1(B(0, R), p) \geq \frac{1}{R^{2p} K(p, n)}.$$

However, much as in 9.4, sharper results can be obtained by use of a Picone-type identity, this time in the form given in [112]. Benedikt and Drábek [21] adopt this procedure: a special case of their results is that if $\Omega = B(0, R)$, then

$$\lambda_1(B(0, R), p) \geq \left(\frac{2n}{R^2}\right)^p \frac{1}{\frac{\sqrt{\pi}\Gamma(p)}{\Gamma(p+\frac{1}{2})} - \frac{1}{p}},$$

from which it follows that

$$\lim_{p\to\infty} \lambda_1(B(0, R), p) = \infty \text{ if } R \leq \sqrt{2n}.$$

For other work on $p-$biharmonic problems, see [21], [50] and [112].

10.4 The $p-$Laplacian

Given any $p \in (1, \infty)$, the $p-$Laplacian Δ_p is defined by

$$\Delta_p u = \sum_{j=1}^n D_j\left(|\nabla u|^{p-2} D_j u\right) \tag{10.4.1}$$

whenever u belongs to some suitable space of functions on $\Omega \subset \mathbb{R}^n$. As in the last section we begin with the one-dimensional problem, but this time there is a difference between the real and complex case and for simplicity we deal with only the real case.

Proposition 10.4.1 *Let $-\infty < a < b \leq \infty$. Then for all real $u \in C^2_0(a, b)$ and all $p \in (1, \infty)$,*

$$\int_a^b \frac{|u(t)|^p}{\delta(t)^{p+p'}} dt \leq \left(\frac{p}{pp'-1}\right)^p \int_a^b |\Delta_p u(t)|^{p'} dt. \tag{10.4.2}$$

Proof. If $\Delta_p u \notin L_{p'}(a,b)$, the right-hand side of (10.4.2) is infinite and so the inequality holds in a trivial sense. Suppose $\Delta_p u \in L_{p'}(a,b)$ and let $\alpha > 1$. As before, put $c = (a+b)/2$ if $b < \infty$, and $c = \infty$ otherwise. Using the fact that $\Delta_p u = (p-1)|u'|^{p-2} u''$ we have

$$\int_a^c \frac{|u'(t)|^p}{(t-a)^\alpha} dt \leq \int_a^c (t-a)^{-\alpha} \left(\int_a^t (|u'(s)|^p)' ds \right) dt$$

$$= \int_a^c (|u'(s)|^p)' \left(\int_s^c (t-a)^{-\alpha} dt \right) ds$$

$$= (\alpha-1)^{-1} \int_a^c \left\{ p |u'(s)|^{p-2} u'(s) u''(s) \right\}$$

$$\times \left\{ (s-a)^{-\alpha+1} - (c-a)^{-\alpha+1} \right\} ds$$

$$\leq (\alpha-1)^{-1} p \int_a^c |u'(s)| |\Delta_p u(s)| (s-a)^{-\alpha+1} ds.$$

The choice $\alpha = p'$ gives

$$\int_a^c \frac{|u'(t)|^p}{(t-a)^{p'}} dt \leq p \left(\int_a^c \frac{|u'(t)|^p}{(t-a)^{p'}} dt \right)^{1/p} \left(\int_a^c |\Delta_p u(t)|^{p'} dt \right)^{1/p'},$$

and so

$$\int_a^c \frac{|u'(t)|^p}{(t-a)^{p'}} dt \leq p^{p'} \int_a^c |\Delta_p u(t)|^{p'} dt.$$

Together with a similar inequality over (c,b) this gives

$$\int_a^b \frac{|u'(t)|^p}{\delta(t)^{p'}} dt \leq p^{p'} \int_a^b |\Delta_p u(t)|^{p'} dt. \tag{10.4.3}$$

Now let $\beta > 1 + p'/p$. Then

$$\int_a^c \frac{|u(t)|^p}{(t-a)^\beta} dt = \int_a^c (|u(s)|^p)' \left(\int_s^c (t-a)^{-\beta} dt \right) ds$$

$$= \frac{p}{\beta-1} \int_a^c |u(s)|^{p-2} u(s) u'(s) (s-a)^{-\beta+1} ds$$

$$\leq \frac{p}{\beta-1} \left(\int_a^c \frac{|u'(s)|^p}{(s-a)^{p'}} ds \right)^{1/p} \left(\int_a^c \frac{|u(s)|^p}{(s-a)^{p'(\beta-1-p'/p)}} ds \right)^{1/p'}.$$

Taking $\beta = pp'$, so that $p'(\beta-1-p'/p) = \beta$, we obtain

10.4 The p–Laplacian

$$\int_a^c \frac{|u(t)|^p}{(t-a)^{pp'}} dt \le \frac{p}{pp'-1} \left(\int_a^c \frac{|u'(s)|^p}{(s-a)^{p'}} ds \right)^{1/p} \left(\int_a^c \frac{|u(s)|^p}{(s-a)^{pp'}} ds \right)^{1/p'},$$

and so

$$\int_a^c \frac{|u(t)|^p}{(t-a)^{pp'}} dt \le \left(\frac{p}{pp'-1} \right)^p \int_a^c \frac{|u'(s)|^p}{(s-a)^{p'}} ds.$$

A similar inequality holds on (c, b) and we thus have

$$\int_a^b \frac{|u(t)|^p}{\delta(t)^{pp'}} dt \le \left(\frac{p}{pp'-1} \right)^p \int_a^b \frac{|u'(t)|^p}{\delta(t)^{p'}} dt.$$

Together with (10.4.3) this gives the Proposition. ∎

Note that when $p = 2$, Proposition 10.4.1 asserts that

$$\int_a^b \frac{|u(t)|^2}{\delta(t)^4} dt \le \frac{16}{9} \int_a^b |u''(t)|^2 dt.$$

A sharpening of (10.4.2) may be obtained by using the following result, which appears as Lemma 3 in [74].

Lemma 10.4.2 *Let $d > 0$, $p \in (1, \infty)$, $s \le p - 1$ and suppose that $u \in C_0^2(0, 2d)$. Then*

$$\int_0^{2d} \delta(t)^s |u'(t)|^p dt \ge \left(\frac{p-s-1}{p} \right)^p \int_0^{2d} \left\{ \delta(t)^{s-p} + (p-1)d^{s-p} \right\} |u(t)|^p dt.$$

The combination of this with (10.4.3) produces

Proposition 10.4.3 *Let $d > 0$, $p \in (1, \infty)$ and suppose that $u \in C_0^2(0, 2d)$. Then*

$$\int_0^{2d} |\Delta_p u(t)|^{p'} dt \ge p^{-p'} \left(\frac{pp'-1}{p} \right)^p \int_0^{2d} \left\{ \delta(t)^{-pp'} + (p-1)d^{-pp'} \right\} |u(t)|^p dt.$$

Extension of these one-dimensional results to higher dimensions presents problems that have not so far been satisfactorily resolved: we refer to [57] for details of what has been accomplished.

Chapter 11
More Properties of Sobolev Embeddings

In this chapter we draw attention to various recent developments and an outstanding problem in the theory of Sobolev spaces.

11.1 The Distance Function

Let Ω be a non-empty, proper open subset of $\mathbb{R}^n$ and let $p \in (1, \infty)$; by the distance function we mean the function δ defined by $\delta(x) = \text{dist}(x, \partial\Omega)$ for each $x \in \mathbb{R}^n$. In view of the importance of the subspace $\overset{0}{W}{}^1_p(\Omega)$ of $W^1_p(\Omega)$ because of its connection with the Dirichlet problem for second-order elliptic operators, useful characterisations of it are desirable. One of the best known results of this kind is that (see [54], Theorem V.3.4)

$$\text{if } u \in W^1_p(\Omega) \text{ and } u/\delta \in L_p(\Omega), \text{ then } u \in \overset{0}{W}{}^1_p(\Omega).$$

Kinnunen and Martio [128] showed that the same holds if instead of the condition $u/\delta \in L^p(\Omega)$ it is merely required that u/δ should belong to weak-$L_p(\Omega)$, here denoted as usual by $L_{p,\infty}(\Omega)$. An inequality of Hardy type shows that, under mild conditions on Ω, the converse of the first of these results is true: $u/\delta \in L^p(\Omega)$ whenever $u \in \overset{0}{W}{}^1_p(\Omega)$: see [54], [117] and [200].

Here we give an account of a particular case of the work of [67] (in which spaces of Lebesgue type with variable exponent are considered), and show that the Kinnunen-Martio result can be improved when Ω has a certain regularity property that, for example, rules out inward-pointing cusps.

Definition 11.1.1 Let Ω be bounded. We say that Ω is regular if there are positive constants b, r_0 such that for all $z \in \partial\Omega$ and all $r \in (0, r_0]$,

$$|B(z,r) \cap \Omega^c| \geq b |B(z,r)|.$$

This is similar to the condition that $\Omega^c = \mathbb{R}^n \setminus \Omega$ be interior regular in the sense of [69], p. 59.

Lemma 11.1.2 *Let Ω be bounded and regular. Then given any $R > 0$, there exists $\tilde{b} > 0$ such that for all $z \in \partial \Omega$ and all $r \in (0, R)$,*

$$|B(z,r) \cap \Omega^c| \geq \tilde{b} |B(z,r)|.$$

Proof. Let b, r_0 be as in Definition 11.1.1. We suppose that $R > r_0$ as otherwise there is nothing to prove, and let $r \in (r_0, R)$. Then

$$\frac{|B(z,r) \cap \Omega^c|}{|B(x,r)|} \geq \frac{|B(z,r_0) \cap \Omega^c|}{|B(x,r)|} = \left(\frac{r_0}{r}\right)^N \frac{|B(z,r_0) \cap \Omega^c|}{|B(x,r_0)|} \geq \left(\frac{r_0}{r}\right)^N b := \tilde{b}.$$

∎

Next we remind the reader of the outer cone property. Let $r, h > 0$, put $x' = (x_1, \ldots, x_{n-1}) \in \mathbb{R}^{n-1}$ and define a cone C with vertex at the origin by

$$C = \left\{ (x', x_n) \in \mathbb{R}^n : (h/r) |x'| \leq x_n \leq h \right\}.$$

We say that Ω has the **outer cone property** if there is a fixed cone C such that for each $x \in \partial \Omega$ there exists a cone C_x with vertex x and congruent to C such that $C_x \cap \Omega = \emptyset$. Plainly every bounded open set with the outer cone property is regular; the converse is false (see [67] for a specific example).

The maximal function defined next will play an important role in the subsequent analysis.

Definition 11.1.3 *Suppose that $0 < R < \infty$. The maximal function $M_R u$ is defined by*

$$M_R u(x) = \sup_{0 < r < R} \frac{1}{|B(x,r)|} \int_{B(x,r)} |u(y)| \, dy$$

for all locally integrable functions u on $\mathbb{R}^n$. When $R = \infty$ the corresponding operator is denoted by M.

For each $r > 0$ set

$$\Omega(r) = \{x \in \Omega : \delta(x) < r\}.$$

Given any $x \in \mathbb{R}^n$, let $\bar{x} \in \partial \Omega$ be such that $\delta(x) = |x - \bar{x}|$. Note that $\bar{x}$ need not be unique: it is enough to choose one such point for each x.

Lemma 11.1.4 *Let Ω be regular and let b, r_0 be the constants appearing in Definition 11.1.1. Then there is a positive constant $C(b)$ such that*

11.1 The Distance Function

$$|u(x)|/\delta(x) \leq CM_{2\delta(x)}\left(|\nabla u|\,\chi_{B(\overline{x},\delta(x))}\right)(x) \tag{11.1.1}$$

for all $u \in \overset{0}{W}{}^1_1(\Omega)$ and all $x \in \Omega(r_0)$.

Proof. This result was proved for $u \in C_0^\infty(\Omega)$ in Proposition 1 of [101]. For $u \in \overset{0}{W}{}^1_1(\Omega)$ we extend u by zero outside Ω. This extension, again denoted by u, belongs to $W^1_1(\mathbb{R}^n)$: see, for example, [54], p. 252. Let (u_m) be a sequence in $C_0^\infty(\Omega)$ such that $\|u - u_m\|_{1,1,\Omega} \to 0$. Then

$$\int_\Omega |u_m(x) - u(x)|\,dx \to 0 \text{ and } \int_\Omega |\nabla u_m(x) - \nabla u(x)|\,dx \to 0.$$

Let (u_{m_k}) be a subsequence such that $u_{m_k}(x) \to u(x)$ and $\nabla u_{m_k}(x) \to \nabla u(x)$ a.e., and let $A \subset \Omega$ be a set of full measure in Ω (that is, the measure of A equals that of Ω) such that

$$u_{m_k}(x) \to u(x) \text{ and } \nabla u_{m_k}(x) \to \nabla u(x) \tag{11.1.2}$$

for all $x \in A$. Fix $x \in A$ and let $0 < r \leq \min\{2\delta(x), r_0\}$. By (11.1.2),

$$|u_{m_k}(x)|/\delta(x) \to |u(x)|/\delta(x). \tag{11.1.3}$$

Moreover

$$\frac{1}{|B(x,r)|}\int_{B(x,r)} \left|\nabla u_{m_k}\cdot\chi_{B(\overline{x},\delta(x))}(t)\right|dt \to \frac{1}{|B(x,r)|}\int_{B(x,r)} \left|\nabla u\cdot\chi_{B(\overline{x},\delta(x))}(t)\right|dt.$$

Let $\varepsilon > 0$. Then there exists $k_0 \in \mathbb{N}$ such that for all $k \geq k_0$,

$$\frac{1}{|B(x,r)|}\int_{B(x,r)} \left|\nabla u_{m_k}\cdot\chi_{B(\overline{x},\delta(x))}(t)\right|dt \leq \frac{1}{|B(x,r)|}\int_{B(x,r)} \left|\nabla u\cdot\chi_{B(\overline{x},\delta(x))}(t)\right|dt$$
$$+ \varepsilon$$
$$\leq M_{2\delta(x)}\left(|\nabla u|\,\chi_{B(\overline{x},\delta(x))}\right)(x) + \varepsilon,$$

so that

$$\limsup_{k\to\infty} M_{2\delta(x)}\left(|\nabla u_{m_k}|\,\chi_{B(\overline{x},\delta(x))}\right)(x) \leq M_{2\delta(x)}\left(|\nabla u|\,\chi_{B(\overline{x},\delta(x))}\right)(x) + \varepsilon.$$

Since ε may be chosen arbitrarily small we obtain

$$\limsup_{k\to\infty} M_{2\delta(x)}\left(|\nabla u_{m_k}|\,\chi_{B(\overline{x},\delta(x))}\right)(x) \leq M_{2d(x)}\left(|\nabla u|\,\chi_{B(\overline{x},\delta(x))}\right)(x).$$

Together with (11.1.3) this gives the desired result. ∎

Lemma 11.1.5 *Suppose that Ω is bounded and regular, and let $p \in (1, \infty)$. Assume further that $|\nabla u| \in L_p(\Omega)$ and $u/\delta \in L_1(\Omega)$. Then $u/\delta \in L_p(\Omega)$.*

Proof. Since $|\nabla u| \in L_p(\Omega)$ it follows that $|\nabla u| \in L_1(\Omega)$ which, together with [54], Theorem V.3.4, implies that $u \in \overset{0}{W}{}^1_1(\Omega)$. Let Q be a cube such that $\operatorname{dist}\left(\overline{\Omega}, Q^c\right) > 0$. Extend u by zero outside Ω: then $u \in W^1_1(\mathbb{R}^n)$.

Choose $R > \operatorname{diam} \Omega$ and let $\widetilde{b}$ be the number corresponding to it in Lemma 11.1.2. By Lemma 11.1.4,

$$|u(x)|/\delta(x) \le C M_{2\delta(x)}\left(|\nabla u| \chi_{B(\overline{x},\delta(x))}\right)(x) \qquad (11.1.4)$$

for all $x \in \Omega(R)$. Since $\Omega_R = \Omega$, it follows that (11.1.4) holds for all $x \in \Omega$. As

$$M_{2\delta(x)}\left(|\nabla u| \chi_{B(\overline{x},\delta(x))}\right)(x) \le M\left(|\nabla u| \chi_{B(\overline{x},\delta(x))}\right)(x) \le M\left(|\nabla u|\right)(x),$$

we see that

$$|u(x)|/\delta(x) \le C M(|\nabla u|)(x) \quad (x \in \Omega),$$

and because u is zero outside Ω, this inequality holds for all $x \in \mathbb{R}^n$. Using the boundedness of M on $L_p(\mathbb{R}^n)$ we obtain

$$\|u/\delta\|_{p,\Omega} = \|u/\delta\|_{p,\mathbb{R}^n} \le C \|M(|\nabla u|)\|_{p,\mathbb{R}^n} \le \widetilde{C} \||\nabla u|\|_{p,\mathbb{R}^n}$$
$$= \widetilde{C} \||\nabla u|\|_{p,\Omega},$$

which completes the proof. ■

As an immediate consequence of this and [54], Theorem V.3.4 we have

Theorem 11.1.6 *Let Ω be bounded and regular, and let $p \in (1, \infty)$. Then $u \in \overset{0}{W}{}^1_p(\Omega)$ if and only if*

$$|\nabla u| \in L_p(\Omega) \text{ and } u/\delta \in L_1(\Omega).$$

Thus when Ω is bounded and regular, $u \in \overset{0}{W}{}^1_p(\Omega)$ if and only if $u \in W^1_p(\Omega)$ and $u/\delta \in L^1(\Omega)$. Since $L_{p,\infty}(\Omega) \subset L_1(\Omega)$, it follows that for such a set Ω we have an extension of the result of Kinnunen and Martio [128] in which the assumption that $u/\delta \in L_{p,\infty}(\Omega)$ is made. That the theorem is false if the hypothesis of regularity is dropped is shown by means of an example in [67].

Sobolev spaces weighted by means of powers of the distance function arise naturally in view of the discussion above. Here we report on results contained in [56], where further references to work on this topic may be found. Let Ω be a (possibly unbounded) open subset of $\mathbb{R}^n$ such that

$$\delta(\Omega) := \sup_{x \in \Omega} \delta(x) < \infty. \qquad (11.1.5)$$

11.1 The Distance Function

Let $p \in [1, \infty)$ and $\mu, \theta \in [0, \infty)$. By $W_p^1(\Omega; \mu, \theta)$ we shall mean the space of all functions f on Ω such that $d^\mu \nabla f$ and $d^{-\theta} f$ belong to $L_p(\Omega)$. Endowed with the norm

$$\left\| u | W_p^1(\Omega; \mu, \theta) \right\| := \left\{ \left\| \delta^\mu \nabla u \right\|_p^p + \left\| \delta^{-\theta} u \right\|_p^p \right\}^{1/p}$$

it is a Banach space which is continuously embedded in $L_p(\Omega)$ since

$$\| u \|_p \leq \delta(\Omega)^\theta \left\| u | W_p^1(\Omega; \mu, \theta) \right\|.$$

Moreover,

$$W_p^1(\Omega; \mu_1, \theta_1) \hookrightarrow W_p^1(\Omega; \mu_2, \theta_2) \text{ if } 0 \leq \mu_1 \leq \mu_2 \text{ and } \theta_1 \geq \theta_2 \geq 0.$$

These spaces are contained in the general family of spaces given in [204], 3.2; see also [140], section 9 and [151], 2.6.3. Of course, $W_p^1(\Omega; 0, 0)$ is just the classical Sobolev space $W_p^1(\Omega)$.

Further analysis is facilitated by use of Whitney decompositions of Ω. We recall that by such a decomposition is meant a countable family $\mathcal{F}$ of closed cubes Q_j with pairwise disjoint interiors, sides parallel to the coordinate axes and such that $\Omega = \cup_{j=1}^\infty Q_j$; with $\delta_m(Q_j)$ denoting the diameter of Q_j, the cubes are so chosen that

(i) the ratio of the side lengths of any two of them is 2^k for some integer k;
(ii) $\delta_m(Q_j) \leq \text{dist}(Q_j, \partial\Omega) \leq 4\delta_m(Q_j)$ for each j;
(iii) if $\varepsilon \in (0, 1/4)$ and Q_j^* is the cube concentric with Q_j, with sides parallel to and of length $(1 + \varepsilon)$ times those of Q_j, then each point of Ω is contained in at most 12^n of the Q_j^*.

These properties imply that there are positive constants c_1, c_2 such that for all $Q \in \mathcal{F}$,

$$c_1 \delta_m(Q) \leq \text{dist}(Q, \partial\Omega) \leq \delta(x) \leq \text{dist}(Q, \partial\Omega) + \delta_m(Q) \leq c_2 \delta_m(Q). \quad (11.1.6)$$

For later convenience we put $b_1 = c_1\sqrt{n}$, $b_2 = c_2\sqrt{n}$.

For details of such decompositions we refer to [194]. Now let

$$E_{\mu,\theta} : W_p^1(\Omega; \mu, \theta) \to L_p(\Omega)$$

be the natural embedding.

Theorem 11.1.7 *Suppose that (11.1.5) holds, let F be a Whitney covering of Ω and suppose that for all $\varepsilon > 0$,*

$$D(\varepsilon) := |\{ x \in \Omega : \delta(x) > \varepsilon \}| < \infty. \quad (11.1.7)$$

Define a map $P : W_p^1(\Omega; \mu, \theta) \to L_p(\Omega)$ by

$$Pf = \sum_{\mathcal{F}} f_Q \chi_Q,$$

where χ_Q is the characteristic function of Q and $f_Q = |Q|^{-1} \int_Q f(x)dx$. Then if $\mu < 1$, the embedding $E_{\mu,\theta}$ is compact if and only if P is compact. If $\theta > 0$, then P is compact.

Proof. Given $\varepsilon > 0$ set

$$\mathcal{F}_\varepsilon = \{Q \in \mathcal{F} : \delta_m(Q) \leq \varepsilon\}, \ \Omega_\varepsilon = \cup\{Q : Q \in \mathcal{F}_\varepsilon\}$$

and

$$P_\varepsilon f = \sum_{\mathcal{F}_\varepsilon} f_Q \chi_Q.$$

Because of the dyadic method of construction of the cubes in the Whitney decomposition, there is a sequence $\{\varepsilon_k\}$ of positive numbers that converges to zero and is such that for all $k \in \mathbb{N}$ the cubes in $\mathcal{F}\backslash\mathcal{F}_{\varepsilon_k}$ may be decomposed into congruent cubes of diameter ε_k. For notational convenience we restrict our attention to positive numbers of the form ε_k, so that, for example, the phrase 'for all $\varepsilon > 0$' is in this context to be interpreted as 'for all $\varepsilon_k > 0$'. With this understanding we denote by Q_ε the congruent cubes of diameter ε arising from the decomposition of the cubes in $\mathcal{F}\backslash\mathcal{F}_\varepsilon$; let $\mathcal{C}_\varepsilon$ be the family of all such Q_ε, so that

$$\Omega\backslash\Omega_\varepsilon = \cup\{Q_\varepsilon : Q_\varepsilon \in \mathcal{C}_\varepsilon\}.$$

Since every $Q_\varepsilon \in \mathcal{C}_\varepsilon$ lies in some $Q \in \mathcal{F}$ with $\delta_m(Q) \geq \varepsilon$, it follows from (11.1.6) that, for $x \in Q_\varepsilon \in \mathcal{C}_\varepsilon$,

$$c_1\delta_m(Q_\varepsilon) \leq c_1\delta_m(Q) \leq \delta(x) \leq c_2\delta_m(Q) \lesssim \delta_m(\Omega). \qquad (11.1.8)$$

Given $f \in W_p^1(\Omega; \mu, \theta)$ and $\varepsilon > 0$, let

$$R_\varepsilon f = \sum_{Q_\varepsilon \in \mathcal{C}_\varepsilon} f_{Q_\varepsilon} \chi_{Q_\varepsilon}.$$

From the Poincaré inequality

$$\|f - f_Q\|_{p,Q} \lesssim \delta_m(Q) \|\nabla f\|_{p,Q}$$

and (11.1.6) we have, for all $Q \in \mathcal{F}$,

$$\|f - f_Q\|_{p,Q} \lesssim \delta_m(Q)^{1-\mu} \|\delta^\mu \nabla f\|_{p,Q}.$$

Another use of the Poincaré inequality together with (11.1.8) gives, for all $Q_\varepsilon \in \mathcal{C}_\varepsilon$,

$$\|f - f_{Q_\varepsilon}\|_{p,Q_\varepsilon} \lesssim \delta_m(Q_\varepsilon)^{1-\mu} \|\delta^\mu \nabla f\|_{p,Q_\varepsilon} = \varepsilon^{1-\mu} \|\delta^\mu \nabla f\|_{p,Q_\varepsilon}.$$

11.1 The Distance Function

Thus if $\mu \le 1$,

$$\|E_{\mu,\theta}f - P_\varepsilon f - R_\varepsilon f\|_{p,\Omega}^p = \sum_{Q \in \mathcal{F}_\varepsilon} \|f - f_Q\|_{p,Q}^p + \sum_{Q_\varepsilon \in \mathcal{C}_\varepsilon} \|f - f_{Q_\varepsilon}\|_{p,Q_\varepsilon}^p$$
$$\lesssim \varepsilon^{(1-\mu)p} \|\delta^\mu \nabla f\|_{p,Q}^p.$$

In view of (11.1.7), $\sharp\{Q \in \mathcal{F} : \delta_m(Q) > \varepsilon\} < \infty$ for all $\varepsilon > 0$. Hence $P - P_\varepsilon : W_p^1(\Omega; \mu, \theta) \to L_p(\Omega)$ has finite rank, and is bounded since

$$\|(P - P_\varepsilon)f\|_{p,\Omega}^p = \sum_{Q \in \mathcal{F}, \delta_m(Q) > \varepsilon} \|f_Q\|_{p,Q}^p \le \|f\|_{p,\Omega \setminus \Omega_\varepsilon}^p$$
$$\le \delta_m(\Omega)^{\theta p} \|\delta^{-\theta} f\|_{p,\Omega}^p \le \delta_m(\Omega)^{\theta p} \|f | W_p^1(\Omega; \mu, \theta)\|^p.$$

As $P - P_\varepsilon - R_\varepsilon$ has finite rank, it follows from (11.1.8) that if $\mu < 1$, then $E_{\mu,\theta}$ is compact if and only if P is compact. Moreover, since

$$|f_Q| \le |Q|^{-1/p} \|f\|_{p,Q} \text{ and } \|f_Q\|_{p,Q} \le \|f\|_{p,Q}$$

we have

$$\|P_\varepsilon f\|_{p,Q}^p = \sum_{Q \in \mathcal{F}_\varepsilon} \|f_Q\|_{p,Q}^p \lesssim \sum_{Q \in \mathcal{F}_\varepsilon} \delta_m(Q)^{\theta p} \|\delta^{-\theta} f\|_{p,Q}^p$$
$$\lesssim \varepsilon^{\theta p} \|f | W_p^1(\Omega; \mu, \theta)\|^p,$$

from which the rest of the theorem follows. ∎

Remark 11.1.8

Note that under the assumption (11.1.7), P is compact if

$$\lim_{\varepsilon \to 0} \|P_\varepsilon | W_p^1(\Omega; \mu, \theta) \to L_p(\Omega)\| = 0. \tag{11.1.9}$$

Thus if if $\mu < 1$, $E_{\mu,\theta}$ is compact if (11.1.7) and (11.1.9) hold.

Remark 11.1.9

From the Poincaré inequality and (11.1.6),

$$\|E_{\mu,\theta}f - P_\varepsilon f\|_{p,\Omega_\varepsilon}^p = \sum_{Q \in \mathcal{F}_\varepsilon} \|f - f_Q\|_{p,Q}^p \lesssim \sum_{Q \in \mathcal{F}_\varepsilon} \delta_m(Q)^{(1-\mu)p} \|d^\mu \nabla f\|_{p,Q}^p$$
$$\lesssim \varepsilon^{(1-\mu)p} \|f | W_p^1(\Omega; \mu, \theta)\|^p.$$

Thus if $\mu < 1$, (11.1.9) is equivalent to

$$\sup \{\|f\|_{p,\Omega_\varepsilon} : \|f | W_p^1(\Omega; \mu, \theta)\| = 1\} \to 0 \text{ as } \varepsilon \to 0.$$

When $\mu = \theta = 0$ this reduces to Amick's necessary and sufficient condition for the compactness of $E_{0,0} : W_p^1(\Omega) \to L_p(\Omega)$ (see [54], V.5).

The arguments above illustrate how useful Whitney decompositions can be. This technique is further deployed in [204] to show that if $\theta \in [0, 1]$, then $C_0^\infty(\Omega)$ is dense in $W_p^1(\Omega; 1 - \theta, \theta)$ and hence $W_p^1(\Omega; 0, 1) \subset \overset{0}{W}{}_p^1(\Omega)$, thus recovering the result mentioned at the beginning of this section. Note that in [204], 3.2.4 Triebel gives the more general result that $C_0^\infty(\Omega)$ is dense in $W_p^1(\Omega; \mu, \theta)$ if $\mu \geq 1 - \theta$, while Theorem 9.7 of [140] gives the result of [56] when $\theta \neq 1/p$. For additional information about the embedding of $W_p^1(\Omega; \mu, \theta)$ in $L_p(\Omega)$, including upper and lower estimates of its approximation numbers, we refer to [56].

Next we study the case $p = 2$ and make the assumption throughout that

$$0 \leq \mu < 1, \theta > 0 \text{ and } D(\varepsilon) := |\{x \in \Omega : d(x) > \varepsilon\}| < \infty \text{ for all } \varepsilon > 0.$$

Let $T_{\mu,\theta}(\Omega)$ be the self-adjoint map acting in $L_2(\Omega)$ defined by the quadratic form

$$\|\delta^\mu \nabla u\|_2^2 + \|\delta^{-\theta} u\|_2^2$$

with domain $W_2^1(\Omega; \mu, \theta)$. This is just the Neumann operator generated by the degenerate elliptic operator

$$-\mathrm{div}\left(\delta^{2\mu}\mathrm{grad}\right) + \delta^{-2\theta}.$$

From Remark 11.1.8 it follows that $T_{\mu,\theta}(\Omega)$ has a compact resolvent and hence a discrete spectrum. Denote by $N(\lambda, T_{\mu,\theta}(\Omega))$ the number of eigenvalues of $T_{\mu,\theta}(\Omega)$ less than λ. In [56] it is shown that

$$N(\lambda, T_{\mu,\theta}(\Omega)) \lesssim \left\{\lambda^{n(1+\mu/\theta)/2} + \lambda^{n/(2\theta)}\right\} D\left(b_1 \lambda^{-1/(2\theta)}/b_2\right);$$

a lower bound of greater complexity can also be obtained. The arguments to establish these results require some technical sophistication and accordingly are not given here: to encourage the reader to look at them we observe that in particular, when $\mu = 0$ and $\theta = 1$, we have, with a suitable constant c,

$$\sum_{l > c\lambda^{-1/2}} 1 \lesssim N\left(\lambda, T_{0,1}(\Omega)\right) \lesssim \lambda^{n/2} D\left(b_1 \lambda^{-1/2}/b_2\right),$$

where the summation is over all those cubes in $\mathcal{F}$ of side length $l > c\lambda^{-1/2}$.

We conclude this section with a simple example in which $\Omega = B(0, 1) \subset \mathbb{R}^n$ and $n < p < \infty$. The distance function δ is then given by $\delta(x) = 1 - |x|$ $(x \in \Omega)$.

Theorem 11.1.10 *Let $u \in W_p^1(\Omega)$. Suppose that $\theta \in (n/p, 1)$ and assume that $u/\delta^\theta \in L_p(\Omega)$. Then $u \in \overset{0}{W}{}_p^1(\Omega)$.*

11.1 The Distance Function

Proof. By Theorem 1.3.4 (ii), $u \in C^\lambda(\overline{\Omega})$, where $\lambda = 1 - n/p$. Suppose there exists $x_0 \in \partial\Omega$ such that $u(x_0) \neq 0$, $|u(x_0)| = \eta > 0$, say. Then there exists $\varepsilon \in (0, 1)$ such that $|u(x)| > \eta/2$ if $|x - x_0| < \varepsilon$, $x \in \overline{\Omega}$. Hence

$$\int_\Omega \left|\frac{u(x)}{\delta^\theta(x)}\right|^p dx \geq \int_{B(x_0,\varepsilon)\cap\Omega} \left|\frac{u(x)}{\delta^\theta(x)}\right|^p dx \geq (\eta/2)^p \int_{B(x_0,\varepsilon)\cap\Omega} \frac{1}{(1-|x|)^{\theta p}} dx.$$

Put $y = x - x_0$ and observe that $1 - |x| = |x_0| - |y + x_0|$ and thus $|1 - |x|| \leq |y|$ ($x \in \Omega$). Change of coordinates and use of the condition $\theta p > n$ hence gives

$$\int_\Omega \left|\frac{u(x)}{\delta^\theta(x)}\right|^p dx \geq (\eta/2)^p \int_{|y|<\varepsilon, |y+x_0|<1} |y|^{-\theta p} dy \geq C(\eta/2)^p \int_0^\varepsilon r^{-\theta p + n - 1} dr = \infty,$$

and we have a contradiction. Thus $u = 0$ on $\partial\Omega$, so that by Theorem 1.3.16, $u \in \overset{0}{W}{}^1_p(\Omega)$. ∎

Note that since $\theta < 1$, the condition $u/\delta \in L_p(\Omega)$ implies that $u/\delta^\theta \in L_p(\Omega)$. In view of the results mentioned at the beginning of this section and Theorem 11.1.10, it follows that

$$u/\delta \in L_p(\Omega) \implies u/\delta^\theta \in L_p(\Omega) \implies u \in \overset{0}{W}{}^1_p(\Omega) \Leftrightarrow u/\delta \in L_p(\Omega).$$

Thus for a function $u \in W^1_p(\Omega)$ and under the very special conditions that $n < p < \infty$ and $\Omega = B(0, 1)$, the conditions $u/\delta \in L_p(\Omega)$ and $u/\delta^\theta \in L_p(\Omega)$ are equivalent. The hypothesis that $\Omega = B(0, 1)$ is made for simplicity of exposition only; the assumption that Ω has boundary of class C^2 would be sufficient.

11.2 Nuclear Maps

In this section we describe some recent work concerning nuclearity and embeddings of function spaces. For a bounded linear map T between Banach spaces it is shown that if the Bernstein numbers of T do not decay sufficiently quickly, then T is not nuclear. At a more concrete level, we consider the embedding id of one Besov space (on a bounded Lipschitz domain) in another such space. Necessary and sufficient conditions are given for id to be nuclear. This characterisation also holds for Lizorkin–Triebel spaces and therefore for (fractional) Sobolev spaces.

Given Banach spaces X and Y, a map $T \in B(X, Y)$ is said to be *nuclear* if it can be written in the form

$$T = \sum_{j=1}^\infty \tau_j e_j^* \otimes f_j, \tag{11.2.1}$$

where for each j, e_j^* belongs to the unit sphere S_{X^*} of X^* and $f_j \in S_Y$; and $\{\tau_j\} \in l_1$. Each τ_j can be chosen to be non-negative; we assume this choice to have been made below. The family of all nuclear maps is denoted by $\mathcal{N}(X, Y)$, or by $\mathcal{N}(X)$ if $X = Y$. The nuclear norm of $T \in \mathcal{N}(X, Y)$ is defined to be

$$\|T\|_{\mathcal{N}(X,Y)} = \inf \sum_{j=1}^{\infty} \tau_j,$$

where the infimum is taken over all possible representations (11.2.1) of T. If the context is clear, we simply denote this norm by $\|T\|_{\mathcal{N}}$. Since every $T \in \mathcal{N}(X, Y)$ is the limit (in $B(X, Y)$) of finite-rank operators it is compact.

There are connections between the nuclear behaviour of a map and that of its approximation numbers: recall that for each $k \in \mathbb{N}$, the kth approximation number $a_k(T)$ of $T \in B(X, Y)$ is defined by

$$a_k(T) = \inf \|T - F\|,$$

where the infimum is taken over all $F \in B(X, Y)$ with rank $F < k$. The family of all maps $T \in B(X, Y)$ with $\{a_k(T)\} \in l_1$ is denoted by $\mathcal{A}(X, Y)$, and by $\mathcal{A}(X)$ if $X = Y$.

Lemma 11.2.1 *For all Banach spaces X and Y,*

$$\mathcal{A}(X, Y) \subset \mathcal{N}(X, Y).$$

Moreover, $A(X) = N(X)$ if and only if X is isomorphic to a Hilbert space.

The inclusion is established in [167]; for the second statement, see [131], Corollary 4.b.12.

Some background information about nuclearity may be helpful. Nuclear maps were introduced by Grothendieck in the 1950s, initially in connection with an abstract form of Schwartz's kernel theorem. Later, he and also Ruston [179] extended Fredholm's determinant theory to operators acting in Banach spaces: this hinged on the availability of the notion of the *trace* of an operator for a sufficiently wide class of operators. We recall that if $T \in B(X)$ has finite rank, written $T \in F(X)$, so that a representation

$$T = \sum_{j=1}^{m} x_j^* \otimes x_j$$

is possible, then the trace of T is defined to be

$$\text{tr} T = \sum_{j=1}^{m} \langle x_j, x_j^* \rangle_X$$

and is independent of the particular representation of T. This trace is often referred to as the *matrix trace*, and it is classical that it coincides with the *spectral trace* $\sum \lambda_j(T)$, where the $\lambda_j(T)$ are the eigenvalues of T. The obvious attempt to extend this idea to infinite-dimensional nuclear maps represented by

11.2 Nuclear Maps

$$T = \sum_{j=1}^{\infty} x_j^* \otimes x_j, \text{ with } \sum_{j=1}^{\infty} \|x_j^*\|_{X^*} \|x_j\|_X < \infty,$$

quickly runs into trouble, for while it is plain that

$$\sum_{j=1}^{\infty} \langle x_j, x_j^* \rangle_X$$

is absolutely convergent, its value may well depend on the particular representation of T. To explain this in more detail we need the notion of the *approximation property* (AP). A Banach space X is said to have the AP if, given any compact set $K \subset X$ and any $\varepsilon > 0$, there exists $A \in F(X)$ such that $\|Ax - x\| < \varepsilon$ for all $x \in K$. In a famous paper, Enflo [70] showed that not every separable Banach space has the AP. Grothendieck [95] proved that X has the AP if and only if

$$|\operatorname{tr} T| \le \|T\|_{\mathcal{N}(X)} \text{ for all } T \in F(X).$$

Since $F(X)$ is dense in $(\mathcal{N}(X), \|\cdot\|_{\mathcal{N}(X)})$, it follows that if X has the AP, then the functional tr has a unique continuous extension from $F(X)$ to $(\mathcal{N}(X), \|\cdot\|_{\mathcal{N}(X)})$. Denoting this extension again by tr it follows that if $T \in \mathcal{N}(X)$, $T = \sum_{j \in \mathbb{N}} x_j^* \otimes x_j$ say, then

$$\operatorname{tr} T = \sum_{j \in \mathbb{N}} \langle x_j, x_j^* \rangle$$

gives a well-defined trace for every nuclear operator T, independent of the particular representation of T, provided that X has the AP. It is natural to hope that this trace coincides with the sum of the eigenvalues of T, as it does when $T \in F(X)$. As pointed out by König [131], p. 219, this hope is doomed: there exists $T \in \mathcal{N}(c_0)$ for which tr $T = 1$ but all its eigenvalues are zero. However, if H is a Hilbert space, then it has the AP and $A(H) = \mathcal{N}(H)$, by Lemma 11.2.1. An inequality of Weyl (see [131], p. 35) asserts that if $S \in B(H)$, then

$$\sum_j |\lambda_j(S)| \le \sum_j a_j(S) < \infty,$$

and so $\sum_j \lambda_j(S)$ is well-defined; by a theorem of Lidskii [146],

$$\operatorname{tr} S = \sum_j \lambda_j(S),$$

so that the spectral and matrix versions of the trace coincide. Lidskii's theorem was extended by König [131] to arbitrary Banach spaces X for maps $T \in A(X) \subset \mathcal{N}(X)$. For an interesting discussion of this circle of ideas see [168], 5.7.4 and 6.5.1.

Given any Hilbert space H and any $p \in [1, \infty)$, the family of all those maps $T \in B(H)$ such that $(a_k(T)) \in l_p$ is denoted by $S_p(H)$ and is called the *Schatten p-class* (of H). Evidently $S_1(H) = \mathcal{N}(H)$, the nuclear maps on H. Moreover, the Hilbert–Schmidt maps are precisely those in $S_2(H)$; and it follows immediately from the multiplicative properties of the approximation numbers that if $T_1, T_2 \in S_2(H)$, then $T_1 \circ T_2 \in \mathcal{N}(H)$: the composition of Hilbert–Schmidt maps is nuclear.

Returning to the development of the theory that we shall need, the following three elementary lemmas are proved in [111], pp. 17–18.

Lemma 11.2.2 *Let $k \in \mathbb{N}$, suppose that X_k is a $k-$ dimensional Banach space and let id: $X_k \to X_k$ be the identity map. Then $\|id\|_{\mathcal{N}(X_k)} = k$.*

Lemma 11.2.3 *Let W, X, Y, Z be Banach spaces and suppose that $R \in B(W, X)$, $S \in B(X, Y)$ and $T \in B(Y, Z)$. Then*

$$\|T \circ S \circ R\| \leq \|T\| \, \|S\|_{\mathcal{N}(X,Y)} \, \|R\| \, .$$

Lemma 11.2.4 *Let X, Y be Banach spaces, let Z be a subspace of X, suppose that $T \in B(X, Y)$ and denote by T_Z the restriction of T to Z. Then*

$$\|T_Z\|_{\mathcal{N}(Z,Y)} \leq \|T\|_{\mathcal{N}(X,Y)} \, .$$

This last lemma deals with change of domain space; consequences of certain changes in the target space are given in the next lemma.

Lemma 11.2.5 *Let X, Y be Banach spaces, let Y_m be an $m-$ dimensional subspace of Y and suppose that $T \in B(X, Y)$. Then there is a projection $P : Y \to Y_m$ such that*

$$\|P \circ T\|_{\mathcal{N}(X,Y_m)} \leq \sqrt{m} \, \|T\|_{\mathcal{N}(X,Y)} \, .$$

Proof. By a result of Kadets and Snobar (see, for example, [168], 6.1.1.7), there is a projection $P : Y \to Y_m$ such that

$$\|P\| \leq \sqrt{m}. \tag{11.2.2}$$

Now use Lemmas 11.2.3 and 11.2.4:

$$\|P \circ T\|_{\mathcal{N}(X,Y_m)} \leq \|P\| \, \|T\|_{\mathcal{N}(X,Y)} \leq \sqrt{m} \, \|T\|_{\mathcal{N}(X,Y)} \, .$$

∎

We next give a condition sufficient for a map to be not nuclear. This involves the notion of the *Bernstein numbers* of a map $T \in B(X, Y)$, where X and Y are Banach spaces. Given any $k \in \mathbb{N}$, the kth Bernstein number of T is

$$b_k(T) := \sup_{X_k} \inf_{x \in X_k \setminus \{0\}} \frac{\|Tx\|_Y}{\|x\|_X},$$

11.2 Nuclear Maps

where the supremum is taken over all k-dimensional subspaces X_k of X. Further details of these numbers are given in [63], Chapter 5 where, in particular, it is shown that they are dominated by the approximation numbers of T :

$$b_k(T) \le a_k(T) \text{ for all } k \in \mathbb{N}.$$

Theorem 11.2.6 *Theorem 11.2.6 Let X and Y be Banach spaces, let $T \in B(X, Y)$ and suppose there exists $\gamma > 0$ such that $b_k(T) \ge \gamma/\sqrt{k}$ for all $k \in \mathbb{N}$. Then T is not nuclear.*

Proof. Suppose that T is nuclear and let $\eta \in (0, \|T\|_\mathcal{N})$. Then T has a representation

$$T = \sum_{j=1}^{\infty} \tau_j e_j^* \otimes y_j,$$

where the $e_j^* \in S_{X^*}$, $y_j \in S_Y$, $\tau_j \ge 0$ and

$$0 < \sum_{j=1}^{\infty} \tau_j - \eta \le \|T\|_\mathcal{N} \le \sum_{j=1}^{\infty} \tau_j.$$

Let $\varepsilon \in (0, \eta)$. Then there exists $m \in \mathbb{N}$ such that $\sum_{j=m+1}^{\infty} \tau_j < \varepsilon$. Put

$$T_{(m)} = \sum_{j=1}^{m} \tau_j e_j^* \otimes y_j, \quad T^{(m)} = \sum_{j=m+1}^{\infty} \tau_j,$$

and let $\mathcal{E}_m = \text{sp}\{e_1^*, ..., e_m^*\} \subset X^*$. The polar set of $\mathcal{E}_m$ is

$$\mathcal{E}_m^0 = \{g \in X : \langle g, e^* \rangle_X = 0 \text{ for all } e^* \in \mathcal{E}_m\} = \cap_{j=1}^{m} \{e_j^*\}^0 \subset X.$$

Plainly $\mathcal{E}_m^0 \subset \ker T_{(m)}$, and so the restriction $T \restriction_{\mathcal{E}_m^0}$ of T to $\mathcal{E}_m^0$ coincides with $T^{(m)} \restriction_{\mathcal{E}_m^0}$. Fix $k \in \mathbb{N}$ such that $k > m$. From the definition of $b_k(T)$ it follows that there is a k-dimensional subspace $X_k = \text{sp}\{f_1, ..., f_k\}$ of X such that

$$\inf_{x \in X_k \setminus \{0\}} \frac{\|Tx\|_Y}{\|x\|_X} \ge \frac{1}{2} b_k(T) \ge \frac{\gamma}{2\sqrt{k}} > 0.$$

Hence $T_k = T \restriction_{X_k}$ has inverse T_k^{-1} and, with $Y_k := T(X_k)$,

$$\|T_k^{-1} \restriction_{Y_k}\| \le 2\sqrt{k}/\gamma. \tag{11.2.3}$$

Since

$$k \ge \dim(X_k \cap \mathcal{E}_m^0) \ge k - m > 0,$$

there exist at least $k - m$ linearly independent vectors in $X_k \cap \mathcal{E}_m^0$; denote the first $k - m$ of these by $a_1, ..., a_{k-m}$. As each a_j can be expressed as a linear combination of $f_1, ..., f_k$, there is a $(k-m) \times k$ matrix $\mathbf{A} = (a_{ij})$, with rank $k - m$, for which

$\mathbf{a}^t = \mathbf{A}\mathbf{f}^t$, where $\mathbf{a} = (a_1, ..., a_{k-m})$ and $\mathbf{f} = (f_1, ..., f_k)$.

Without loss of generality we may suppose that the first $k - m$ columns of $\mathbf{A}$ are linearly independent, for if not, the order of the f_i may be changed to achieve this. Denote by $\mathbf{B}$ the $(k - m) \times (k - m)$ matrix given by the first $k - m$ columns of $\mathbf{A}$, so that $\mathbf{B} = (a_{ij})_{i=1, j=1}^{k-m, k-m}$. Then $\mathbf{B}^{-1}\mathbf{A}$ is a matrix with (i, j) – th entry equal to the Kronecker delta δ_{ij} when $1 \le i, j \le k - m$. Define $\mathbf{c} = (c_1, ..., c_{k-m})$ by

$$\mathbf{c}^t = \mathbf{B}^{-1}\mathbf{A}\mathbf{f}^t.$$

Then $c_1, ..., c_{k-m}$ form a set of $k - m$ linearly independent vectors in $X_k \cap \mathcal{E}_m^0$ which may be expressed as a linear combination of $f_1, ..., f_k$ in such a way that

$$c_i = \sum_{j=1}^{k} c_{ij} f_j, \text{ where } c_{ij} = \delta_{ij} \text{ for } 1 \le i, j \le k - m.$$

Now put

$$X_{k-m} = \text{sp}\{f_1, ..., f_{k-m}\}, Y_{k-m} = \text{sp}\{Tf_1, ..., Tf_{k-m}\},$$

where the f_i are the functions that generate X_k (perhaps not in order due to the possible rearrangement of these functions). Of course, T maps X_{k-m} onto Y_{k-m}. Use of Lemmas 11.2.3 and 11.2.4, together with (11.2.3) gives

$$k - m = \|\text{id}\|_{\mathcal{N}(X_{k-m})} = \left\|\left(T_k^{-1}\restriction_{Y_{k-m}}\right) \circ \left(T_k\restriction_{X_{k-m}}\right)\right\|_{\mathcal{N}(X_{k-m})}$$

$$\le \left\|T_k\restriction_{X_{k-m}}\right\|_{\mathcal{N}(X_{k-m}, Y_{k-m})} \left\|T_k^{-1}\restriction_{Y_k}\right\| \le \frac{2\sqrt{k}}{\gamma} \left\|T_k\restriction_{X_k}\right\|_{\mathcal{N}(X_{k-m}, Y_{k-m})}.$$
(11.2.4)

With the help of Lemmas 11.2.4 and 11.2.5 we thus have

$$k - m \le \frac{2\sqrt{k(k-m)}}{\gamma} \left\|T_k\restriction_{X_{k-m}}\right\|_{\mathcal{N}(X_{k-m}, Y)}$$

$$\le \frac{2\sqrt{k(k-m)}}{\gamma} \left\|T\restriction_{X_{k-m} \cap \mathcal{E}_m^0}\right\|_{\mathcal{N}(X_{k-m} \cap \mathcal{E}_m^0, Y)},$$
(11.2.5)

and so

$$\left\|T\restriction_{X_{k-m} \cap \mathcal{E}_m^0}\right\|_{\mathcal{N}(X_{k-m} \cap \mathcal{E}_m^0, Y)} \ge \frac{\gamma\sqrt{k-m}}{2\sqrt{k}}.$$
(11.2.6)

But

$$\left\|T\restriction_{X_{k-m} \cap \mathcal{E}_m^0}\right\|_{\mathcal{N}(X_{k-m} \cap \mathcal{E}_m^0, Y)} \le \left\|T\restriction_{\mathcal{E}_m^0}\right\|_{\mathcal{N}(\mathcal{E}_m^0, Y)} \le \left\|T^{(m)}\right\|_{\mathcal{N}(X, Y)} < \varepsilon.$$

11.2 Nuclear Maps

Since k may be chosen arbitrarily large we have a contradiction and the proof is complete. ∎

We now turn to the situation in which T is an embedding map between Besov spaces. Let Ω be a bounded open subset of $\mathbb{R}^n$ with C^∞ boundary and suppose that $s > 0$ and $p, q \in [1, \infty)$. The *Besov space* $B_{p,q}^s(\Omega)$ may be defined in various equivalent ways: see, for example, [69], [102], [103], [204] and [206]. For definiteness we choose the approach via dyadic resolutions of unity which begins with the Schwartz space $\mathcal{S}$ of rapidly decreasing functions on $\mathbb{R}^n$ and its dual $\mathcal{S}'$, the space of tempered distributions. Let $\phi = \phi_0 \in \mathcal{S}$ be such that

$$\operatorname{supp} \phi \subset \{y \in \mathbb{R}^n : |y| < 2\} \text{ and } \phi(x) = 1 \text{ if } |x| \leq 1,$$

and for each $j \in \mathbb{N}$ let $\phi_j(x) = \phi(2^{-j}x) - \phi(2^{-j+1}x)$. Given any $f \in \mathcal{S}'$, we denote by $\mathcal{F}f$ and $\mathcal{F}^{-1}f$ its Fourier transform and inverse Fourier transform respectively. The Besov space $B_{p,q}^s(\mathbb{R}^n)$ is defined to be the family of all $f \in \mathcal{S}'$ such that

$$\|f\|_{B_{p,q}^s(\mathbb{R}^n)} = \left(\sum_{j=0}^\infty 2^{jsq} \left\|\mathcal{F}^{-1}(\phi_j \mathcal{F}f)\right\|_{p,\mathbb{R}^n}^q\right)^{1/q} < \infty.$$

Endowed with the norm $\|\cdot\|_{B_{p,q}^s(\mathbb{R}^n)}$ this is a Banach space. The space $B_{p,q}^s(\Omega)$ is defined by restriction:

$$B_{p,q}^s(\Omega) := \left\{f \in \mathcal{D}'(\Omega) : f = g \upharpoonright_\Omega \text{ for some } g \in B_{p,q}^s(\mathbb{R}^n)\right\},$$

where $\mathcal{D}'(\Omega)$ is the space of distributions on Ω; it is normed by

$$\|f\|_{B_{p,q}^s(\Omega)} = \inf_g \|g\|_{B_{p,q}^s(\mathbb{R}^n)},$$

where the infimum is taken over all $g \in B_{p,q}^s(\mathbb{R}^n)$ with $f = g \upharpoonright_\Omega$. The spaces obtained by this procedure coincide (up to equivalence of norms) with those derived by the use of moduli of smoothness in [161], for example. The *Lizorkin–Triebel spaces* $F_{p,q}^s$ are defined by means of a similar process, starting from

$$\|f\|_{F_{p,q}^s(\mathbb{R}^n)} = \left\|\left(\sum_{j=0}^\infty 2^{jsq} \left|\mathcal{F}^{-1}(\phi_j \mathcal{F}f)\right|^q\right)^{1/q}\right\|_{p,\mathbb{R}^n} < \infty.$$

In [59] the embedding $id : B_{p_1,q}^s(U) \to L_{p_2}(U)$ is considered, under certain conditions on the parameters s, p_1, p_2, q, and where U is the open unit ball in $\mathbb{R}^n$. It is shown that the analogue of Theorem 11.2.6 for id holds under the weaker assumption that $b_k(id) \geq \gamma/k$. Together with sharp upper and lower estimates of $b_k(id)$ this enables necessary and sufficient conditions to be given for id to be nuclear. More recently, Triebel [207] has given a characterisation of nuclearity that holds for a wider range of parameters than in [59], and also allows the target space to be of Besov type. His result is the following:

Theorem 11.2.7 *Let Ω be a bounded Lipschitz domain in $\mathbb{R}^n$ ($n \in \mathbb{N}$), suppose that $p_1, p_2, q_1, q_2 \in (1, \infty)$, let $s_1, s_2 \in \mathbb{R}$ and assume that*

$$s_1 - s_2 > n \left(\frac{1}{p_2} - \frac{1}{p_1} \right)_+ . \tag{11.2.7}$$

(This condition is necessary and sufficient for $id : B_{p_1,q_1}^{s_1}(\Omega) \to B_{p_2,q_2}^{s_2}(\Omega)$ to be compact.) Then id is nuclear if and only if

$$s_1 - s_2 > n - n \left(\frac{1}{p_2} - \frac{1}{p_1} \right)_+ . \tag{11.2.8}$$

The proof involves wavelet techniques. Note that the assertions of the theorem are independent of the q-parameters. From the inclusions

$$B_{p,\min(p,q)}^{s}(\Omega) \hookrightarrow F_{p,q}^{s}(\Omega) \hookrightarrow B_{p,\max(p,q)}^{s}(\Omega)$$

(see [69], p. 106) it thus follows that the theorem also holds for Lizorkin–Triebel spaces. Since $F_{p,2}^{s}(\Omega)$ coincides with the (fractional) Sobolev space $H_p^s(\Omega)$ (which is just $W_p^s(\Omega)$ if $s \in \mathbb{N}$) the result is available for $H_p^s(\Omega)$. In particular, the embedding of $W_p^{s_1}(\Omega)$ in $W_p^{s_2}(\Omega)$ (with $s_1, s_2 \in \mathbb{N}$) is nuclear if and only if

$$s_1 - s_2 > n.$$

Standard procedures show that all these results hold for the spaces $\overset{0}{B}{}_{p,q}^{s}(\Omega)$ and $\overset{0}{F}{}_{p,q}^{s}(\Omega)$.

Some comparison of the above theorem with known results is desirable. In [222], p. 279 it is shown that the embedding of $\overset{0}{W}{}_{2}^{k}(\Omega)$ in $\overset{0}{W}{}_{2}^{j}(\Omega)$ is nuclear if $k - j > n$ ($k, j \in \mathbb{N}_0$); a similar result is given in [154], p. 336. These results are covered by Theorem 11.2.7; note that in each case the underlying setting is that of L_2. The if-part of the theorem is essentially covered by Pietsch's paper [166] (following on from [169]) combined with such technicalities as restrictions, extensions and lifts. In more detail, Pietsch proved that id is nuclear if

$$s_1 - s_2 > n - n \left(\frac{1}{p_2} - \frac{1}{p_1} \right)_+ ,$$

and that it is not nuclear if

$$s_1 - s_2 < n - n \left(\frac{1}{p_2} - \frac{1}{p_1} \right)_+ .$$

11.2 Nuclear Maps

This left open the case in which

$$s_1 - s_2 = n - n\left(\frac{1}{p_2} - \frac{1}{p_1}\right)_+.$$

A first step to seal this gap was taken in [59], and the whole matter was settled in [207].

11.3 Asymptotic Formulae for Approximation Numbers of Sobolev Embeddings

Let Ω be a bounded open subset of $\mathbb{R}^n$ and suppose that $p \in (1, \infty)$ and $m \in \mathbb{N}$; let $\mathrm{id}_{m,p} \colon \overset{0}{W}{}_p^m(\Omega) \to L_p(\Omega)$ be the natural embedding. Then (see [69], Chapter 3) $\mathrm{id}_{m,p}$ is compact and its approximation numbers $a_k(\mathrm{id}_{m,p})$ decay like $k^{-m/n}$. More precisely, there are positive constants c_1, c_2 such that for all $k \in \mathbb{N}$,

$$c_1 \le k^{m/n} a_k(\mathrm{id}_{m,p}) \le c_2.$$

This leaves open the question as to whether or not

$$\lim_{k \to \infty} k^{m/n} a_k(\mathrm{id}_{m,p}) \text{ exists.}$$

Of course, the same question arises for other s–numbers of $\mathrm{id}_{m,p}$.

When $p = 2$ there is a positive answer to this question. For example, suppose that $m = 1$ and for simplicity write id instead of $\mathrm{id}_{1,2}$. Then (see, for example, Theorem II.5.10 of [54])

$$a_k(\mathrm{id}) = \lambda_k^{-1/2},$$

where λ_k is the kth eigenvalue of the Dirichlet Laplacian. The asymptotic behaviour of these eigenvalues is known [181]:

$$\lim_{k \to \infty} k^{-2/n} \lambda_k = 4\pi^2 \left(\omega_n |\Omega|\right)^{-2/n},$$

where ω_n is the volume of the unit ball in $\mathbb{R}^n$. Hence

$$\lim_{k \to \infty} k^{1/n} a_k(\mathrm{id}) = \frac{|\Omega|^{1/n}}{2\sqrt{\pi} \left(\Gamma\left(1 + n/2\right)\right)^{1/n}}.$$

Similar results hold for arbitrary values of $m \in \mathbb{N}$, always supposing that $p = 2$.

If $p \ne 2$, it appears that only when $n = 1$, $\Omega = (a, b) \subset \mathbb{R}$ and $m = 1$ or 2 has anything been established about the existence or otherwise of

$$\lim_{k \to \infty} k^{m/n} a_k(id_{m,p}).$$

In fact, when $m = n = 1$ and $\Omega = (a, b)$, the approximation numbers of $\mathrm{id}_{1,p}$ can be calculated precisely for all $p \in (1, \infty)$:

$$a_k\left(\mathrm{id}_{1,p}\right) = \gamma_p (b-a) k^{-1} \quad (k \in \mathbb{N}),$$

where

$$\gamma_p = \frac{1}{2\pi} \left(p'\right)^{1/p} p^{1/p'} \sin(\pi/p).$$

This was proved in [61] (see also [63]) by means of techniques involving generalised trigonometric functions. For the case $m = 2$ no formula is known for the individual approximation numbers, but in [65] it is shown, again supposing $\Omega = (a, b)$ and $p \in (1, \infty)$, that

$$\lim_{k \to \infty} k^2 a_k(\mathrm{id}_{2,p}) = C(b-a)^2,$$

where C depends only on p. The techniques used to obtain this result are more sophisticated than those when $m = 1$: they involve eigenvalues of the p-biharmonic operator, sharp estimates from above of the approximation numbers of $\mathrm{id}_{2,p}$ and sharp estimates from below of the corresponding Bernstein numbers. It seems possible that with greater effort similar results may be established when $m > 2$.

The outstanding open question is thus the following: does

$$\lim_{k \to \infty} k^{m/n} a_k(\mathrm{id}_{m,p}) \text{ exist when } p \neq 2 \text{ and } n > 1?$$

The same problem arises for other embeddings of Sobolev type and indeed for embeddings involving Besov and Lizorkin–Triebel spaces for which sharp upper and lower bounds are known for the approximation numbers. Of course it is tempting to conjecture that such limits really do exist. Indeed, the optimist may well hope to establish the existence not only of the limit but also of a second term in the asymptotic expansion of $a_k(\mathrm{id}_{m,p})$ for certain Ω, as is the case when $p = 2$. Those of a more melancholy disposition will surely point to the many examples in which the siren voice of a special case acts like Lorelei and leads to the mathematical rocks.

11.4 Spaces with Variable Exponent

These spaces result from the replacement of the exponent p in the Lebesgue spaces L_p by a function. A powerful reason for wishing to do this is that the resulting spaces are peculiarly adapted to problems stemming from certain physical situations (see, for example, [180]) and to variational problems involving integrands with non-standard growth (see [223]). Moreover, they are special cases of Musielak spaces (see

11.4 Spaces with Variable Exponent

[157]), which continue to enjoy a good deal of attention. The literature on spaces with variable exponent is now very large, but there are still genuinely interesting problems to be solved, and we hope that this short section will give the reader some idea of current developments and help to stimulate interest in the area. First some of the more important results that have been obtained are described: as such matters are well covered in the literature we simply give references rather than proofs. Given this body of facts, we draw attention to some topics that we believe are of especial interest. It is clearly important to know how replacement of the function p by approximants affects the outcome: in practical situations it may well be that p is not precisely known and that only approximations to it are available. Moreover, properties of Sobolev embedding maps, such as the behaviour of their approximation numbers, that are well known and useful in a classical context deserve to be known in the setting of spaces with variable exponent. We indicate some recent developments in these directions together with some suggestions for future study.

Let Ω be an open subset of $\mathbb{R}^n$ with positive Lebesgue n–measure $|\Omega|_n$ (written as $|\Omega|$ if the context is clear), let $\mathcal{M}(\Omega)$ be the family of all extended scalar-valued measurable functions on Ω and denote by $\mathcal{P}(\Omega)$ the subset of $\mathcal{M}(\Omega)$ consisting of all those functions p that map Ω to $[1, \infty)$. For each $p \in \mathcal{P}(\Omega)$ let

$$\Omega_0 := \{x \in \Omega : 1 < p(x) < \infty\}, \Omega_1 := \{x \in \Omega : p(x) = 1\}$$

and set

$$p_- := \operatorname*{ess\,inf}_{x \in \Omega_0} p(x), \; p_+ := \operatorname*{ess\,sup}_{x \in \Omega_0} p(x) \text{ if } |\Omega_0|_n > 0,$$

$$p_- := p_+ := 1 \text{ otherwise.}$$

The function p' conjugate to p is defined by

$$p'(x) = \begin{cases} p(x)/(p(x) - 1) & \text{if } x \in \Omega_0, \\ \infty & \text{if } x \in \Omega_1. \end{cases}$$

For every $f \in \mathcal{M}(\Omega)$ and $p \in \mathcal{P}(\Omega)$ define

$$\rho_p(f) = \rho_{p,\Omega}(f) = \int_\Omega |f(x)|^{p(x)} \, dx$$

and

$$\|f\|_{p,\Omega} = \inf \left\{ \lambda > 0 : \rho_p(f/\lambda) \le 1 \right\}.$$

By $L_p(\Omega)$ is meant the set of all f with $\|f\|_{p,\Omega} < \infty$. In the literature the notation $\|f\|_{p(\cdot),\Omega}$ and $L_{p(\cdot)}(\Omega)$ is often used to make it clear that the case of variable p is being considered, but here we adopt the simpler notation in the belief that it will be clear from the context what is intended. As in the classical case, we may write $\|f\|_p$ instead of $\|f\|_{p,\Omega}$ if no ambiguity is possible.

We now give some of the basic properties of $L_p(\Omega)$; proofs can be found in [62], [49] or [135].

Theorem 11.4.1 *Let $p \in \mathcal{P}(\Omega)$. Then*

(i) *$L_p(\Omega)$ is a Banach function space (see 1.3) that coincides with the classical Lebesgue space $L_p(\Omega)$ when p is constant in Ω.*

(ii) *Let $f \in L_p(\Omega)$. Then $\|f\|_{p,\Omega} \leq 1$ if and only if $\rho_p(f) \leq 1$; if $\|f\|_{p,\Omega} \leq 1$, then $\rho_p(f) \leq \|f\|_{p,\Omega}$.*

(iii) *Suppose that $1 < p_- \leq p_+ < \infty$. Then for all $f \in L_p(\Omega)$,*

$$\min\left\{\|f\|_{p,\Omega}^{p_-}, \|f\|_{p,\Omega}^{p_+}\right\} \leq \rho_p(f) \leq \max\left\{\|f\|_{p,\Omega}^{p_-}, \|f\|_{p,\Omega}^{p_+}\right\}.$$

(iv) *Let $(f_k)_{k\in\mathbb{N}}$ be a sequence in $L_p(\Omega)$ that converges to f in $L_p(\Omega)$. Then there is a subsequence of $(f_k)_{k\in\mathbb{N}}$ that converges pointwise a.e. in Ω to f.*

(v) *Suppose that $p_+ < \infty$. Then $\rho_p(f_k) \to 0$ if and only if $\|f_k\|_{p,\Omega} \to 0$; if $\|f_k\|_{p,\Omega} \to 0$ then $f_k \to 0$ in measure.*

(vi) *Suppose that $p_+ < \infty$ and let $f, f_k \in L_p(\Omega)$ $(k \in \mathbb{N})$. Then $f_k \to f$ in $L_p(\Omega)$ if and only if*

(f_k) converges to f in measure on Ω and $\rho_p(f_k) \to \rho_p(f)$.

(vii) *For all $f \in L_p(\Omega)$ and $g \in L_{p'}(\Omega)$,*

$$\int_\Omega |f(x)g(x)|\,dx \leq r_p \|f\|_{p,\Omega} \|g\|_{p',\Omega},$$

where

$$r_p = c_p + 1/p_- - 1/p_+ \text{ and } c_p = \|\chi_{\Omega_1}\|_{\infty,\Omega} + \|\chi_{\Omega_0}\|_{\infty,\Omega}.$$

(viii) *$L_p(\Omega)$ is separable if and only if $p_+ < \infty$, and this holds if and only if $L_p(\Omega)$ has absolutely continuous norm (see 1.3).*

(ix) *If $p_+ < \infty$, the dual of $L_p(\Omega)$ is isomorphic to $L_{p'}(\Omega)$.*

(x) *The following statements are equivalent:*

(a) *$L_p(\Omega)$ is reflexive;*
(b) *$L_p(\Omega)$ and $L_{p'}(\Omega)$ have absolutely continuous norms;*
(c) *$L_p(\Omega)$ is uniformly convex;*
(d) *$1 < p_- \leq p_+ < \infty$.*

These properties give a distinct impression that the spaces $L_p(\Omega)$ are not so very different from their classical counterparts, especially if $1 < p_- \leq p_+ < \infty$. However, there are various flies in this soothing ointment. The first arises from the notion of p−mean continuity. When p is a constant in the interval $[1, \infty)$ and Ω is a bounded open subset of $\mathbb{R}^n$, it is a familiar fact that every function $f \in L_p(\Omega)$,

11.4 Spaces with Variable Exponent

extended by 0 outside Ω, is p—mean continuous in the sense that given any $\varepsilon > 0$, there exists $\delta > 0$ such that

$$\left(\int_\Omega |f(x+h) - f(x)|^p\right)^{1/p} < \varepsilon \text{ if } |h| < \delta.$$

By analogy with this, when $p \in \mathcal{P}(\Omega)$, a function $f \in L_{p(\cdot)}(\Omega)$ is said to be p—mean continuous if $\rho_p(f - \tau_h f) < \varepsilon$ whenever $|h| < \delta$. Here $\tau_h f := f(\cdot + h)$. In general, elements of $L_{p(\cdot)}(\Omega)$ are not p—mean continuous: it is shown in [137] that if there is a ball contained in Ω on which p is continuous and non-constant, then there is a function $f \in L_p(\Omega)$ that is not p—mean continuous. Other difficulties arise in connection with convolutions (Young's inequality does not hold, in general) and, most importantly, with the Hardy-Littlewood maximal operator M, defined for all $f \in L_{1,loc}(\Omega)$, by

$$(Mf)(x) = \sup |B|^{-1} \int_{B \cap \Omega} |f(y)|\,dy \ (x \in \Omega),$$

where the supremum is taken over all balls B that contain x and for which $|B \cap \Omega| > 0$. When p is a constant in the interval $(1, \infty)$, M maps $L_p(\Omega)$ boundedly to itself. This is not so for all $p \in \mathcal{P}(\Omega)$: see [49]. All is well when a suitable additional condition is imposed on p. Thus let Ω be bounded and suppose that $p \in \mathcal{P}(\Omega)$ has the property that there is a constant C such that for all $x, y \in \Omega$ with $0 < |x - y| < 1/2$,

$$|p(x) - p(y)| \log \frac{1}{|x-y|} \le C.$$

The family of all such $p \in \mathcal{P}(\Omega)$ is denoted by $\mathcal{P}_l(\Omega)$.

Theorem 11.4.2 *Suppose that Ω is bounded and that $p \in \mathcal{P}_l(\Omega)$. Then there is a constant $c = c(p, \Omega)$ such that for all $f \in L_p(\Omega)$,*

$$\|Mf\|_{p,\Omega} \le c \|f\|_{p,\Omega}.$$

Originally due to Diening, this result is thoroughly discussed in [49]. We now turn to the ordering between different variable exponent spaces, always supposing that Ω is bounded. When p and q are constant and belong to $(1, \infty)$, it is clear that $L_q(\Omega) \hookrightarrow L_p(\Omega)$ if $p \le q$. A similar result holds when $p, q \in \mathcal{P}(\Omega)$: in fact, $L_q(\Omega) \hookrightarrow L_p(\Omega)$ if $p(x) \le q(x)$ for a.e. $x \in \Omega$, and the corresponding embedding map id satisfies

$$\|id\| \le 1 + |\Omega|.$$

This was proved in [135], where it is also shown that for the embedding $L_q(\Omega) \hookrightarrow L_p(\Omega)$ to exist it is necessary that $p(x) \le q(x)$ for a.e. $x \in \Omega$. For various purposes it is desirable to have sharper estimates for $\|id\|$ when p and q are close together, and we next address this matter, following [66].

Lemma 11.4.3 *Let $\varepsilon \in [0, 1/2]$ and suppose that $p, q \in \mathcal{P}(\Omega)$ with $1 \leq p(x) \leq q(x) \leq p(x)/(1-\varepsilon)$ for all $x \in \Omega$. If $\rho_q(f) \leq 1$, then*

$$\rho_p(f) \leq 1 + |\Omega| K\varepsilon, \text{ where } K = \sup_{0 < \alpha \leq 1} \alpha^{1/2} |\log \alpha|. \qquad (11.4.1)$$

Proof. Suppose $\rho_q(f) \leq 1$ and set

$$\Omega_1 = \{x \in \Omega : |f(x)| \leq 1\}, \Omega_2 = \{x \in \Omega : |f(x)| > 1\}.$$

Then

$$\rho_p(f) = \rho_{p,\Omega_1}(f) + \rho_{p,\Omega_2}(f)$$
$$\leq \int_{\Omega_1} |f(x)|^{(1-\varepsilon)q(x)} dx + \int_{\Omega_2} |f(x)|^{q(x)} dx := I_1 + I_2.$$

Evidently

$$I_1 \leq \rho_{q,\Omega_1}^{1-\varepsilon}(f) |\Omega_1|^\varepsilon \text{ and } I_2 \leq 1 - \rho_{q,\Omega_1}(f).$$

Assume that $|\Omega_1| > 0$ and put $\alpha = |\Omega_1|^{-1} \rho_{q,\Omega_1}(f)$. Then

$$\rho_p(f) \leq 1 + |\Omega_1| \left\{ \left(\frac{1}{|\Omega_1|} \int_{\Omega_1} |f(x)|^{q(x)} dx \right)^{1-\varepsilon} - \frac{1}{|\Omega_1|} \int_{\Omega_1} |f(x)|^{q(x)} dx \right\}$$
$$\leq 1 + |\Omega| \left(\alpha^{1-\varepsilon} - \alpha \right).$$

The desired result now follows from the inequality $\alpha^{1-\varepsilon} - \alpha \leq K\varepsilon$ that is established by an elementary mean value argument. If $|\Omega_1| = 0$ the result is obvious, for then $\rho_p(f) \leq I_2 \leq 1$. ∎

Proposition 11.4.4 *Let $\varepsilon \in [0, 1/2]$ and suppose that $p, q \in \mathcal{P}(\Omega)$ with $1 \leq p(x) \leq q(x) \leq p(x) + \varepsilon$ for all $x \in \Omega$. Then $L_q(\Omega) \hookrightarrow L_p(\Omega)$ and the norm of the corresponding embedding map id satisfies*

$$\|id\| \leq 1 + |\Omega| K\varepsilon,$$

where K is given by (11.4.1).

Proof. Let $\delta = \varepsilon/(1+\varepsilon)$; note that $0 \leq \delta \leq \varepsilon \leq 1/2$ and

$$1 \leq p(x) \leq q(x) \leq p(x) + \delta/(1+\delta) \leq p(x)\left(1 + \frac{\delta}{1-\delta}\right) = \frac{p(x)}{1-\delta} \quad (x \in \Omega).$$

Suppose that $\|f\|_q \leq 1$. Then $\rho_q(f) \leq 1$, and hence, by Lemma 11.4.3, $\rho_p(f) \leq 1 + K\delta |\Omega|$. Hence

11.4 Spaces with Variable Exponent

$$\int_\Omega \left(\frac{|f(x)|}{(1+K\delta|\Omega|)^{1/p_-}}\right)^{p(x)} dx \leq \int_\Omega \left(\frac{|f(x)|}{(1+K\delta|\Omega|)^{1/p(x)}}\right)^{p(x)} dx$$
$$= \int_\Omega \frac{|f(x)|^{p(x)}}{1+K\delta|\Omega|} dx \leq 1.$$

Thus
$$\|f\|_p \leq (1+K\delta|\Omega|)^{1/p_-} \leq 1+K\delta|\Omega| \leq 1+|\Omega|K\varepsilon.$$

∎

As for an estimate of $\|id\|$ from below, we have the following:

Proposition 11.4.5 *Let $\varepsilon \geq 0$ and suppose that $p, q \in \mathcal{P}(\Omega)$ with $1 \leq p(x) \leq q(x) \leq p(x) + \varepsilon < p(x) + \varepsilon_0$ for all $x \in \Omega$, where*

$$\varepsilon_0 = \begin{cases} 1 & \text{if } |\Omega| \geq 1, \\ \frac{1}{\log(1/|\Omega|)} & \text{if } 0 < |\Omega| < 1. \end{cases}$$

Then there exists $L \in [0, 1/\varepsilon_0]$ such that for the embedding id of $L_q(\Omega)$ in $L_p(\Omega)$ we have

$$\|id\| \geq 1 - \varepsilon L.$$

Proof. Let f be the function defined by $f(x) = |\Omega|^{-1/q(x)}$ $(x \in \Omega)$. Since $\rho_q(f) = 1$, it follows that $\|f\|_q = 1$. If $|\Omega| \geq 1$,

$$\rho_p(f) = \int_\Omega |\Omega|^{-p(x)/q(x)} dx \geq \int_\Omega |\Omega|^{-1} dx = 1,$$

so that $\|f\|_p \geq 1$ and hence $\|id\| \geq 1$.

If $|\Omega| < 1$,

$$\rho_p(f) = \int_\Omega |\Omega|^{-p(x)/q(x)} dx \geq \int_\Omega |\Omega|^{\varepsilon/q(x)-1} dx \geq \int_\Omega |\Omega|^{\varepsilon-1} dx = |\Omega|^\varepsilon.$$

Let $\lambda \in \left(0, |\Omega|^{\varepsilon/p_-}\right)$. Then

$$\int_\Omega \left(\frac{|f(x)|}{\lambda}\right)^{p(x)} dx \geq \int_\Omega \left(\frac{|f(x)|}{|\Omega|^{\varepsilon/p_-}}\right)^{p(x)} dx \geq \int_\Omega \left(\frac{|f(x)|}{|\Omega|^{\varepsilon/p(x)}}\right)^{p(x)} dx \geq 1,$$

and so
$$\|f\|_p \geq |\Omega|^{\varepsilon/p_-} \geq |\Omega|^\varepsilon.$$

Thus $\|id\| \geq |\Omega|^\varepsilon \geq 1 - \varepsilon \log(1/|\Omega|)$ and the result follows. ∎

As an immediate consequence of these last two Propositions we have

Corollary 11.4.6 *Let $p \in \mathcal{P}(\Omega)$ and suppose that for each $k \in \mathbb{N}$, $q_k \in \mathcal{P}(\Omega)$ and $\varepsilon_k > 0$, where $\lim_{k \to \infty} \varepsilon_k = 0$. Assume further that for all $k \in \mathbb{N}$ and all $x \in \Omega$,*

$$p(x) \le q_k(x) \le p(x) + \varepsilon_k.$$

Denote by id_k the natural embedding of $L_{q_k}(\Omega)$ in $L_p(\Omega)$. Then

$$\lim_{k \to \infty} \|id_k\| = 1.$$

Sobolev spaces based on Lebesgue spaces with variable exponent are defined in the obvious way. For our purposes it is sufficient to deal with first-order spaces. Thus given $p \in \mathcal{P}(\Omega)$, the associated Sobolev space $W_p^1(\Omega)$ is defined to be the set of all $u \in L_p(\Omega)$ such that each distributional derivative $D_j u$ ($j = 1, ..., n$) also belongs to $L_p(\Omega)$; this space is endowed with the norm given by

$$\|u\|_{1,p} := \|u\|_p + \||\nabla u|\|_p.$$

The closure of $C_0^\infty(\Omega)$ in $W_p^1(\Omega)$ is denoted by $\overset{0}{W}{}_p^1(\Omega)$. Basic properties of these spaces are summarised in the next theorem; for proofs, see [62], [49] and [135].

Theorem 11.4.7 *Let $p \in \mathcal{P}(\Omega)$. Then:*

(i) $W_p^1(\Omega)$ is a Banach space; if $p_+ < \infty$, the space is separable; if in addition $p_- > 1$, it is reflexive.

(ii) If Ω is bounded with Lipschitz boundary and $p \in \mathcal{P}_l(\Omega)$ with $1 < p_- \le p_+ < \infty$, then both $C^\infty(\overline{\Omega})$ and $C^\infty(\Omega) \cap W_p^1(\Omega)$ are dense in $W_p^1(\Omega)$.

(iii) Suppose Ω is bounded and $p \in \mathcal{P}_l(\Omega)$. Then there is a constant c such that for all $u \in W_p^1(\Omega)$,

$$\left\| u - \frac{1}{|\Omega|} \int_\Omega u(x) dx \right\|_p \le c \||\nabla u|\|_p.$$

Moreover, there is a constant C such that for all $u \in \overset{0}{W}{}_p^1(\Omega)$,

$$\|u\|_p \le C \||\nabla u|\|_p.$$

(iv) Suppose that Ω is bounded and $p \in \mathcal{P}_l(\Omega)$, with $p_+ < \infty$. Then $\overset{0}{W}{}_p^1(\Omega)$ is compactly embedded in $L_p(\Omega)$. The same holds if instead of the condition $p \in \mathcal{P}_l(\Omega)$ it is assumed that $p \in C(\overline{\Omega})$, $p_- > 1$ and $n > 1$.

(v) Let Ω be bounded and suppose that $q \in \mathcal{P}(\Omega)$, with $p(x) \le q(x)$ a.e. in Ω. Then $W_q^1(\Omega)$ is continuously embedded in $W_p^1(\Omega)$, and the norm of the embedding map is bounded above by $1 + |\Omega|$. The same holds for the embedding of $\overset{0}{W}{}_q^1(\Omega)$ in $\overset{0}{W}{}_p^1(\Omega)$.

11.4 Spaces with Variable Exponent

When Ω is bounded and p is constant, the embedding id of $\overset{0}{W}{}^1_p(\Omega)$ in $L_p(\Omega)$ is compact. The behaviour of the approximation numbers of id is known: $a_k(id) \asymp k^{-1/n}$. The literature for this result and corresponding estimates for more general embeddings between function spaces is now quite large: see [69] and [206] in particular. However, nothing has been available when spaces of variable exponent are involved until quite recently. Here we outline the work of [66] in which similar estimates are obtained when $p \in \mathcal{P}_l(\Omega)$. The idea is to use a covering of $\overline{\Omega}$ by cubes together with accurate local estimates. Throughout it is supposed that Ω is bounded.

Given any $S \subset \mathbb{R}^n$ with positive, finite Lebesgue $n-$ measure and any measurable function u on S we write

$$u_S = \frac{1}{|S|} \int_S u(x) dx.$$

Lemma 11.4.8 *Let $Q \subset R^n$ be a cube and suppose that p, q are constants such that*

$$1 \leq p \leq q < \infty \text{ and } h := 1/n - (1/p - 1/q) > 0.$$

Then

(i) there is a constant c such that for all $u \in W^1_p(Q)$,

$$\|u - u_Q\|_{q,Q} \leq c \left(2^{-n} |Q|\right)^h \|\nabla u\|_{p,Q};$$

(ii) with $\sigma_n := 2\pi^{n/2}/\Gamma(n/2)$,

$$\sup \frac{\|u\|_{q,Q}}{\|\nabla u\|_{p,Q}} \geq 2^{-2-2n/q+n/p} \sigma_n^{1/q-1/p} n^{1/p-1/q} |Q|^h,$$

where the supremum is taken over all $u \in \overset{0}{W}{}^1_p(Q) \setminus \{0\}$.

Proof. (i) When Q is the unit cube this follows immediately from Theorem V.3.24 of [54]; now use scaling.

(ii) Let Q have centre x_0 and side length $2s$; let $B = \{x \in R^n : |x - x_0| < s\}$ and define a function u by

$$u(x) = \max\{s - |x - x_0|, 0\} \quad (x \in Q).$$

Then $u \in \overset{0}{W}{}^1_p(Q)$ since it is Lipschitz-continuous and is zero on ∂Q; and $|\nabla u(x)| = \chi_B(x)$. Also, $u(x) \geq s/2$ if $|x - x_0| < s/2$. It now follows easily that

$$\|u\|_{q,Q} \geq 2^{-1-n/q} (\sigma_n/n)^{1/q} s^{1+n/q} \text{ and } \|\nabla u\|_{p,Q} = (\sigma_n/n)^{1/p} s^{n/p}.$$

Now use the fact that $s = |Q|^{1/n}/2$. ∎

Lemma 11.4.9 *There is a constant $A_\Omega \geq 1$ such that given any $k \in \mathbb{N}$, there is a covering of $\overline{\Omega}$ by a collection $\{Q_i\}_{i=1}^m$ of non-overlapping congruent cubes with*

$$k \leq \sharp\{j : Q_j \subset \Omega\} \leq m \leq A_\Omega k.$$

Details of the proof are given in [66].

Next we note that the condition of log-Hölder continuity imposed on p in the definition of $\mathcal{P}_l(\Omega)$ may be written as

$$|p(x) - p(y)| \leq C\omega(|x - y|),$$

where ω is the modulus of continuity defined by

$$\omega(t) = \begin{cases} 0, & t = 0, \\ \left(\log \frac{\sqrt{n}}{t}\right)^{-1}, & t \in (0, \sqrt{n}e^{-2}), \\ \frac{1}{2}, & t \in [\sqrt{n}e^{-2}, \infty). \end{cases}$$

After this preparation we can give the promised result concerning approximation numbers.

Theorem 11.4.10 *Let Ω be bounded and $p \in \mathcal{P}_l(\Omega)$; let $id : \overset{0}{W}{}^1_p(\Omega) \to L_p(\Omega)$ be the natural embedding. Then there are positive constants c_1, c_2 such that for all $k \in \mathbb{N}$,*

$$c_1 k^{-1/n} \leq a_k(id) \leq c_2 k^{-1/n}.$$

Proof. Let Q be a cube containing Ω. The function p may be extended to Q with preservation of its modulus of continuity ω (see, for example, Lemma 2.2 of [66]); denote this extension again by p. Given $u \in \overset{0}{W}{}^1_p(\Omega)$, extend it by zero outside Ω, denote this extension again by u and observe that $u \in \overset{0}{W}{}^1_p(Q)$. Now fix $k \in \mathbb{N}$: by Lemma 11.4.9, there is a covering of $\overline{\Omega}$ by cubes $\{Q_i\}_{i=1}^N$ with $k \leq \sharp\{i : Q_i \subset \Omega\} \leq A_\Omega k$ and

$$k \leq N \leq A_\Omega k, \, k|Q_i| \leq |\overline{\Omega}| \leq N|Q_i| \leq A_\Omega k|Q_i|$$

for each i.

Define an operator P_N of rank N on $\overset{0}{W}{}^1_p(\Omega)$ by

$$P_N u(x) = \sum_{i=1}^N u_{Q_i} \chi_{Q_i}(x);$$

for every i let

$$p_i^+ = \sup_{x \in Q_i} p(x), \quad p_i^- = \inf_{x \in Q_i} p(x)$$

11.4 Spaces with Variable Exponent

and set
$$p^+(x) = \sum_{i=1}^{N} p_i^+ \chi_{Q_i}(x), \quad p^-(x) = \sum_{i=1}^{N} p_i^- \chi_{Q_i}(x).$$

Put
$$M = \max_i \sup_{u \in \overset{0}{W}{}^1_p(\Omega)\setminus\{0\}} \frac{\|u - u_{Q_i}\|_{p_i^+, Q_i}}{\||\nabla u|\|_{p_i^-, Q_i}} \quad \text{and} \quad \lambda_0 = M \||\nabla u|\|_{p^-, \Omega}.$$

Then
$$1 = \int_\Omega \left|\frac{\nabla u(x)}{\lambda_0/M}\right|^{p^-(x)} dx = \sum_{i=1}^{N} \int_{Q_i} \left|\frac{\nabla u(x)}{\lambda_0/M}\right|^{p_i^-} dx$$
$$\geq \sum_{i=1}^{N} \left(\int_{Q_i} \left|\frac{\nabla u(x)}{\lambda_0/M}\right|^{p_i^-} dx\right)^{p_i^+/p_i^-}$$
$$= \sum_{i=1}^{N} \lambda_0^{-p_i^+} \left\{ M \left(\int_{Q_i} |\nabla u(x)|^{p_i^-} dx\right)^{1/p_i^-} \right\}^{p_i^+}$$
$$\geq \sum_{i=1}^{N} \lambda_0^{-p_i^+} \left\{ \left(\int_{Q_i} |u(x) - u_{Q_i}|^{p_i^+} dx\right)^{1/p_i^+} \right\}^{p_i^+}$$
$$= \sum_{i=1}^{N} \int_{Q_i} \left|\frac{u(x) - u_{Q_i}}{\lambda_0}\right|^{p_i^+} dx$$
$$= \sum_{i=1}^{N} \int_\Omega \left|\frac{u(x) - u_{Q_i}}{\lambda_0}\right|^{p_i^+} \chi_{Q_i}(x) dx = \int_\Omega \left|\frac{u(x) - P_N u}{\lambda_0}\right|^{p^+} dx.$$

Hence
$$\|u - P_N u\|_{p^+, \Omega} \leq \lambda_0 = M \||\nabla u|\|_{p^-, \Omega}.$$

Thus for the natural embedding $id_{-,+}$ of $\overset{0}{W}{}^1_{p^-}(\Omega)$ in $L_{p^+}(\Omega)$ we have
$$a_{N+1}(id_{-,+}) \leq \sup\{\|u - P_N u\|_{p^+, \Omega} : \||\nabla u|\|_{p^-, \Omega} \leq 1\} \leq M.$$

By Lemma 11.4.8 (i) we see that for each i,
$$\sup_{u \neq 0} \frac{\|u - u_{Q_i}\|_{p_i^+, Q_i}}{\||\nabla u|\|_{p_i^-, Q_i}} \leq \tilde{c} \left(\frac{|\Omega|}{k}\right)^{1/n} |Q_i|^{-(1/p_i^- - 1/p_i^+)}.$$

For a given i, fix $\tilde{x}, \tilde{y} \in Q_i$ with $p_i^+ = p(\tilde{x})$ and $p_i^- = p(\tilde{y})$, and observe that since $p \in \mathcal{P}_l(\Omega)$,

$$-(1/p_i^- - 1/p_i^+)\log|Q_i| \leq (p_i^+ - p_i^-)\log(1/|Q_i|)$$
$$\leq (p(\tilde{x}) - p(\tilde{y}))\log\left(\frac{\sqrt{n}}{|\tilde{x}-\tilde{y}|}\right)^n \leq nC.$$

Hence
$$|Q_i|^{-(1/p_i^- - 1/p_i^+)} = \exp\left(-(1/p_i^- - 1/p_i^+)\right)\log|Q_i| \leq \exp(nC),$$

and so there is a constant $L > 0$ such that
$$\sup_{u \in \overset{0}{W}{}^1_p(\Omega)\setminus\{0\}} \frac{\|u - u_{Q_i}\|_{p_i^+, Q_i}}{\|\nabla u\|_{p_i^-, Q_i}} \leq L\left(\frac{|\Omega|}{k}\right)^{1/n},$$

which implies that $M \leq L\left(\frac{|\Omega|}{k}\right)^{1/n}$. It follows that
$$a_{N+1}(id_{-,+}) \leq L\left(\frac{|\Omega|}{k}\right)^{1/n}.$$

To obtain the desired result concerning $id : \overset{0}{W}{}^1_p(\Omega) \to L_p(\Omega)$ we note that $id = id_2 \circ id_{-,+} \circ id_1$, where $id_1 : \overset{0}{W}{}^1_p(\Omega) \to \overset{0}{W}{}^1_{p^-}(\Omega)$ and $id_2 : L_{p^+}(\Omega) \to L_p(\Omega)$ are the natural embeddings. Given any $x \in \Omega$, there exists i such that $x \in Q_i$, and since $|p^+(x) - p^-(x)|\log(1/|Q_i|) \leq c$ we have
$$|p^+(x) - p^-(x)| \leq \frac{c}{\log(N/|\Omega|)}.$$

By Lemma 11.4.3, there is a positive constant R such that
$$\|id_1\|, \|id_2\| \leq 1 + R|\Omega|c/\log(N/|\Omega|) := \widetilde{R}.$$

Thus
$$a_{N+1}(id) \leq \|id_1\| a_{N+1}(id_{-,+}) \|id_2\| \leq \widetilde{R}^2 L\left(\frac{|\Omega|}{k}\right)^{1/n},$$

and so
$$a_{kA_\Omega+1}(id) \leq a_{N+1}(id) \leq \widetilde{R}^2 L\left(\frac{|\Omega|}{k}\right)^{1/n} \leq 2\widetilde{R}^2 L\left(\frac{A_\Omega|\Omega|}{A_\Omega k + 1}\right)^{1/n},$$

from which the upper estimate of the theorem follows.

11.4 Spaces with Variable Exponent

For the lower estimate our strategy is to obtain a lower bound for the Bernstein numbers of id: we recall that they are defined in 11.2 and that they are dominated by the approximation numbers. To do this consider

$$\mu := \min_i \sup \frac{\|u\|_{p_i^-, Q_i}}{\||\nabla u|\|_{p_i^+, Q_i}},$$

where the minimum is taken over all $i \in \{j : Q_j \subset \Omega\}$ and the supremum is taken over all non-zero elements of $\overset{0}{W}{}^1_{p_i^+}(Q_i)$. Set

$$2\lambda_0 = \mu \||\nabla u|\|_{p^+, \Omega}.$$

For each $i \in \{j : Q_j \subset \Omega\}$ let $u_i \in \overset{0}{W}{}^1_{p_i^+}(Q_i)$ be such that

$$\frac{\|u\|_{p_i^-, Q_i}}{\||\nabla u|\|_{p_i^+, Q_i}} \geq \mu/2,$$

and denote by X_μ the linear span of these u_i: evidently X_μ is a linear subspace of $\overset{0}{W}{}^1_{p^+}(\Omega)$ and $k \leq \dim X_\mu \leq A_\Omega k$. Let $u \in X_\mu$, $u = \sum_{i \in \{j: Q_j \subset \Omega\}} \alpha_i u_i$, say. Then

$$1 = \int_\Omega \left| \frac{\nabla u(x)}{2\lambda_0/\mu} \right|^{p^+(x)} dx = \sum_{i \in \{j:Q_j \subset \Omega\}} \int_{Q_i} \left| \frac{\alpha_i \nabla u_i(x)}{2\lambda_0/\mu} \right|^{p_i^+} dx$$

$$\leq \sum_{i \in \{j:Q_j \subset \Omega\}} \left(\int_{Q_i} \left| \frac{\alpha_i \nabla u_i(x)}{2\lambda_0/\mu} \right|^{p_i^+} dx \right)^{p_i^-/p_i^+}$$

$$= \sum_{i \in \{j:Q_j \subset \Omega\}} \lambda_0^{-p_i^-} \left\{ \frac{\mu}{2} \left(\int_{Q_i} |\alpha_i \nabla u_i(x)|^{p_i^+} dx \right)^{1/p_i^+} \right\}^{p_i^-}$$

$$\leq \sum_{i \in \{j:Q_j \subset \Omega\}} \lambda_0^{-p_i^-} \left\{ \left(\int_{Q_i} |\alpha_i u_i(x)|^{p_i^-} dx \right)^{1/p_i^-} \right\}^{p_i^-}$$

$$= \sum_{i \in \{j:Q_j \subset \Omega\}} \int_{Q_i} \left| \frac{\alpha_i u_i(x)}{\lambda_0} \right|^{p_i^-} dx$$

$$= \int_\Omega \left| \frac{u(x)}{\lambda_0} \right|^{p^-(x)} dx.$$

It follows that for the embedding $id_{+,-}$ of $\overset{0}{W}{}^1_{p^+}(\Omega)$ in $L_{p^-}(\Omega)$ we have

$$b_k(id_{+,-}) \geq \mu/2.$$

Using Lemma 11.4.8 (ii) we see that

$$\sup \frac{\|u\|_{p_i^-, Q_i}}{\|\nabla u\|_{p_i^+, Q_i}} \geq D := 2^{-2-2n/p^- + n/p^+} \min(1, \sigma_n)^{1/p^- - 1/p^+} n^{1/p^+ - 1/p^-},$$

and so for each i,

$$\mu \geq D |Q_i|^{1/n - (1/p_i^+ - 1/p_i^-)} = D |Q_i|^{1/n} |Q_i|^{-(1/p_i^+ - 1/p_i^-)}.$$

Since

$$|Q_i|^{1/p_i^+ - 1/p_i^-} \leq \exp(nC)$$

we have, for each $i \in \{j : Q_j \subset \Omega\}$,

$$\mu/2 \geq D |Q_i|^{1/n} /2 \exp(-nC) := B |Q_i|^{1/n}.$$

Thus

$$b_k(id_{+,-}) \geq B |Q_i|^{1/n}$$

for each $i \in \{j : Q_j \subset \Omega\}$. With obvious notation,

$$\overset{0}{W}/{}^1_{p^+}(\Omega) \overset{id_3}{\hookrightarrow} \overset{0}{W}/{}^1_{p}(\Omega) \overset{id}{\hookrightarrow} L_p(\Omega) \hookrightarrow L_{p^-}(\Omega).$$

As before, there is a positive constant $\widetilde{S}$ such that $\|id_3\|, \|id_4\| \leq \widetilde{S}$. Hence

$$B \left(\frac{|\Omega|}{A_\Omega k} \right)^{1/n} \leq B |Q_i|^{1/n} \leq b_k(id_{+,-}) \leq \|id_3\| b_k(id) \|id_4\| \leq \widetilde{S}^2 b_k(id),$$

and so $b_k(id) \geq Ck^{-1/n}$, as claimed. Because $b_k(id) \leq a_k(id)$ the result follows. ∎

Sharp upper and lower estimates are known for the approximation numbers of compact embeddings between Sobolev spaces of arbitrary order modelled on classical Lebesgue spaces: indeed the same is true for compact embeddings between spaces that are of Besov or Lizorkin–Triebel type. Much remains to be done to acheive anything like comparability when spaces of variable exponent are involved. Even the embedding of $\overset{0}{W}{}^1_p(\Omega)$ in $L_q(\Omega)$ when $p, q \in \mathcal{P}_l(\Omega)$ and $p \neq q$ presents a challenge, for it seems likely that the techniques used in the last theorem will simply give upper and lower bounds with different rates of decay. Real progress in classical settings was made possible by the use of more sophisticated techniques involving wavelets, atomic decompositions, etc. Perhaps a similar revolution is needed to deal effectively with the variable exponent situation.

11.5 Notes

1. Extension of the results of 11.1 to higher-order Sobolev spaces is carried out in [68]. There it is shown that if $m \in \mathbb{N}$, $p \in (1, \infty)$ and Ω is a bounded, open, regular subset of $\mathbb{R}^n$, then $u \in \overset{0}{W}{}_p^m(\Omega)$ if and only if all distributional derivatives of u of order m belong to $L_p(\Omega)$ and $u/\delta \in L_1(\Omega)$. In fact what is proved is the analogous result when the Sobolev space is based on a member of a class of Banach function spaces that includes both $L_p(\Omega)$ and $L_{q(\cdot)}(\Omega)$, the Lebesgue space with variable exponent q that satisfies a log-Hölder condition.

2. Operators of Hardy type occur naturally in connection with embeddings of Sobolev spaces when the underlying space domain Ω is a generalised ridged domain in the sense of Evans and Harris ([73]; see also [55]). These operators act between Lebesgue spaces based on trees arising from the generalised ridge. The simplest case arises when the generalised ridge is a bounded interval $(a, b) = I \subset \mathbb{R}$; then the operator of Hardy type is $T : L_p(I) \to L_q(I)$, where $p, q \in [1, \infty]$ and

$$Tf(x) = v(x) \int_a^x u(t) f(t) dt,$$

u and v being real-valued functions determined by the geometry of Ω, with $u \in L_{p'}(a, X)$ and $v \in L_q(X, b)$ for all $X \in I$. For a detailed study of such maps T from the standpoint of approximation numbers (and other s–numbers), see [55] and [63], together with the references given in these books: in certain cases it is possible to obtain asymptotic formulae for the approximation numbers of T, even with remainder terms. When $u = v = 1$ and $p = q$, the map T is not nuclear (see [64]). For information about certain operators $S_m : L_p(I) \to L_p(I)$ of Volterra type given by

$$S_m f(x) = \int_a^x (x-t)^{m-1} f(t) dt,$$

where $m \in \mathbb{N}$, see [63] and [65], where in particular it is shown that if $1 < p < \infty$, then

$$a_k(S_1) = \frac{\gamma_p |I|}{k - 1/2} \quad (k \in \mathbb{N})$$

and

$$\lim_{k \to \infty} k^2 a_k(S_2) = C(p) |I|^2.$$

3. The spaces with variable exponent discussed in 11.4 can help to illuminate classical questions. For example, it is well known that $L_1(\mathbb{T})$ (where $\mathbb{T}$ is the one-dimensional torus) differs from every $L_p(\mathbb{T})$ with $p > 1$ in that it contains a function with Fourier series that is almost everywhere divergent. The first to show this was Kolmogorov: the function he constructed belongs to $L \log \log L(\mathbb{T})$, a space that is slightly smaller than $L_1(\mathbb{T})$, and has partial Fourier sums that diverge unboundedly

a.e. Later, Marcinkiewicz gave an example of a function in $L_1(\mathbb{T})$ with a.e. divergent Fourier series and bounded partial sums. Thus from this point of view there is a big gulf between $L_1(\mathbb{T})$ and $\cup_{p>1} L_p(\mathbb{T})$. The position is quite different if we consider Lebesgue spaces with variable exponent, for then

$$L_1(\mathbb{T}) = \cup L_p(\mathbb{T}),$$

where the union is taken over all measurable $p : \mathbb{T} \to [1, \infty)$ with $p(x) > 1$ a.e. Moreover, there is a variable exponent space $L_p(\mathbb{T})$, with $1 < p(x) < \infty$ a.e., which has in common with $L_\infty(\mathbb{T})$ the property that the space $C(\mathbb{T})$ of continuous functions on $\mathbb{T}$ is a closed linear subspace in it, while both the Kolmogorov and Marcinkiewicz functions belong to its dual $L_q(\mathbb{T})$, where $1/q(x) = 1 - 1/p(x)$. For details of these assertions we refer to [58].

Chapter 12
The Dirac Operator

12.1 Preamble

The "free" Dirac expression $\mathbb{D}'$ is the Dirac expression shorn of its electric and magnetic potentials, and in appropriate units, is such that $\mathbb{D}'^2 = (-\Delta + 1)I_4$, where I_4 is the identity on $\mathbb{C}^4$. For any domain $\Omega \subseteq \mathbb{R}^3$, $\mathbb{D}'$ defines a symmetric operator $\mathbb{D}$ on $C_0^\infty(\Omega, \mathbb{C}^4)$ in $L_2(\Omega, \mathbb{C}^4)$ which is unbounded below and above, ands so the KVB theory of Chapter 6 does not apply to determine its self-adjoint extensions. Various approximations have been investigated to model the relativistic properties that the Dirac operator was designed to describe, and which define symmetric operators which are strictly positive. The two considered in this chapter have free forms $\mathbb{B}$ and $\mathbb{H}$, where $\mathbb{B}$ is the Brown–Ravenhall operator, and $\mathbb{H} = \sqrt{-\Delta + 1}$. The Brown–Ravenhall operator is defined in terms of the projection onto the positive spectral subspace of $\mathbb{D}$, and is strictly positive, as is $\mathbb{H}$; details will follow in section 12.3 below.

Sections 12.1–12.4 will catalogue results established in the last twenty years or so, on self-adjoint realisations of perturbations of $\mathbb{D}$, $\mathbb{B}$ and $\mathbb{H}$ by Coulomb potentials $-\gamma/|\mathbf{x}|$, when $\Omega = \mathbb{R}^3$. In section 12.5, the self-adjoint realisations of the free operators are discussed in the case $\Omega \subsetneq \mathbb{R}^3$.

12.2 The Dirac Equation

The time-independent Dirac equation describing the motion of an electron or positron in $\mathbb{R}^3$ under the influence of an electrical potential V and a magnetic potential $\mathbf{A} = (A_1, A_2, A_3)$ is given by

$$\left\{ \sum_{j=1}^3 \alpha_j \left(\frac{\hbar}{i} D_j + A_j \right) + mc^2 \beta + V \right\} u = \lambda u, \quad (12.2.1)$$

where $2\pi\hbar$ is Planck's constant, c the velocity of light, m the mass of the particle, $\alpha_1, \alpha_2, \alpha_3, \beta$ are Hermitian 4×4 matrices which satisfy

$$\alpha_j \alpha_k + \alpha_k \alpha_j = \delta_{jk} I_4 \ (j, k = 1, 2, 3, 4), \quad \alpha_4 = \beta, \tag{12.2.2}$$

and $\lambda \in \mathbb{C}$ is a parameter which describes the energy; in (12.2.2), δ_{jk} is the Kronecker delta. In the standard representation, the *Dirac matrices* α_j, $j = 1, 2, 3, 4$ are given by

$$\alpha_j = \begin{pmatrix} 0_2 & \sigma_j \\ \sigma_j & 0_2 \end{pmatrix} \ (j = 1, 2, 3), \quad \alpha_4 = \beta = \begin{pmatrix} I_2 & 0_2 \\ 0_2 & -I_2 \end{pmatrix}, \tag{12.2.3}$$

where $\sigma_1, \sigma_2, \sigma_3$ are the *Pauli matrices*

$$\sigma_1 = \begin{pmatrix} 0 & 1 \\ 1 & 0 \end{pmatrix}, \quad \sigma_2 = \begin{pmatrix} 0 & -i \\ i & 0 \end{pmatrix}, \quad \sigma_3 = \begin{pmatrix} 1 & 0 \\ 0 & -1 \end{pmatrix}. \tag{12.2.4}$$

The function u to be determined in (12.2.1) is $\mathbb{C}^4$-valued. The equation (12.2.1) has a natural interpretation in terms of an operator defined by the expression

$$\left\{ \alpha \cdot \left(\frac{\hbar}{i} \nabla + \mathbf{A} \right) + mc^2 \beta + V \right\}$$

in the Hilbert space $L_2(\mathbb{R}^3; \mathbb{C}^4)$ of $\mathbb{C}^4$-valued functions whose components lie in $L_2(\mathbb{R}^3)$; λ in (12.2.1) is then the spectral parameter of this operator. To simplify the notation we take $m = c = \hbar = 1$ which is adequate for our purpose.

12.3 The Free Dirac Operator

When there are no external forces; (12.2.1) is called the *free Dirac equation* and the closure $\mathbb{D}$ in $L_2(\mathbb{R}^3, \mathbb{C}^4)$ of the operator defined by the expression

$$\mathbb{D}' := \alpha \cdot \left(\frac{1}{i} \nabla \right) + \beta$$

on $C_0^\infty(\mathbb{R}^3, \mathbb{C}^4)$ is called the *free Dirac operator*. The identities (12.2.2) imply that $\mathbb{D}'^2 = (-\Delta + 1) I_4$.

Our first task is to determine $\mathbb{D}$. For this we recall some notation from sections 1.2 and 1.3.2. The inner product and norm on $L_2(\mathbb{R}^3, \mathbb{C}^4)$ are defined as follows: for

$$u = \begin{pmatrix} u_1 \\ u_2 \\ u_3 \\ u_4 \end{pmatrix}, \quad v = \begin{pmatrix} v_1 \\ v_2 \\ v_3 \\ v_4 \end{pmatrix},$$

12.3 The Free Dirac Operator

we have

$$(u, v)_{L_2(\mathbb{R}^3, \mathbb{C}^4)} := \int_{\mathbb{R}^3} \langle u(\mathbf{x}), v(\mathbf{x}) \rangle_{\mathbb{C}^4} \, d\mathbf{x}, \quad \|u\|^2_{L_2(\mathbb{R}^3, \mathbb{C}^4)} := \int_{\mathbb{R}^3} |u(\mathbf{x})|^2_{\mathbb{C}^4} d\mathbf{x},$$

where

$$\langle u(\mathbf{x}), v(\mathbf{x}) \rangle_{\mathbb{C}^4} = \sum_{j=1}^{4} u_j(\mathbf{x}) \overline{v_j(\mathbf{x})}, \quad |u(\mathbf{x})|_{\mathbb{C}^4} = \left(\sum_{j=1}^{4} |u_j(\mathbf{x})|^2 \right)^{1/2},$$

are respectively the inner-product and norm on $\mathbb{C}^4$. The Sobolev space $H^1(\mathbb{R}^3, \mathbb{C}^4) = W_2^1(\mathbb{R}, \mathbb{C}^4)$ is the Hilbert space of $\mathbb{C}^4$-valued functions (*spinors*) with components in $H^1(\mathbb{R}^3)$, endowed with the inner-product

$$(u, v)_{H^1(\mathbb{R}^3, \mathbb{C}^4)} = (\nabla u, \nabla v)_{L_2(\mathbb{R}^3, \mathbb{C}^4)} + (u, v)_{L_2(\mathbb{R}^3, \mathbb{C}^4)},$$

where ∇ is the gradient, defined in the weak sense, and

$$(\nabla u, \nabla v)_{L_2(\mathbb{R}^3, \mathbb{C}^4)} := \sum_{j=1}^{3} (D_j u, D_j v)_{L_2(\mathbb{R}^3, \mathbb{C}^4)}.$$

We denote the closure of $C_0^\infty(\mathbb{R}^3, \mathbb{C}^4)$ in $H^1(\mathbb{R}^3, \mathbb{C}^4)$ by $\overset{0}{H}{}^1(\mathbb{R}^3, \mathbb{C}^4)$. Since $\mathbb{D}'^2 = (-\Delta + 1)I_4$ it follows that for $u, v \in C_0^\infty(\mathbb{R}^3, \mathbb{C}^4)$,

$$(u, v)_{H^1(\mathbb{R}^3, \mathbb{C}^4)} = (\mathbb{D}'u, \mathbb{D}'v)_{L_2(\mathbb{R}^3)}, \quad (12.3.1)$$

and hence that the closure of $\mathbb{D}'$ is the operator

$$\mathbb{D} := \alpha \cdot \left(\frac{1}{i} \nabla \right) + \alpha_4, \quad \mathcal{D}(\mathbb{D}) = \overset{0}{H}{}^1(\mathbb{R}^3, \mathbb{C}^4).$$

To simplify notation, we shall denote $L_2(\mathbb{R}^3, \mathbb{C}^4)$ and $H^1(\mathbb{R}^3, \mathbb{C}^4)$ by H, H^1 respectively, and their inner products by $(\cdot, \cdot)$, $(\cdot, \cdot)_1$, respectively, with $\|\cdot\|$, $\|\cdot\|_1$ the associated norms.

Lemma 12.3.1 *The free Dirac operator $\mathbb{D}$ is self-adjoint and $\mathbb{D}^2 \geq 1$.*

Proof. For any $u \in \overset{0}{H}{}^1(\mathbb{R}^3, \mathbb{C}^4)$ and $\varphi \in C_0^\infty(\mathbb{R}^3, \mathbb{C}^4)$, the fact that the Dirac matrices α_j, $j = 1, 2, 3, 4$, are Hermitian implies that

$$(\mathbb{D}'\varphi, u) = (\varphi, \mathbb{D}u)$$

and this yields
$$\mathbb{D}^* = (\mathbb{D}')^* = \mathbb{D},$$
which establishes the self-adjointness of $\mathbb{D}$. From (12.3.1), for $u \in \mathcal{D}(\mathbb{D}^2)$,
$$(\mathbb{D}^2 u, u) = (\mathbb{D}u, \mathbb{D}u) = \|u\|_1^2 \geq \|u\|^2$$
and hence $\mathbb{D}^2 \geq 1$. ∎

The free Dirac operator is most easily analysed by means of the Fourier transform
$$\hat{u}(\mathbf{p}) := (\mathbb{F}u)(\mathbf{p}) := \frac{1}{(2\pi)^{3/2}} \int_{\mathbb{R}^3} e^{-i\mathbf{p}\cdot\mathbf{x}} u(\mathbf{x}) d\mathbf{x}, \qquad (12.3.2)$$
defined on $C_0^\infty(\mathbb{R}^3, \mathbb{C}^4)$; it will be convenient in this chapter to take the slightly different definition of the Fourier transform in (12.3.2) rather than its analogue on $C_0^\infty(\mathbb{R}^3)$ defined in Chapter 1. We shall need the following properties of $\mathbb{F}$ which are familiar for scalar-valued functions u and readily extended to $\mathbb{C}^4$-valued u:

1. $\mathbb{F}$ is a linear bijection of the Schwarz space $\mathcal{S}(\mathbb{R}^3, \mathbb{C}^4)$ onto itself and its inverse is given by
$$(\mathbb{F}^{-1} u)(\mathbf{x}) = \frac{1}{(2\pi)^{3/2}} \int_{\mathbb{R}^3} e^{i\mathbf{x}\cdot\mathbf{p}} u(\mathbf{p}) d\mathbf{p}, \qquad (12.3.3)$$
$(\mathbb{F}^{-1})(\mathbf{x}) = (\mathbb{F})(-\mathbf{x})$ and $\mathbb{F}^4 = I_4$, the identity on $\mathcal{S}(\mathbb{R}^3, \mathbb{C}^4)$;
2. for $u \in \mathcal{S}(\mathbb{R}^3, \mathbb{C}^4)$ and $\kappa \in \mathbb{N}^n$,
$$(\mathbb{F} D^\kappa \mathbb{F}^{-1}) u(\mathbf{p}) = (i\mathbf{p})^\kappa u(\mathbf{p}); \qquad (12.3.4)$$
3. $\mathbb{F}$ can be uniquely extended to a unitary operator on H which we shall continue to call $\mathbb{F}$. Thus
$$\mathbb{F}^{-1} = \mathbb{F}^*. \qquad (12.3.5)$$

When describing the motion of a particle, $\mathbf{x}$ represents its position and $\mathbf{p}$ its momentum, and $\mathbb{F}$ is described as a mapping of the $\mathbf{x}$-space (the *configuration space*) onto the $\mathbf{p}$-space (the *momentum space*). The configuration space is $H = L_2(\mathbb{R}^3, \mathbb{C}^4)$, and we denote the momentum space $\mathbb{F}H$ by $\hat{H}$.

For $u \in C_0^\infty(\mathbb{R}^3, \mathbb{C}^4)$ and setting $\hat{u} = \mathbb{F}u$, we have from (12.3.4) that
$$(\mathbb{F}\mathbb{D}'\mathbb{F}^{-1})\hat{u}(\mathbf{p}) = \mathbb{F}\left[\alpha \cdot \left(\frac{1}{i}\nabla\right) + \beta\right] u = (M\hat{u})(\mathbf{p}),$$
where M is the matrix multiplication
$$(M\hat{u})(\mathbf{p}) = (\alpha \cdot \mathbf{p} + \beta)\hat{u}(\mathbf{p}). \qquad (12.3.6)$$

12.3 The Free Dirac Operator

It follows that $\mathbb{D} = \mathbb{F}^{-1}\mathbb{M}\mathbb{F}$ where

$$\mathbb{M}\hat{u} = M\hat{u}, \quad \mathcal{D}(\mathbb{M}) = \left\{\hat{u} : \int_{\mathbb{R}^3} (|\mathbf{p}|^2 + 1)|\hat{u}(\mathbf{p})|^2 d\mathbf{p}\right\} < \infty. \tag{12.3.7}$$

From (12.3.6), $\mathbb{M}$ is the operator of multiplication by $\sqrt{p^2+1}\Lambda(\mathbf{p})$, where

$$\Lambda(\mathbf{p}) = \frac{\alpha \cdot \mathbf{p} + \beta}{\sqrt{p^2+1}}. \tag{12.3.8}$$

The 4×4 matrix $\Lambda(\mathbf{p})$ is a unitary involution, that is, $\Lambda(\mathbf{p})^2 = I_4$, and hence has double eigenvalues at ± 1. The projections $\Lambda_{\pm}(\mathbf{p})$ of $\mathbb{C}^4$ onto the eigenspaces at ± 1 are given by

$$\Lambda_{\pm}(\mathbf{p}) = \frac{1}{2}\left\{I_4 \pm \frac{\alpha \cdot \mathbf{p} + \beta}{\sqrt{p^2+1}}\right\}. \tag{12.3.9}$$

Denote by $\Lambda_{\pm}$ the operators in $\hat{H}$ of multiplication by $\Lambda_{\pm}(\mathbf{p})$ and set $L_{\pm} = \mathbb{F}^{-1}\Lambda_{\pm}\mathbb{F}$. For $u \in H$, and $u_{\pm} := L_{\pm}u$, we have by Parseval's formula,

$$2(u_+, u_-) = 2(\hat{u}_+, \hat{u}_-) = (\hat{u}, \hat{u}) + (\hat{u}, \Lambda\hat{u}) - (\Lambda\hat{u}, \hat{u}) - (\Lambda\hat{u}, \Lambda\hat{u}) = 0,$$

since Λ is symmetric and $\Lambda^2 = I$, the identity on $\hat{H}$. The projections $L_{\pm}$ are therefore orthogonal in H and we have the orthogonal sum decomposition

$$H = H_+ \oplus H_-, \tag{12.3.10}$$

where

$$H_{\pm} = L_{\pm}H.$$

The subspaces H_+, H_- are called the *positive* and *negative* subspaces respectively: if $E(\mathbf{p}) = \sqrt{p^2+1}$ and $u_+ \in H_+ \cap \mathcal{D}(\mathbb{D})$, then $\hat{u}_+ \in \mathcal{D}(\mathbb{M})$ and for $u_+ \neq 0$,

$$(\mathbb{D}u_+, u_+) = (\mathbb{M}\hat{u}_+, \hat{u}_+) = (E\hat{u}_+, \hat{u}_+) > 0;$$

similarly $\mathbb{D}$ is negative on H_-.

The matrix $\Lambda(\mathbf{p})$ is diagonalised by the unitary matrix

$$U(\mathbf{p}) = \phi_+(p)I_4 + \phi_-(p)\beta\frac{(\alpha \cdot \mathbf{p})}{p},$$

where

$$\phi_{\pm}(\mathbf{p}) = \frac{1}{\sqrt{2}}\sqrt{1 \pm 1/E(\mathbf{p})}.$$

$$U(\mathbf{p})^{-1} = \phi_+(\mathbf{p})I_4 - \phi_-(\mathbf{p})\beta\frac{(\alpha \cdot \mathbf{p})}{p}$$

and we have
$$U\Lambda U^{-1} = \beta.$$

Hence $UMU^{-1} = UE\Lambda U^{-1} = E\beta$ and

$$\left(\mathbb{F}^{-1}U\mathbb{F}\right)\left(\mathbb{F}^{-1}M\mathbb{F}\right)\left(\mathbb{F}^{-1}U^{-1}\mathbb{F}\right) = \mathbb{F}^{-1}UMU^{-1}\mathbb{F} = \left(\mathbb{F}^{-1}E\mathbb{F}\right)\beta.$$

On defining
$$\mathbb{H} = \sqrt{-\Delta + 1} := \mathbb{F}^{-1}E\mathbb{F} \qquad (12.3.11)$$

we see that $U_{FW} := \mathbb{F}^{-1}U\mathbb{F}$ transforms $\mathbb{D}$ into the following 2×2-block form in H:

$$U_{FW}\mathbb{D}U_{FW}^{-1} = \begin{pmatrix} \sqrt{-\Delta + 1} & 0_2 \\ 0_2 & -\sqrt{-\Delta + 1} \end{pmatrix}. \qquad (12.3.12)$$

The operator $\mathbb{H}$ is called the *quasi-relativistic Schrödinger operator* and U_{FW} the *Fouldy–Wouthuysen transformation*.

12.4 The Brown–Ravenhall Operator

The operator $L_+ = \chi_{(0,\infty)}(\mathbb{D})$, where $\chi_{(0,\infty)}$ is the characteristic function on $(0, \infty)$, is the projection of H onto the positive spectral subspace H_+ of $\mathbb{D}$. In terms of the notation in the previous section, $L_+ = \mathbb{F}^{-1}\Lambda_+\mathbb{F}$, where Λ_+ is the operator of multiplication by $\Lambda_+(\mathbf{p})$ in $\hat{H}$. The matrix $\Lambda_+(\mathbf{p})$ can be written in terms of the Pauli matrices $\sigma_1, \sigma_2, \sigma_3$ as

$$\Lambda_+(\mathbf{p}) = \frac{1}{2E(\mathbf{p})}\begin{pmatrix} E(\mathbf{p}) + 1 & \sigma \cdot \mathbf{p} \\ \sigma \cdot \mathbf{p} & E(\mathbf{p}) - 1 \end{pmatrix}. \qquad (12.4.1)$$

Corresponding to its eigenvalue 1, $\Lambda_+(\mathbf{p})$ has the two orthonormal eigenvectors

$$\frac{1}{N(\mathbf{p})}\begin{pmatrix} [E(\mathbf{p}) + 1]\hat{\varphi}_j \\ (\sigma \cdot \mathbf{p})\hat{\varphi}_j \end{pmatrix} \quad (j = 1, 2), \quad N(\mathbf{p}) = \sqrt{2E(\mathbf{p})[E(\mathbf{p}) + 1]}, \qquad (12.4.2)$$

where $\hat{\varphi}_1$ and $\hat{\varphi}_2$ are orthonormal vectors in $\mathbb{C}^2$. Thus any eigenvector of $\Lambda_+(\mathbf{p})$ is of the form

$$\hat{\psi}(\mathbf{p}) = \frac{1}{N(\mathbf{p})}\begin{pmatrix} [E(\mathbf{p}) + 1]\hat{\varphi}(\mathbf{p}) \\ (\sigma \cdot \mathbf{p})\hat{\varphi}(\mathbf{p}) \end{pmatrix}, \qquad (12.4.3)$$

12.4 The Brown–Ravenhall Operator

where $\hat{\varphi}(\mathbf{p}) \in \mathbb{C}^2$; the $\mathbb{C}^2$-valued function $\hat{\varphi}$ is called a 2-spinor or *Pauli spinor*. Moreover,

$$\int_{\mathbb{R}^3} |\hat{\psi}(\mathbf{p})|^2_{\mathbb{C}^4} d\mathbf{p} = \int_{\mathbb{R}^3} |\hat{\varphi}(\mathbf{p})|^2_{\mathbb{C}^2} d\mathbf{p}$$

and if $\hat{\psi}^{(j)}(\mathbf{p})$ is of the form (12.4.3) for $\hat{\varphi} = \hat{\varphi}^j$, $j = 1, 2$, then

$$\langle \hat{\psi}^{(1)}(\mathbf{p}), \hat{\psi}^{(2)}(\mathbf{p}) \rangle_{\mathbb{C}^4} = \langle \hat{\varphi}^{(1)}(\mathbf{p}), \hat{\varphi}^{(2)}(\mathbf{p}) \rangle_{\mathbb{C}^2},$$

where the angle brackets are inner products on $\mathbb{C}^4$ and $\mathbb{C}^2$.

The Brown–Ravenhall operator is given formally as

$$\mathbb{B} = L_+ \mathbb{D} L_+, \quad L_+ = \mathbb{F}^{-1} \Lambda_+ \mathbb{F}, \tag{12.4.4}$$

acting in $L_2(\mathbb{R}^3, \mathbb{C}^4)$. On recalling that $\mathbb{D} = \mathbb{F}^{-1} \mathbb{M} \mathbb{F}$ and $\mathbb{M} = E\Lambda$, we also note that

$$\begin{aligned} \mathbb{D} L_+ &= \mathbb{F}^{-1} \mathbb{M} \mathbb{F} L_+ \\ &= \mathbb{F}^{-1} E \Lambda \mathbb{F} L_+ \\ &= \left(\mathbb{F}^{-1} E \mathbb{F} \right) \left(\mathbb{F}^{-1} \Lambda \Lambda_+ \mathbb{F} \right) \\ &= \sqrt{-\Delta + 1} I_4 L_+ \end{aligned}$$

since $\Lambda \Lambda_+ = \Lambda_+$. Hence

$$\mathbb{B} = L_+ (\mathbb{H} I_4) L_+. \tag{12.4.5}$$

12.5 Sums of Operators and Coulomb Potentials

Let V_γ denote the Coulomb potential

$$V_\gamma(\mathbf{x}) = -\gamma/|\mathbf{x}|, \quad \gamma \in \mathbb{R}, \tag{12.5.1}$$

and let A stand for one of the operators $\mathbb{D}$, $\mathbb{H}$ or $\mathbb{B}$. In this section, we shall consider the problem of defining a self-adjoint operator which can be regarded, in some sense, as the sum of A and V_γ for a range of values of γ. In the case of $A = \mathbb{D}$, the restriction of $A + V_\gamma$ to $C_0^\infty(\mathbb{R}^3, \mathbb{C}^4)$ is essentially self-adjoint, while for $A = \mathbb{H}$ or $\mathbb{B}$ the self-adjoint operator is a form sum. The account that follows is a brief survey of some known results, and is by no means exhaustive. Proofs are not given, but any necessary background material and precise references are provided.

12.5.1 The Case $A = \mathbb{D}$

Let
$$\mathbb{D}'_\gamma := \mathbb{D}' + V_\gamma, \quad \mathcal{D}(\mathbb{D}'_\gamma) = C_0^\infty(\mathbb{R}^3, \mathbb{C}^4),$$

and let $\mathbf{x} = (x_1, x_2, x_3) = (r, \theta, \varphi)$ denote the spherical polar co-ordinates

$$x_1 = r \sin\theta \cos\varphi, \quad x_2 = r \sin\theta \sin\varphi, \quad x_3 = r \cos\theta,$$

where
$$r = |\mathbf{x}| \in (0, \infty), \quad \theta = \cos^{-1}(x_3/|\mathbf{x}|) \in [0, \pi), \quad \varphi \in [-\pi, \pi).$$

The Dirac operator $\mathbb{D}'_\gamma$ can then be expressed in the form

$$\mathbb{D}'_\gamma \cong \bigoplus_{l \in \mathbb{N}_0} \bigoplus_{m=\pm(l+1/2)} \bigoplus_{s=\pm 1/2} (\mathbb{D}'_{0,l,s} + V_\gamma), \qquad (12.5.2)$$

where $\cong$ indicates unitary equivalence and the $\mathbb{D}'_{0,l,s}$ are the so-called *partial wave operators* defined on $C_0^\infty(\mathbb{R}_+, \mathbb{C}^2)$ by

$$\mathbb{D}'_{0,l,s} = \begin{pmatrix} 1 & -\frac{d}{dr} - \frac{\kappa_{l,s}}{r} \\ \frac{d}{dr} - \frac{\kappa_{l,s}}{r} & -1 \end{pmatrix}$$
$$= -i\sigma_2 \frac{d}{dr} - \frac{\kappa_{l,s}}{r}\sigma_1 + \sigma_3 \qquad (12.5.3)$$

where $\sigma_1, \sigma_2, \sigma_3$ are the Pauli matrices of (12.2.4) and $\kappa_{l,s} = (2sl + s + 1/2)$. The unitary equivalence in (12.5.2) is given by a natural unitary isomorphism between the Hilbert spaces $L_2(\mathbb{R}^3, \mathbb{C}^4)$ and $L_2(\mathbb{R}_+, r^2 dr) \otimes L_2(\mathbb{S}^2, \mathbb{C}^4)$, and $\{(l, m, s) : l \in \mathbb{N}_0, m = \pm(l + 1/2), s = \pm 1/2\}$ are admissible indices in an orthogonal decomposition of $L_2(\mathbb{S}^2, \mathbb{C}^4)$ in terms of *spherical spinors*. It is shown in [15], section 2.1.2, that $\mathbb{D}'_\gamma$ is essentially self-adjoint if and only if, for all values of l and s, $\mathbb{D}'_{0,l,s} + V_\gamma$ is in the limit-point case at 0, and that this is so for

$$\gamma^2 \leq \kappa_{l,s}^2 - 1/4.$$

Since $l \geq 1$ when $s = -1/2$, we have that $\kappa_{l,s}^2 = (2sl + s + 1/2)^2 \geq 1$ and hence $\mathbb{D}'_\gamma$ is essentially self-adjoint if and only if $|\gamma| \leq \sqrt{3}/2$.

As already noted, proofs of the above results are given in [15], but the technique described and results stated are well-known; the problems are analysed in a similar way in [199]. We now enumerate related results of interest established by other authors.

1. $\mathbb{D}'_\gamma$ is essentially self adjoint if and only if $|\gamma| \leq \sqrt{3}/2$, and the closure of $\mathbb{D}'_\gamma$ is the operator sum $\mathbb{D}_\gamma = \mathbb{D} + V_\gamma$ with domain $\mathcal{D}(\mathbb{D}_\gamma) = \mathcal{D}(\mathbb{D}) = H^1(\mathbb{R}^3, \mathbb{C}^4)$

if $|\gamma| < \sqrt{3}/2$; see Theorem 2.1.6 and Remark 2.1.8 in [15], and section 4.3.3 in [199]. The essential self-adjointness of $\mathbb{D}' + V_\gamma$ depends only on the local behaviour of V_γ, this being a property established by Jörgens in [114] and Chernoff in [36], [37]; see also Remark 2.1.10 in [15].
2. If $\sqrt{3}/2 < |\gamma| < 1$, $\mathbb{D}'_\gamma$ is no longer essentially self-adjoint, but a distinguished self-adjoint extension exists which is a certain limit of Dirac operators with cut-off potentials, and is characterised by its domain being in $\mathcal{D}((\mathbb{D}'_\gamma)^*) \cap \mathcal{D}(|\mathbf{x}|^{-1})$. Proofs of different aspects of this result, under the more general condition on a potential V,

$$\sup_{\mathbf{x} \neq 0} |\mathbf{x}||V(x)| < 1,$$

are given in [217], [218], [219], [189] and [129].
3. If $|\gamma| \leq \sqrt{3}/2$, the essential spectrum of the self-adjoint operator $\mathbb{D}_\gamma$ is $(-\infty, -1] \cup [1, \infty)$. In fact this is true for any self-adjoint operator $\mathbb{D} + V$ with a potential V which satisfies $V(\mathbf{x}) \to 0$ as $|\mathbf{x}| \to \infty$; see [199], section 4.3.4.
4. If $\sqrt{3}/2 < |\gamma| < 1$, Schechter proved in [187] that the spectrum of the distinguished self-adjoint extension of item 2 above is contained in
$\mathbb{R} \setminus (-\sqrt{1-\eta^2}, \sqrt{1-\eta^2})$ for any $\eta \in (|\gamma|, 1)$.

12.5.2 The Case $A = \mathbb{H}$

A comprehensive study of the spectral properties of the natural self-adjoint extension $\mathbb{H}_\gamma$ of $\mathbb{H}' + V_\gamma$ was made by Herbst in [107] and Weder in [214]. An important ingredient in the analysis of this case is Kato's inequality, that for all $u \in \mathcal{S}(\mathbb{R}^3)$,

$$\int_{\mathbb{R}^3} |\mathbf{x}|^{-1}|u(\mathbf{x})|^2 d\mathbf{x} \leq \frac{\pi}{2} \int_{\mathbb{R}^3} |\mathbf{p}||\hat{u}(\mathbf{p})|^2 d\mathbf{p}, \qquad (12.5.4)$$

where the constant $\pi/2$ is sharp; see [15], Theorem 2.2.4 for a proof. It follows that with $E(\mathbf{p}) = \sqrt{p^2 + 1}$,

$$\|V_\gamma u\|^2 \leq \frac{\pi|\gamma|^2}{2} \int_{\mathbb{R}^3} |E(\mathbf{p})||\hat{u}(\mathbf{p})|^2 d\mathbf{p} = \frac{\pi|\gamma|^2}{2} \|\mathbb{H}^{1/2} u\|^2. \qquad (12.5.5)$$

The quadratic form

$$h[u] := \int_{\mathbb{R}^3} \left(|(\mathbb{H}^{1/2})(\mathbf{x})|^2 + V_\gamma(\mathbf{x})|u(\mathbf{x})|^2 \right) d\mathbf{x}, \qquad (12.5.6)$$

is therefore non-negative on $H^{1/2}(\mathbb{R}^3)$ for $|\gamma| \leq 2/\pi$ and, on applying Theorem 6.3.2, $\mathbb{H}_\gamma$ is defined as the associated operator with form domain $H^{1/2}(\mathbb{R}^3)$;

it is the Friedrichs extension of the restriction of $\mathbb{H} + V_\gamma$ to $\mathcal{D}(\mathbb{H}) = H^1(\mathbb{R}^3)$. If $|\gamma| < 2/\pi$, $\mathbb{H}_\gamma$ coincides with the form sum $\mathbb{H} \dotplus V_\gamma$.

If $|\gamma| < 1/2$, V_γ is a relatively bounded perturbation of $\mathbb{H}$ with relative bound less than 1, since by Hardy's inequality,

$$\|V_\gamma u\| \le (2|\gamma|)\|\mathbb{H}\|, \quad u \in H^1(\mathbb{R}^3);$$

see [15], section 1.7. It follows that $\mathbb{H} + V_\gamma$ is self-adjoint with domain $\mathcal{D}(\mathbb{H})$ and is the operator sum. If $\gamma = 1/2$, $\mathbb{H}' + V_\gamma$ is essentially self-adjoint by a theorem of Wüst in [217].

In [107], Theorem 2.5, Herbst determines the norm of the operator $C_\alpha := |\mathbf{x}|^{-\alpha}|\mathbf{p}|^{-\alpha}$ as a map on $L_q(\mathbb{R}^n)$ under optimal conditions on α, n and q, and thereby obtains an inequality which includes both the Hardy and Kato inequalities. This inequality enables analogues of the above results for $h[\cdot]$ (and $\mathbb{H}_\gamma$ in the case $n = 3$) to be proved for all $n \ge 1$.

Further interesting properties in the range $1/2 < \gamma < 2/\pi$ have been determined by following a similar path to that in Sect. 12.4 of decomposing the operator (or the associated quadratic form in this case) into partial sums. From the result

$$\mathbb{F}\left(|\cdot|^{-1}\right)(\mathbf{p}) = \frac{1}{(2\pi)^{3/2}} \int_{\mathbb{R}^3} e^{-i\mathbf{p}\cdot\mathbf{x}}|\mathbf{x}|^{-1}d\mathbf{x} = \sqrt{2/\pi}|\mathbf{p}|^{-2},$$

it follows from (1.5.8) and (1.5.9) in [15] that

$$\mathbb{F}\left(\frac{1}{|\cdot|}f\right)(\mathbf{p}) = \left(\mathbb{F}\frac{1}{|\cdot|} * \hat{f}\right)(\mathbf{p})$$
$$= \frac{1}{2\pi^2} \int_{\mathbb{R}^3} \frac{1}{|\mathbf{p}-\mathbf{p}'|^2}\hat{f}(\mathbf{p}')d\mathbf{p}'$$

and thus, for all $f \in H^1(\mathbb{R}^3)$,

$$\mathbb{F}\left(\mathbb{H}_\gamma f\right)(\mathbf{p}) = \sqrt{p^2+1}f(\mathbf{p}) - \frac{\gamma}{2\pi^2} \int_{\mathbb{R}^3} \frac{1}{|\mathbf{p}-\mathbf{p}'|^2}\hat{f}(\mathbf{p}')d\mathbf{p}'.$$

The functions $\hat{f}$ can be expanded in terms of the spherical harmonics $Y_{l,m}$ as

$$\hat{f}(\mathbf{p}) = \sum_{l,m} p^{-1}c_{l,m}(p)Y_{l,m}(\omega_\mathbf{p})$$

where

$$p^{-1}c_{l,m}(p) = \int_{\mathbb{R}^3} \hat{f}(p\omega_\mathbf{p})Y_{l,m}(\omega_\mathbf{p})d\omega_\mathbf{p}$$

and the index ranges are $l \in \mathbb{N}_0$ and $m = \pm(l+1/2)$. On using the identity

12.5 Sums of Operators and Coulomb Potentials

$$\frac{1}{|\mathbf{p}-\mathbf{p}'|^2} = \frac{4\pi}{2pp'} \sum_{k=0}^{\infty} Q_k(z) \sum_{t=-k}^{k} \bar{Y}_{k,t}(\omega_\mathbf{p}) Y_{k,t}(\omega_{\mathbf{p}'}),$$

where the Q_k are Legendre functions of the second kind, given in terms of the Legendre polynomials P_l by

$$Q_l(z) = \frac{1}{2} \int_{-1}^{1} \frac{P_l(t)}{z-t} dt,$$

we obtain

$$\mathbb{F}(\mathbb{H}_\gamma f)(\mathbf{p}) = \sum_{l,m} \left\{ \sqrt{p^2+1}\, p^{-1} c_{l,m}(p) \right\}$$
$$- \sum_{l,m} \left\{ \frac{\gamma}{\pi} \int_0^\infty p^{-1} c_{l,m}(p') Q_l\left(\frac{1}{2}\left[\frac{p}{p'}+\frac{p'}{p}\right]\right) dp' \right\} Y_{l,m}(\omega_\mathbf{p}).$$
(12.5.7)

For details of these calculations see [15], section 2.2. It follows that for $f \in H^1(\mathbb{R}^3)$,

$$(f, \mathbb{H}_\gamma f) = \sum_{l,m} \left\{ \int_0^\infty \sqrt{p^2+1}\, |c_{l,m}(p)|^2 \right\} dp$$
$$- \sum_{l,m} \left\{ \frac{\gamma}{\pi} \int_0^\infty \int_0^\infty Q_l\left(\frac{1}{2}\left[\frac{p}{p'}+\frac{p'}{p}\right]\right) c_{l,m}(p') \overline{c_{l,m}(p)} dp dp' \right\}$$
$$= \sum_{l,m} (c_{l,m}, h_l c_{l,m}),$$
(12.5.8)

where $h_l = E - K_l$, with E the operation of multiplication by $\sqrt{p^2+1}$ and

$$(K_l u)(p) = \frac{\gamma}{\pi} \int_0^\infty Q_l\left(\frac{1}{2}\left[\frac{p}{p'}+\frac{p'}{p}\right]\right) u(p') dp'.$$
(12.5.9)

The operators h_l are shown in [15], section 2.2 to have the following properties:

1. for $\gamma < 1/2$, h_0 is self-adjoint with domain $L_2(\mathbb{R}_+; (p^2+1)dp)$;
2. for $\gamma = 1/2$, h_0 on $L_2(\mathbb{R}_+; (p^2+1)dp)$ is essentially self-adjoint;
3. for $l \geq 1$, h_l, with domain $L_2(\mathbb{R}_+; (p^2+1)dp)$ is self-adjoint for all $\gamma < 3/2$, and thus throughout the critical range $\gamma \leq 2/\pi$.

The operator $\mathbb{H}_\gamma$ is self-adjoint or essentially self-adjoint if and only if the same is true for all the operators h_l. Thus the result noted earlier, that $\mathbb{H}_\gamma$ is self-adjoint for $|\gamma| < 1/2$ and essentially self-adjoint if $|\gamma| = 1/2$, is recovered.

For γ in the range $1/2 < \gamma < 2/\pi$, the Mellin transform $\mathcal{M}$ is used in [221], sub-section 2.2.3, to investigate the operator h_0. The Mellin transform is a unitary map from $L_2(\mathbb{R}_+)$ onto $L_2(\mathbb{R})$ and is defined by

$$\mathcal{M}\psi(s) := \frac{1}{\sqrt{2\pi}} \int_0^\infty p^{-\frac{1}{2}-is}\psi(p)dp.$$

For γ in the range $1/2 < \gamma \geq 2/\pi$ a number of interesting properties of the operator $\mathcal{M}h_0\mathcal{M}^{-1}$ are established in [221], and a particular consequence of one of these is that h_0 is a closed symmetric operator in $L_2(\mathbb{R}_+)$ with deficiency indices $(1, 1)$.

Let $0 \leq \gamma \leq 2/\pi$. The essential spectrum of $\mathbb{H}_\gamma$ coincides with that of $\mathbb{H}$, namely $[1, \infty)$; see [15], Theorem 3.2.1 for $\gamma < 2/\pi$ and [221], Lemma 4 for $\gamma = 2/\pi$. In $(0, 1)$ the spectrum of $\mathbb{H}_\gamma$ consists of an infinite number of isolated eigenvalues of finite multiplicity which accumulate only at 1. If $\gamma < 2/\pi$, the smallest eigenvalue λ_0 satisfies $\lambda_0 \geq (1 - [\pi\gamma/2]^2)^{1/2}$, which follows from the inequality for $\mathbb{H}_\gamma$ proved by Herbst in [107], Theorem 2.2. If $\gamma = 2/\pi$, λ_0 is still positive as it was proved in [174] that $\lambda_0 > 0.4825$. A virial theorem for $\mathbb{H}_\gamma$ is established in [107], Theorem 2.4, and this implies that in $(1, \infty)$, $\mathbb{H}_\gamma$ has no embedded eigenvalues, the spectrum in $[1, \infty)$ being absolutely continuous.

12.5.3 The Case $A = \mathbb{B}$

A brief description is given in this subsection of some of the very interesting properties of the Brown–Ravenhall operator concerning its definition as a self-adjoint operator and its spectral theory. Our intention is to communicate the rich flavour of the topic, and refer to [15] and references therein for details and further reading.

Let $\psi \in H_+ \cap \mathcal{S}(\mathbb{R}^3, \mathbb{C}^4)$. By Parseval's formula and (12.4.5),

$$\begin{aligned}
(\psi, \mathbb{B}_\gamma \psi) &= \int_{\mathbb{R}^3} \langle \hat{\psi}, \mathbb{F}[L_+ \mathbb{H}_\gamma I_4 L_+ \psi] \rangle (\mathbf{p}) d\mathbf{p} \\
&= \int_{\mathbb{R}^3} E(\mathbf{p}) |\hat{\psi}(\mathbf{p})|^2 d\mathbf{p} + \int_{\mathbb{R}^3} \langle \hat{\psi}, \mathbb{F}[L_+ V_\gamma L_+ \psi] \rangle (\mathbf{p}) d\mathbf{p} \\
&= \int_{\mathbb{R}^3} E(\mathbf{p}) |\hat{\psi}(\mathbf{p})|^2 d\mathbf{p} + \int_{\mathbb{R}^3} \langle \hat{\psi}, \Lambda_+ \mathbb{F}[V_\gamma \psi] \rangle (\mathbf{p}) d\mathbf{p}, \quad (12.5.10)
\end{aligned}$$

since $\mathbb{F}L_+\mathbb{F}^{-1} = \Lambda_+$ and $L_+\psi = \psi$. Moreover

$$\begin{aligned}
\mathbb{F}[V_\gamma \psi](\mathbf{p}) &= (\hat{V}_\gamma * \hat{\psi})(\mathbf{p}) \\
&= \frac{\gamma}{2\pi^2} \int_{\mathbb{R}^3} |\mathbf{p} - \mathbf{p}'|^{-2} \hat{\psi}(\mathbf{p}') d\mathbf{p}'.
\end{aligned}$$

On substituting in (12.5.10) and taking (12.4.3) into account, we obtain

12.5 Sums of Operators and Coulomb Potentials

$$(\psi, \mathbb{B}_\gamma \psi) = \int_{\mathbb{R}^3} E(\mathbf{p})|\hat{\varphi}(\mathbf{p})|^2 d\mathbf{p} - \frac{\gamma}{2\pi^2} \int_{\mathbb{R}^3} \int_{\mathbb{R}^3} \hat{\varphi}(\mathbf{p})^* K(\mathbf{p}', \mathbf{p}) \hat{\varphi}(\mathbf{p}) d\mathbf{p} d\mathbf{p}'$$
$$=: (\hat{\varphi}, b_\gamma \hat{\varphi}) \quad (12.5.11)$$

where * denotes the Hermitian conjugate,

$$K(\mathbf{p}', \mathbf{p}) = \frac{(E(p')+1)(E(p)+1) I_2 + (\mathbf{p}' \cdot \boldsymbol{\sigma})(\mathbf{p} \cdot \boldsymbol{\sigma})}{N(p')|\mathbf{p}' - \mathbf{p}|^2 N(p)},$$

and $(\hat{\varphi}, b_\gamma \hat{\varphi})$ is a quadratic form on the space $\mathcal{S}(\mathbb{R}^3, \mathbb{C}^2)$ of rapidly decreasing Pauli spinors.

In [15], the quadratic form $(\cdot, b_\gamma \cdot)$ is analysed by decomposing it into quadratic forms in the weighted space $L_2(\mathbb{R}_+, e(p)dp)$, with weight $e(p) = E(p) = \sqrt{p^2+1}$. This is achieved by expanding the spinors $\hat{\varphi}$ as infinite series involving *spherical 2-spinors* $\Omega_{(l,m,s)}$ which form an orthonormal basis of $L_2(\mathbb{S}^2, \mathbb{C}^2)$:

$$\hat{\varphi}(\mathbf{p}) = \sum_{(l,m,s)} p^{-1} a_{l,m,s}(p) \Omega_{l,m,s}(\omega).$$

The outcome is that

$$(\hat{\varphi}, b_\gamma \hat{\varphi}) = \sum_{(l,m,s)} \left(\int_0^\infty e(p) |a_{(l,m,s)}(p)|^2 dp \right)$$
$$- \sum_{(l,m,s)} \left(\frac{\gamma}{\pi} \int_0^\infty \int_0^\infty \overline{a_{(l,m,s)}} k_{l,s}(p', p) a_{(l,m,s)}(p) dp' dp \right)$$
$$=: \sum_{(l,m,s)} \left(a_{l,m,s}, b_{l,s}(\gamma) a_{l,m,s} \right), \quad (12.5.12)$$

say, where with $n(p) = N(p) = \sqrt{2e(p)[e(p)+1]}$,

$$k_{l,s}(p'p) = \frac{[e(p')+1] Q_l\left(\frac{1}{2}\left[\frac{p'}{p} + \frac{p}{p'}\right]\right)[e(p)+1] + p' Q_{l+2s}\left(\frac{1}{2}\left[\frac{p'}{p} + \frac{p}{p'}\right]\right) p}{n(p)n(p')},$$
$$(12.5.13)$$

and the operators $b_{l,s}(\gamma)$ are defined by

$$b_{l,s}(\gamma) f(p) = e(p) f(p) - \frac{\gamma}{\pi} \int_0^\infty k_{l,s}(p', p) f(p') dp'.$$

In [15], Theorems 2.3.7 and 2.3.11, the following results are established in terms of $\gamma_c = 2(\pi/2 + 2/\pi)^{-1}$ which is a critical value of γ.

1. If $\gamma \le \gamma_c$, $b_{l,s}(\gamma) \ge (1 - \gamma/\gamma_c)$.

2. If $\gamma = \gamma_c$, $b_{l,s}(\gamma)$ has no eigenvalue at 0. In fact, for $\gamma \leq \gamma_c$, Tix proves in [202] that $b_{l,s}(\gamma) \geq (1 - \gamma_c) > 0.09$.
3. If $\gamma > \gamma_c$, $b_{l,s}(\gamma)$ is unbounded below.
4. The operators $b_{l,s}(\gamma)$ have form domain $L_2(\mathbb{R}_+, e(p)dp)$. For $(l, s) = (0, 1/2)$ or $(1, -1/2)$, $b_{l,s}(\gamma)$ is self-adjoint for $\gamma < 3/4$ and essentially self-adjoint if $\gamma = 3/4$. For other values of l, s, $b_{l,s}(\gamma)$ is self-adjoint if $\gamma < 15/7$ and hence for $\gamma \leq \gamma_c$.

In [203], Tix proves results which are analogous to those in [221] quoted in section 12.5.2. He proves that for $3/4 < \gamma \leq \gamma_c$, $b_{0,1/2}(\gamma)$ and $b_{1,-1/2}(\gamma)$ are closed symmetric operators with deficiency indices $(1, 1)$.

12.6 The Free Dirac Operator on a Bounded Domain

Our main objective in this section is to construct self-adjoint extensions of the symmetric operator defined by the free Dirac expression

$$\mathbb{D}' := \mathbb{D}'_0 + \beta, \quad \mathbb{D}'_0 = \alpha \cdot \left(\frac{1}{i}\nabla\right)$$

on $C_0^\infty(\Omega, \mathbb{C}^4)$, when Ω is a domain in $\mathbb{R}^3$ which is sufficiently smooth for Gauss' Theorem to hold. The rich KVD theory of Chapter 2 is not available because the Dirac operator is not semi-bounded, and we can only expect limited results. These will be achieved through the use of the supersymmetric structure of $\mathbb{D}'$ as was done by Schmidt in [188].

It follows as in section 12.3 that the closure $\mathbb{D}$ of $\mathbb{D}'$ is the operator

$$\mathbb{D} := \mathbb{D}', \quad \mathcal{D}(\mathbb{D}) = \overset{0}{H^1}(\Omega, \mathbb{C}^4),$$

and that the adjoint $\mathbb{D}^*$ is given by

$$\mathbb{D}^* = \mathbb{D}', \quad \mathcal{D}(\mathbb{D}^*) = \{u : u, \mathbb{D}'u \in L_2(\Omega, \mathbb{C}^4)\},$$

where $\mathbb{D}'u$ is understood in the weak (distributional) sense. Clearly $H^1(\Omega, \mathbb{C}^4) \subset \mathcal{D}(\mathbb{D}^*)$ and for $u, v \in H^1(\Omega, \mathbb{C}^4)$,

$$\begin{aligned}(\mathbb{D}'u, v) - (u, \mathbb{D}'v) &= (\mathbb{D}'_0 u, v) - (u, \mathbb{D}'_0 v) \\ &= \int_\Omega \left\{ \left\langle \left[\alpha \cdot (\tfrac{1}{i}\nabla)\right] u(\mathbf{x}), v(\mathbf{x}) \right\rangle_{\mathbb{C}^4} - \left\langle u(\mathbf{x}), \left[\alpha \cdot (\tfrac{1}{i}\nabla)\right] v(\mathbf{x}) \right\rangle_{\mathbb{C}^4} \right\} d\mathbf{x} \\ &= -i \int_{\partial\Omega} \langle (\nu \cdot \alpha)u, v\rangle_{\mathbb{C}^4} \, ds, \end{aligned} \qquad (12.6.1)$$

12.6 The Free Dirac Operator on a Bounded Domain

where ν is the exterior unit normal field of $\partial\Omega$ and ds is the measure defined by the surface area on $\partial\Omega$. On setting

$$u = \begin{pmatrix} u^{(1)} \\ u^{(2)} \end{pmatrix}, \quad u^{(j)} = \begin{pmatrix} u_1^{(j)} \\ u_2^{(j)} \end{pmatrix}, \quad j = 1, 2,$$

we have

$$(\nu \cdot \alpha) u = \sum_{j=1}^{3} \nu_j \begin{pmatrix} 0 & \sigma_j \\ \sigma_j & 0 \end{pmatrix} \begin{pmatrix} u^{(1)} \\ u^{(2)} \end{pmatrix} = \sum_{j=1}^{3} \nu_j \begin{pmatrix} \sigma_j u^{(2)} \\ \sigma_j u^{(1)} \end{pmatrix}.$$

Hence, with

$$v = \begin{pmatrix} v^{(1)} \\ v^{(2)} \end{pmatrix}, \quad v^{(j)} = \begin{pmatrix} v_1^{(j)} \\ v_2^{(j)} \end{pmatrix}, \quad j = 1, 2,$$

it follows that (with the superscript T denoting transpose)

$$\begin{aligned}
& (\mathbb{D}'u, v) - (u, \mathbb{D}'v) \\
&= -i \int_{\partial\Omega} \sum_j \left\{ \nu_j \left([\sigma_j u^{(2)}]^T \overline{v^{(1)}} + [\sigma_j u^{(1)}]^T \overline{v^{(2)}} \right) \right\} ds \\
&= -i \int_{\partial\Omega} \left\{ (\nu_1 - i\nu_2) \left(u_2^{(2)} \overline{v_1^{(1)}} + u_2^{(1)} \overline{v_1^{(2)}} \right) \right\} ds \\
&= -i \int_{\partial\Omega} \left\{ (\nu_1 + i\nu_2) \left(u_1^{(2)} \overline{v_2^{(1)}} + u_1^{(1)} \overline{v_2^{(2)}} \right) \right\} ds \\
& \quad -i \int_{\partial\Omega} \nu_3 \left(u_1^{(2)} \overline{v_1^{(1)}} - u_2^{(2)} \overline{v_2^{(1)}} + u_1^{(1)} \overline{v_1^{(2)}} - u_2^{(1)} \overline{v_2^{(2)}} \right) ds.
\end{aligned} \quad (12.6.2)$$

Therefore,

$$\left\{ u : u \in H^1(\Omega, \mathbb{C}^4), u^{(j)} = 0 \right\}, \quad j = 1, 2,$$

are both domains of symmetric extensions of $\mathbb{D}$.

To construct self-adjoint extensions of $\mathbb{D}$ we follow Schmidt in [188]. We write $L_2(\Omega, \mathbb{C}^4)$ as the orthogonal sum $L_2(\Omega, \mathbb{C}^2) \oplus L_2(\Omega, \mathbb{C}^2)$, and define the operator

$$\mathbb{D}_0 := \begin{pmatrix} 0 & A^* \\ A & 0 \end{pmatrix}, \quad \mathcal{D}(\mathbb{D}_0) = \mathcal{D}(A) \oplus \mathcal{D}(A^*), \quad (12.6.3)$$

where A is to be a closed operator with a dense domain in $L_2(\Omega, \mathbb{C}^2)$. It follows that the adjoint of $\mathbb{D}_0$ is

$$\mathbb{D}_0^* = \begin{pmatrix} 0 & A^* \\ A^{**} & 0 \end{pmatrix} = \mathbb{D}_0$$

since A is closed and hence $A^{**} = A$; see [199], section 5.2 for details. Therefore the operator $\mathbb{D}_0$ in (12.6.3) is self-adjoint in $L_2(\Omega, \mathbb{C}^4)$. On returning to the free Dirac operator, we choose

$$A = \sigma \cdot \left(\frac{1}{i}\nabla\right), \quad \mathcal{D}(A) = \{u^1 : u^1 \in H^1(\Omega, \mathbb{C}^2), u^{(1)} = 0 \text{ on } \partial\Omega\};$$

the operator defined with the boundary condition $u^{(2)} = 0$ on $\partial\Omega$ is an alternative choice. The adjoint of A is then

$$A^* = \sigma \cdot \left(\frac{1}{i}\nabla\right), \quad \mathcal{D}(A^*) = \left\{u^{(2)} : u^{(2)}, \left[\sigma \cdot \left(\frac{1}{i}\nabla\right)\right] u^{(2)} \in L_2(\Omega, \mathbb{C}^2)\right\}.$$

It follows that $\mathbb{D} = \mathbb{D}_0 + \beta$ is self-adjoint in $L_2(\Omega, \mathbb{C}^4)$ with domain

$$\mathcal{D}(\mathbb{D}) = \overset{0}{H}{}^1(\Omega, \mathbb{C}^2) \oplus \left\{u^{(2)} : u^{(2)}, \left[\sigma \cdot \left(\frac{1}{i}\nabla\right)\right] u^{(2)} \in L_2(\Omega, \mathbb{C}^2)\right\}. \quad (12.6.4)$$

The following interesting feature of this self-adjoint operator was observed in [72] for a ball Ω in $\mathbb{R}^3$, and investigated later for a general smooth bounded domain in $\mathbb{R}^2$ in [188].

Proposition 12.6.1 *If $\Omega \subset \mathbb{R}^3$ is bounded, the self-adjoint operator $\mathbb{D} = \mathbb{D}_0 + \beta$, with domain (12.6.4), has an eigenvalue of infinite multiplicity at -1.*

Proof. Let

$$U = \begin{pmatrix} u^{(1)} \\ u^{(2)} \end{pmatrix}, \quad u^{(2)} = \begin{pmatrix} f_1 \\ f_2 \end{pmatrix}$$

where $u^{(1)} = 0$ and f_1, f_2 are yet to be chosen. Then $(\mathbb{D} + 1)U = 0$ if

$$A^* u^{(2)} = \frac{1}{i} \left\{ \begin{pmatrix} D_1 f_2 \\ D_1 f_1 \end{pmatrix} + i \begin{pmatrix} -D_2 f_2 \\ D_2 f_1 \end{pmatrix} + \begin{pmatrix} D_3 f_1 \\ D_3 f_2 \end{pmatrix} \right\} = 0.$$

Let $\mathbf{x} = (x_1, x_2, x_3)$ and choose $f_2(\mathbf{x}) = f(x_1, x_2)$, $f_1(\mathbf{x}) = \overline{f(\mathbf{x})}$, where f satisfies $(D_1 - iD_2)f(x_1, x_2) = 0$; then $A^* u^{(2)} = 0$ and hence $(\mathbb{D} + 1)U = 0$. On identifying $\mathbb{R}^2$ with the complex plane, such functions f can be identified with the set of complex analytic functions in $\mathbb{R}^2$, since $(i/2)(D_1 - iD_2)$ is the Cauchy Riemann operator. The constructed set of functions U is infinite dimensional on a bounded Ω as it contains all monomials in x and y. The proposition therefore follows. ∎

Another noteworthy result in [188] is that $\mathcal{D}(A^*)$ strictly contains $H^1(\Omega, \mathbb{C}^2)$; the following example establishing this is based on one in [188].

Example 12.6.2

Let $B_R := \{z : |z| < R\} \subset \mathbb{C}$, $R \leq 1$, and set $\Omega_R := B_R \times (0, 1)$. Let

12.6 The Free Dirac Operator on a Bounded Domain

$$u^{(2)} = \begin{pmatrix} \overline{f} \\ f \end{pmatrix},$$

where with $\mathbf{x} = (x_1, x_2, x_3)$ and $z = x_1 + ix_2$,

$$f(z) := \sum_{k \in \mathbb{N}_0} \frac{z^k}{\sqrt{k+1}}, \quad z \in B_1.$$

Then using the polar co-ordinates $z = re^{i\theta}$ in B_1, we have

$$\|u^{(2)}\|^2_{L_2(\Omega_1, \mathbb{C}^2)} = 4\pi \sum_{k \in \mathbb{N}_0} \frac{1}{k+1} \int_0^1 r^{2k+1} dr = 2\pi \sum_{k \in \mathbb{N}_0} \frac{1}{(k+1)^2} < \infty.$$

Also,

$$\|u^{(2)}\|^2_{H^1(\Omega_R, \mathbb{C}^2)} = 2 \left\{ \left\| \frac{\partial f}{\partial r} \right\|^2_{L_2(B_R)} + \left\| \frac{1}{r} \frac{\partial f}{\partial \theta} \right\|^2_{L_2(B_R)} + \|f\|^2_{L_2(B_R)} \right\}$$

and

$$\left\| \frac{\partial f}{\partial r} \right\|^2_{L_2(B_R)} + \left\| \frac{1}{r} \frac{\partial f}{\partial \theta} \right\|^2_{L_2(B_R)} = 4\pi \sum_{k \in \mathbb{N}} \frac{k^2}{k+1} \int_0^R r^{2k-1} dr$$

$$= 2\pi \sum_{k \in \mathbb{N}_0} \frac{k}{k+1} R^{2k},$$

which is convergent for $R < 1$ but divergent for $R = 1$. Hence $u^{(2)} \in H^1_{\text{loc}}(\Omega_1, \mathbb{C}^2)$ but not in $H^1(\Omega_1, \mathbb{C}^2)$. Since f is analytic,

$$A^* u^{(2)} = \frac{1}{i}(D_1 - iD_2) u^{(2)} = 0$$

as in the proof of Proposition 12.6.1. Hence $u^{(2)} \in \mathcal{D}(A^*)$

12.7 The Brown–Ravenhall Operator on a Bounded Domain

Motivated by (12.4.5), we define the Brown–Ravenhall operator $\mathbb{B}$ on Ω to be the closure in $L_2(\Omega, \mathbb{C}^4)$ of the operator defined by the expression

$$\mathbb{B}' = \mathbb{F}^{-1} \left(\Lambda_+(\mathbf{p}) \sqrt{p^2 + 1} \Lambda_+(\mathbf{p}) \right) \mathbb{F} \tag{12.7.1}$$

on $C_0^\infty(\Omega, \mathbb{C}^4)$; a function defined on the domain Ω will be extended by zero outside Ω. On using the Parseval theorem,

$$\begin{aligned}\Pi &:= \left\{u : u, B'u \in L_2(\Omega, \mathbb{C}^4)\right\} \\ &= \left\{u : \hat{u}, \left(\Lambda_+(\mathbf{p})\sqrt{p^2+1}\Lambda_+(\mathbf{p})\right)\hat{u} \in L_2(\mathbb{R}^3, \mathbb{C}^4)\right\}.\end{aligned} \quad (12.7.2)$$

Also

$$\|u\|^2 + \|B'u\|^2 = \int_{\mathbb{R}^3} \left\{|\hat{u}|_{\mathbb{C}^4}^2 + (p^2+1)\left|\Lambda_+(\mathbf{p})^2\hat{u}\right|_{\mathbb{C}^4}^2\right\} d\mathbf{p}, \quad u \in C_0^\infty(\Omega, \mathbb{C}^4). \quad (12.7.3)$$

The domain of the closure $\mathbb{B}$ of $\mathbb{B}'$ is therefore the completion of $C_0^\infty(\Omega, \mathbb{C}^4)$ with respect to the norm

$$\left\{\int_{\mathbb{R}^3} \left\{|\hat{u}|_{\mathbb{C}^4}^2 + (p^2+1)\left|\Lambda_+(\mathbf{p})^2\hat{u}\right|_{\mathbb{C}^4}^2\right\} d\mathbf{p}\right\}^{1/2}. \quad (12.7.4)$$

If $u \in \Pi$ then for all $\psi \in \mathcal{D}(\mathbb{B}') = C_0^\infty(\Omega, \mathbb{C}^4)$,

$$\begin{aligned}(\mathbb{B}'\psi, u) &= (\mathbb{F}[\mathbb{B}'\psi], \hat{u}) \quad \text{(Parseval)} \\ &= \int_{\mathbb{R}^3} \hat{\psi}(\mathbf{p}) \overline{\left(\Lambda_+(\mathbf{p})\sqrt{p^2+1}\Lambda_+(\mathbf{p})\right)\hat{u}(\mathbf{p})} d\mathbf{p} \\ &= \int_\Omega \psi(\mathbf{x})\overline{\mathbb{F}^{-1}\left[\left(\Lambda_+(\mathbf{p})\sqrt{p^2+1}\Lambda_+(\mathbf{p})\right)\mathbb{F}u\right](\mathbf{x})} d\mathbf{x} \\ &= (\psi, \mathbb{B}'u)\end{aligned} \quad (12.7.5)$$

Hence, since $B^* = (B')^*$,

$$\Pi \subset \mathcal{D}(\mathbb{B}^*). \quad (12.7.6)$$

In particular, $\mathbb{B}' \subset \mathbb{B}^*$ so that $\mathbb{B}'$ is a positive symmetric operator.

Suppose now that $u \in \mathcal{D}(\mathbb{B}^*)$. Then, for all $\psi \in C_0^\infty(\Omega, \mathbb{C}^4)$,

$$\begin{aligned}(\psi, \mathbb{B}^*u) &= (\mathbb{B}'\psi, u) = (\mathbb{F}[\mathbb{B}'\psi], \hat{u}) \\ &= \int_{\mathbb{R}^3} \hat{\psi}(\mathbf{p})\overline{\left(\Lambda_+(\mathbf{p})\sqrt{p^2+1}\Lambda_+(\mathbf{p})\right)\hat{u}(\mathbf{p})} d\mathbf{p} \\ &= \int_\Omega \psi(\mathbf{x})\overline{\left[\mathbb{F}^{-1}\left(\Lambda_+(\mathbf{p})\sqrt{p^2+1}\Lambda_+(\mathbf{p})\right)\mathbb{F}u\right](\mathbf{x})} d\mathbf{x}\end{aligned}$$

It follows that $\mathbb{B}^*u = \mathbb{B}'u$ and hence

$$\mathcal{D}(\mathbb{B}^*) = \Pi = \left\{u : \hat{u}, \left(\Lambda_+(\mathbf{p})\sqrt{p^2+1}\Lambda_+(\mathbf{p})\right)\hat{u} \in L_2(\mathbb{R}^3, \mathbb{C}^4)\right\}, \quad \mathbb{B}^*u = \mathbb{B}'u. \quad (12.7.7)$$

12.7 The Brown–Ravenhall Operator on a Bounded Domain

The form domain of $\mathbb{B}$ is the completion of $C_0^\infty(\Omega, \mathbb{C}^4)$ with respect to the norm defined by

$$\|u\|^2 + (\mathbb{B}'u, u) = \|\hat{u}\|^2 + (\mathbb{F}[\mathbb{B}'u], \hat{u})$$
$$= \|\hat{u}\|^2 + \int_{\mathbb{R}^3} \overline{\hat{u}(\mathbf{p}) \left(\Lambda_+(\mathbf{p}) \sqrt{p^2 + 1} \Lambda_+(\mathbf{p}) \right) \hat{u}(\mathbf{p})} d\mathbf{p}$$
$$= \int_{\mathbb{R}^3} \left\{ |\hat{u}(\mathbf{p})|_{\mathbb{C}^4}^2 + \sqrt{p^2 + 1} \left| \Lambda_+(\mathbf{p})\hat{u}(\mathbf{p}) \right|_{\mathbb{C}^4}^2 \right\} d\mathbf{p}. \quad (12.7.8)$$

Remark 12.7.1

The Paley–Wiener theorem (see [83]) asserts that a function $U(\zeta)$ which is analytic on $\mathbb{C}^n$ is the Fourier-Laplace transform

$$U(\zeta) = \int_{\mathbb{R}^n} e^{-i\zeta \cdot x} u(x) dx \quad (\zeta \in \mathbb{C}^n),$$

of a function $u \in C^\infty(\mathbb{R}^n)$ with support in $\{x \in \mathbb{R}^n : |x| \leq a\}$, if and only if, there are constants $C_m \geq 0$ ($m \in \mathbb{N}_0$) such that for all $m \in \mathbb{N}_0$ and all $\zeta \in \mathbb{C}^n$

$$|U(\zeta)| \leq C_m (1 + |\zeta|)^{-m} e^{a|\mathrm{im}\,\zeta|}.$$

It follows that the integrals in (12.7.3)–(12.7.8) are finite for all $u \in C_0^\infty(\Omega, \mathbb{C}^4)$.

Remark 12.7.2

For any $v \in \mathbb{C}^4$,

$$|\Lambda_+(\mathbf{p})v|^2 = \langle \Lambda_+(\mathbf{p})v, \Lambda_+(\mathbf{p})v \rangle_{\mathbb{C}^4}$$
$$= \langle \Lambda_+(\mathbf{p})^2 v, v \rangle_{\mathbb{C}^4}$$

Since $\Lambda_+(\mathbf{p}) = 1/2 \{I_4 + \Lambda(\mathbf{p})\}$ and $\Lambda(\mathbf{p})^2 = I_4$ we have

$$\Lambda_+(\mathbf{p})^2 = \frac{1}{2} \{I_4 + \Lambda(\mathbf{p})\}.$$

and therefore
$$0 \leq \Lambda_+(\mathbf{p})^2 \leq I_4. \quad (12.7.9)$$

It follows from (12.7.4) that the domain of $\mathbb{B}$ contains the completion of $C_0^\infty(\Omega, \mathbb{C}^4)$ with respect to the norm

$$\left\{ \int_{\mathbb{R}^3} (p^2 + 1) |\hat{u}(\mathbf{p})|_{\mathbb{C}^4}^2 d\mathbf{p} \right\}^{1/2}, \quad (12.7.10)$$

which is $\overset{0}{H}{}^1(\Omega, \mathbb{C}^4)$, with continuous embedding

$$\overset{0}{H^1}(\Omega, \mathbb{C}^4) \hookrightarrow \mathcal{D}(\mathbb{B}) \tag{12.7.11}$$

and embedding norm ≤ 1. Also from (12.7.8), the form domain $Q(\mathbb{B}_F)$ of the Friedrich extension $\mathbb{B}_F$ of $\mathbb{B}$ contains the completion of $C_0^\infty(\Omega, \mathbb{C}^4)$ with respect to the norm

$$\left\{ \int_{\mathbb{R}^3} \sqrt{p^2 + 1} |\hat{u}(\mathbf{p})|_{\mathbb{C}^4}^2 d\mathbf{p} \right\}^{1/2}, \tag{12.7.12}$$

which is $\overset{0}{H^{1/2}}(\Omega, \mathbb{C}^4)$, and we have the continuous embedding

$$\overset{0}{H^{1/2}}(\Omega, \mathbb{C}^4) \hookrightarrow Q(\mathbb{B}_F) \tag{12.7.13}$$

with embedding norm ≤ 1.

Theorem 12.7.3 *The unique self-adjoint extension of $\mathbb{B}$ is $\mathbb{B}_F$.*

Proof. By the KVB theory, the Krein-von Neumann extension $\mathbb{B}_K$ of $\mathbb{B}$ has the same form domain as $\mathbb{B}_F$. Also by (6.4.10),

$$\mathcal{D}(\mathbb{B}_K) = \mathcal{D}(\mathbb{B}_F) \dotplus \mathcal{N}, \quad \mathcal{N} = \ker \mathbb{B}^*.$$

But since the only eigenvalue of $\Lambda_+(\mathbf{p})$ is at 1, we have

$$\mathbb{B}^* u = 0$$
$$\Leftrightarrow \left(\Lambda_+(\mathbf{p}) \sqrt{p^2 + 1} \Lambda_+(\mathbf{p}) \right) \hat{u}(\mathbf{p}) = 0$$
$$\Leftrightarrow \Lambda_+(\mathbf{p}) \hat{u}(\mathbf{p}) = 0$$
$$\Leftrightarrow \hat{u}(\mathbf{p}) = 0$$
$$\Leftrightarrow u = 0.$$

Hence $\mathcal{N} = \{0\}$ and it follows that $\mathbb{B}_K = \mathbb{B}_F$. Consequently $\mathbb{B}_F$ is the unique self-adjoint extension of $\mathbb{B}$ in view of Krein's result (6.4.3) that all the self-adjoint extensions S of $\mathbb{B}$ satisfy $\mathbb{B}_K \leq S \leq \mathbb{B}_F$ in the quadratic form sense. ∎

Similar considerations apply to the operator defined by

$$\mathbb{H}' := \sqrt{-\Delta + 1} := \mathbb{F}^{-1} \sqrt{p^2 + 1} \mathbb{F} \tag{12.7.14}$$

on $C_0^\infty(\Omega)$ for a bounded Ω. The closure $\mathbb{H}$ has domain

$$\overset{0}{H^1}(\Omega) = \left\{ u : \int_{\mathbb{R}^3} (p^2 + 1) |\hat{u}(\mathbf{p})|^2 d\mathbf{p} < \infty \right\} \tag{12.7.15}$$

and the form domain of $\mathbb{H}$ is

12.7 The Brown–Ravenhall Operator on a Bounded Domain

$$\overset{0}{H}{}^{1/2}(\Omega) = \left\{ u : \int_{\mathbb{R}^3} \sqrt{p^2+1}|\hat{u}(\mathbf{p})|^2 d\mathbf{p} < \infty \right\}. \tag{12.7.16}$$

Moreover, the Friedrichs extension $\mathbb{H}_F$ is the unique self-adjoint extension of $\mathbb{H}$.

Bibliography

1. Abels, H., Grubb, G., Wood, I.G.: Extension theory and Kreĭn-type resolvent formulas for non-smooth boundary value problems. J. Funct. Anal. **266**, 4037–4100 (2014)
2. Agmon, S.: The L^p approach to the Dirichlet problem. Ann. Scuola Norm. Sup. Pisa **13**, 49–92 (1959)
3. Agmon, S., Douglis, A., Nirenberg, L.: Estimates near the boundary for solutions of elliptic partial differential equations satisfying general boundary conditions. I. Comm. Pure Appl. Math. **12**, 623–727 (1959)
4. Allegretto, W., Huang, Y.X.: Picone's identity for the $p-$ Laplacian and applications. Nonlinear Anal. **32**, 819–830 (1998)
5. Alonso, A., Simon, B.: The Birman-Krein-Vishik theory of self-adjoint extensions of semi-bounded operators. J. Operator Th. **4**, 251–270 (1980)
6. Ando, T., Nishio, K.: Positive self-adjoint extensions of positive symmetric operators. Tohóku Math. J. **22**, 65–75 (1970)
7. Arendt, W.: Gaussian estimates and interpolation of the spectrum in L^p,. Diff. Int. Equ. **7**(5), 1153–1168 (1994)
8. Arendt, W., Batty, C.: Absorption semigroups and Dirichlet boundary conditions. Math. Ann. **295**, 427–448 (1993)
9. Arlinskii, Y.M.: Maximal sectorial extensions and associated with them closed forms. Ukrain. Math. Zh. **48**, 723–738 (Russian). English translation in Ukrain. Math. J. **48**(1996), 809–827 (1996)
10. Arlinskii, YuM: Extremal extensions of sectorial linear relations. Math. Stud. **7**, 81–96 (1997)
11. Arlinskii, YuM: Abstract boundary conditions for maximal sectorial extensions of sectorial operators. Math. Nachr. **209**, 5–36 (2000)
12. Arlinskii, YuM, Tsekanovskii, E.: On von Neumann's problem in extension theory of non-negative operators. Proc. Am. Math. Soc. **131**(10), 3143–3154 (2003)
13. Ashbaugh, M.S., Gesztesy, F., Mitrea, M., Shterenberg, R., Teschl, G.: The Krein-von Neumann extension and its connection to an abstract buckling problem. Math. Nachr. **283**, 165–179 (2010)
14. Ashbaugh, M.S., Gesztesy, F., Mitrea, M., Shterenberg, R., Teschl, G.: A survey on the the Krein-von Neumann extension, the corresponding abstract buckling problem, and Weyl-type spectral asymptotics for perturbed Krein Laplacians in non-smooth domains. Mathematical Physics, Spectral Theory and Stochastic Analysis. Oper. Th. Adv. Appl. **232**, 165–179 (2013), Adv. Partial Differ. Equ. pp. 165–179. Birkhaüser/Springer, Basel AG, Basel (2013)

15. Balinsky, A., Evans, W.D.: Spectral Analysis of Relativistic Operators. Imperial College Press, Singapore (2011)
16. Balinsky, A., Evans, W.D., Lewis, R.T.: The Analysis and Geometry of Hardy's Inequality. Springer, New York (2015)
17. Bañuelos, R., Osękowski, A.: Sharp martingale inequalities and applications to Riesz transforms on manifolds, Lie groups and Gauss space. J. Funct. Anal. **269**, 1652–1713 (2015)
18. Bañuelos, R., Wang, G.: Sharp inequalities for martingales with applications to the Beurling-Ahlfors and Riesz transformations. Duke. Math. J. **80**, 575–600 (1995)
19. Barbatis, G.: Improved Rellich inequalities for the polyharmonic operator. Indiana Univ. Math. J. **55**(4), 1401–1422 (2006)
20. Benedikt, J., Drábek, P.: Estimates of the principal eigenvalue of the $p-$ Laplacian. J. Math. Anal. Appl. **393**, 311–315 (2012)
21. Benedikt, J., Drábek, P.: Estimates of the principal eigenvalue of the $p-$biharmonic operator. Nonlinear Anal. **75**, 5374–5379 (2012)
22. Behrndt, J., Langer, M.: Elliptic operators, Dirichlet to Neumann maps and quasi-boundary triples. Lond. Math. Soc. Lect. Note Ser. **404**, 121–160 (2012)
23. Behrndt, J., Micheler, T.: Elliptic differential operators on Lipschitz domains and abstract boundary-value problems. J. Funct. Anal. **267**, 3657–3709 (2014)
24. Birman, MSh: On the theory of self-adjoint extensions of positive definite operators. Mat. Sbornik **38**, 431–450 (1956)
25. Birman, M.Sh.: Perturbations of the continuous spectrum by varying the boundary and boundary conditions. Vestnik Leningrad Univ. **17**(1), 22–55 (1962). English translation in Spectral Theory of Differential Operators, AMS Translation, ser. 2, vol. 225, pp. 75–84. American Mathematical Society, Providence (2000)
26. Bourgain, J.: On Pleijel's nodal domain theorem. Int. Math. Res. Not. **2015**(6), 2841–2855 (2013)
27. Brezis, H.: Functional Analysis. Sobolev Spaces and Partial Differential Equations. Springer, New York (2011)
28. Brothers, J., Ziemer, W.P.: Minimal rearrangements of Sobolev functions. J. Reine Angew. Math. **384**, 153–179 (1988)
29. Brown, B.M., Evans, W.D.: Self-adjoint and m-sectorial extensions of Sturm-Liouville operators. Integral Equ. Oper. Theory **85**(2), 151–166 (2016)
30. Brown, B.M., Marletta, M., Naboko, S., Wood, I.: Boundary triplets and M-functions for non-selfadjoint operators, with applications to elliptic PDEs and block operator matrices. J. Lond. Math. Soc. **2**(77), 700–718 (2008)
31. Brown, B.M., Hinchcliffe, J., Marletta, M., Naboko, S., Wood, I.: The abstract Titchmarsh-Weyl M-function for adjoint pairs and its relation to the spectrum. Integral Equ. Oper. Theory **63**, 297–320 (2009)
32. Bruk, V.M.: On a class of boundary-value problems with a spectral parameter in the boundary condition. Math. Sb. **100**, 210–216 (1976)
33. Calkin, J.W.: Abstract symmetric boundary conditions. Trans. Am. Math. Soc. **45**, 369–442 (1939)
34. Calkin, J.W.: Symmetric transformations in Hilbert space. Duke Math. J. **7**, 504–508 (1940)
35. Chaudhuri, J., Everitt, W.N.: On the spectrum of ordinary second order differential operators, Proc. Roy. Soc. Edinb. **LXVIII**, A, Part II, 95–119 (1967/68)
36. Chernoff, P.R.: Self-adjointness of powers of generators of hyerbolic equations. J. Funct. Anal. **12**, 401–414 (1973)
37. Chernoff, P.R.: Schrödinger and Dirac operators with singular potentials and hyperbolic equations. Pacific J. Math. **72**(2), 361–382 (1977)
38. Cianchi, A., Fusco, N.: Functions of bounded variation and rearrangements. Arch. Rat. Mech. Anal. **165**, 1–40 (2002)
39. Cianchi, A., Pick, L.: An optimal endpoint trace embedding. Ann. Inst. Fourier, Grenoble **60**, 939–951 (2010)

40. Cuesta. M.: On the Fučik spectrum of the Laplacian and the p−Laplacian. In: Drábek, P. (ed.) Proceedings of Seminar Differential Equation, pp. 67-96 (2000). (Univ. of West Bohemia, Pilsen)
41. Cuesta, M., de Figueiredo, D.G., Gossez, J.-P.: A nodal domain property for the p−Laplacian. C. R. Acad. Sci. Paris **330**, 669–673 (2000)
42. Courant, R., Hilbert, D.: Methods of Mathematical Physics. I. Interscience, New York (1953)
43. Davies, E.B.: Some norm bounds and quadratic form inequalities for Schrödinger operators (II). J. Operator Theory **12**, 177–196 (1984)
44. Davies, E.B.: Heat Kernels and Spectral Theory. Cambridge University Press, Cambridge (1989)
45. Davies, E.B., Hinz, A.M.: Explicit constants for Rellich inequalities in $L_p(\Omega)$. Math. Zeit. **227**, 511–523 (1998)
46. Deimling, K.: Nonlinear Functional Analysis. Springer, Berlin (1985)
47. Denk, R., Hieber, M., Prüss, J.: $\mathcal{R}$—Boundedness, Fourier Multipliers and Problems of Elliptic and Parabolic Type, pp. 1–114. Memoirs of the American Mathematical Society (2003)
48. Derkach, V.A., Malamud, M.M.: Generalized resolvents and the boundary value problem for Hermitian operators with gaps. J. Funct. Anal. **95**, 1–95 (1991)
49. Diening, L., Harjulehto, P., Hästö, P., Růžička, M.: Lebesgue and Sobolev Spaces with Variable Exponents. Lecture Notes in Mathematics, vol. 2017. Springer, Berlin (2011)
50. Drábek, P., Ôtani, M.: Global bifurcation result for the p−biharmonic operator. Electron. J. Differ. Equ. **2001**(48), 1–19 (2001)
51. Drábek, P., Robinson, S.B.: On a generalization of the Courant nodal domain theorem. J. Differ. Equ. **181**, 58–71 (2002)
52. Dunford, N., Schwartz, J.T.: Linear operators I. Interscience, New York and London (1958)
53. Eckhardt, J., Gesztesy, F., Nichols, R., Teschl, G.: Weyl-Titchmarsh theory for Sturm-Liouville operators with distributional potentials. Opscula Math. **33**(3), 467–563 (2013)
54. Edmunds, D.E., Evans, W.D.: Spectral theory and differential operators, 2nd edn. Oxford University Press, Oxford (2018)
55. Edmunds, D.E., Evans, W.D.: Hardy Operators. Function Spaces and Embeddings. Springer, Berlin (2004)
56. Edmunds, D.E., Evans, W.D.: Spectral problems on arbitrary open subsets of $\mathbb{R}^n$ involving the distance to the boundary. J. Comput. Appl. Math. **194**, 36–53 (2004)
57. Edmunds, D.E., Evans, W.D.: The Rellich inequality. Rev. Math. Complut. **29**(3), 511–530 (2016)
58. Edmunds, D.E., Gogatishvili, A., Kopalini, T.: Construction of function spaces close to L^∞ with associate space close to L^1. J. Fourier Anal. Appl. (To appear)
59. Edmunds, D.E., Gurka, P., Lang, J.: Nuclearity and non-nuclearity of some Sobolev embeddings on domains. J. Approx. Theory **211**, 94–103 (2016)
60. Edmunds, D.E., Kerman, R., Pick, L.: Optimal Sobolev imbeddings involving rearrangement-invariant quasi-norms. J. Funct. Anal. **170**, 307–355 (2000)
61. Edmunds, D.E., Lang, J.: Behaviour of the approximation numbers of a Sobolev embedding in the one-dimensional case. J. Funct. Anal. **206**, 149–166 (2004)
62. Edmunds, D.E., Lang, J., Méndez, O.: Differential Operators on Spaces of Variable Integrability. World Scientific, Singapore (2014)
63. Edmunds, D.E., Lang, J.: Eigenvalues. Embeddings and Generalised Trigonometric Functions. Springer, Berlin (2011)
64. Edmunds, D.E., Lang, J.: Non-nuclearity of a Sobolev embedding on an interval. J. Approx. Theory **178**, 22–29 (2014)
65. Edmunds, D.E., Lang, J.: Asymptotic formulae for s−numbers of a Sobolev embedding and a Volterra type operator. Rev. Math. Complut. **29**(1), 1–11 (2016)
66. Edmunds, D.E., Lang, J., Nekvinda, A.: Estimates of s−numbers of a Sobolev embedding involving spaces of variable exponent. J. Math. Anal. Appl. **430**, 1088–1101 (2015)
67. Edmunds, D.E., Nekvinda, A.: Characterisation of zero trace functions in variable exponent Sobolev spaces. Math. Nachr. **290**, 2247–2258 (2017)

68. Edmunds, D.E., Nekvinda, A.: Characterisation of zero trace functions in higher-order spaces of Sobolev type. J. Math. Anal. Appl. **459**, 879–892 (2018)
69. Edmunds, D.E., Triebel, H.: Function Spaces. Entropy Numbers. Differential Operators. Cambridge University Press, Cambridge (1996)
70. Enflo, P.: A counterexample to the approximation problem in Banach spaces. Acta Math. **130**, 309–317 (1973)
71. Evans, L.C., Gariepy, R.F.: Measure Theory and Fine Properties of Functions. CRC Press, Boca Raton (1992)
72. Evans, W.D.: The Dirac equation with a spherically symmetric scalar potential. Q. J. Math. Oxford **20**(21), 89–99 (1970)
73. Evans, W.D., Harris, D.J.: Sobolev embeddings for generalised ridged domains. Proc. Lond. Math. Soc. **54**, 141–175 (1987)
74. Evans, W.D., Lewis, R.T.: Hardy and Rellich inequalities with remainders. J. Math. Inequal. **1**(4), 473–490 (2007)
75. Fabes, E., Méndez, O., Mitrea, M.: Boundary layers on Sobolev-Besov spaces and Poisson's equation for the Laplacian in Lipschitz domains. J. Funct. Anal. **159**, 323–368 (1998)
76. Fabian, M., Habala, P., Santalucía, V.M., Pelaut, J., Zizler, V.: Functional Analysis and Infinite-Dimensional Geometry. Springer, Berlin (2001)
77. Faris, W.G.: Self-Adjoint Operators. Lecture Notes in Mathematics, vol. 433. Springer, Berlin (1975)
78. di Fazio, G., Fanciullo, M.S., Zamboni, P.: Harnack inequality and regularity for degenerate quasilinear elliptic equations. Math. Zeit. **264**, 679–695 (2010)
79. Fichera, G.: Alcuni recenti sviluppi della teoria dei problemi al contorno per le equazioni alle derivate parziali lineari. In: Equazione lineari alle derivate parziali. Ediz. Cremonese Roma, pp. 174–227 (1955)
80. Fichera, G.: On a unified theory of boundary-value problems for elliptic-parabolic equations of second order. Boundary Problems in Differential Equations, pp. 97–120. University of Wisconsin Press, Madison (1960)
81. Fraenkel, L.E.: Introduction to Maximum Principles and Symmetry in Elliptic Problems. Cambridge University Press, Cambridge (2000)
82. Franzina, G., Lamberti, P.D.: Existence and uniqueness for a $p-$Laplacian nonlinear eigenvalue problem. Electron. J. Differ. Equ. **26**, 10 (2010)
83. Friedlander, F.G.: Introduction to the Theory of Distributions. Cambridge University Press, Cambridge (1982)
84. Friedlander, L.: Asymptotic behavior of the eigenvalues of the $p-$Laplacian. Commun. Partial Differ. Equ. **14**, 1059–1069 (1989)
85. Friedman, A.: Partial Differential Equations. Holt-Rinehart-Winston, New York (1969)
86. Friedrichs, K.: Spektraltheorie halbbeschränkter Operatoren und Anwendung auf die Spektralzerlegung von Differentialoperatoren I. Math. Ann. **109**, 465–487 (1933–34)
87. Fukagai, N., Ito, M., Narukawa, K.: Limit as $p \to \infty$ of p-Laplace eigenvalue problems and L^∞-inequality of the Poincaré type. Differ. Integral Equ. **12**, 183–206 (1999)
88. Gesztesy, F., Mitrea, M.: A description of all self-adjoint extensions of the Laplacian and Krein-type resolvent formulas on non-smooth domains. J. D'Anal. Math. **113**, 53–129 (2011)
89. Giaquinta, M.: Introduction to Regularity Theory for Nonlinear Elliptic Systems. Birkhäuser, Basel (1993)
90. Gilbarg, D., Trudinger, N.S.: Elliptic Partial Differential Equations of Second Order, 2nd edn., Revised 3rd Printing. Springer, Berlin (1998)
91. Goldstein, J.A.: Semigroups of Linear Operators and Applications. Oxford University Press, Oxford (1985)
92. Gorbachuk, V.I., Gorbachuk, M.L.: Boundary Value Problems for Operator Differential Equations. Kluwer, Dordrecht (1991)
93. Grafakos, L.: Best bounds for the Hilbert transform on $L^p(\mathbb{R}^1)$. Math. Res. Lett. **4**, 469–471 (1997)
94. Grafakos, L.: Classical Fourier Analysis, 2nd edn. Springer, New York (2008)

95. Grothendieck, A.: Produits tensoriels topologiques et espaces nucléaires. Mem. Am. Math. Soc. **16** (1955)
96. Grubb, G.: A characterization of the non-local boundary-value problems associated with an elliptic operator. Ann. Scuola Norm. Sup. Pisa **3**(22), 425–513 (1968)
97. Grubb, G.: Spectral asymptotics for the "soft" self-adjoint extension of a symmetric differential operator. J. Operator Th. **10**, 9–20 (1983)
98. Grubb, G.: Krein resolvent formulas for elliptic boundary problems in non-smooth domains. Rend. Semin. Mat. Univ. Politec. Torino **66**, 13–39 (2008)
99. Hadamard, J.: La théorie des équations aux dérivées partielles. Éditions Scientifiques, Pekin (1964)
100. Han, Qing: Lin, Fanghua: Elliptic Partial Differential Equations. Courant Lecture Notes in Mathematics. American Mathematical Society, Providence (1997)
101. Hajłasz, P.: Pointwise Hardy inequalities. Časopis Pěst Math. **91**, 463–471 (1966)
102. Haroske, D.D., Triebel, H.: Distributions. Sobolev Spaces. Elliptic Equations. European Mathematical Society, Zürich (2008)
103. Haroske, D.D.: Envelopes and Sharp Embeddings of Function Spaces. Chapman and Hall/CRC, Boca Raton (2007)
104. Hartman, P.: Ordinary Differential Equations. John Hopkins Univ, Baltimore (1973)
105. Hayden, T.L.: Representation theorems in reflexive Banach spaces. Math. Zeit. **104**, 405–406 (1968)
106. Helms, L.L.: Introduction to Potential Theory. Wiley, New York (1969)
107. Herbst, I.: Spectral theory of the operator $(p^2 + m^2)^{1/2} - ze^2/r$. Commun. Math. Phys. **53**(3), 285–294 (1977)
108. Hewitt, E., Stromberg, K.: Real and Abstract Analysis. Springer, New York (1965)
109. Ivrii, V.Ya.: On the second term of the spectral asymptotics for the Laplace-Beltrami operator on manifolds with boundary. Funct. Anal. Appl. **14** , 98–106 (1980)
110. Iwaniec, T., Martin, G.: Riesz transforms and related singular integrals. J. Reine Angew. Math. **473**, 25–57 (1996)
111. Jameson, G.J.O.: Summing and Nuclear Norms in Banach Space Theory. Cambridge University Press, Cambridge (1987)
112. Jaroš, J.: Picone's identity for the $p-$biharmonic operator with applications. Electron. J. Differ. Equ. **2011**(122), 1–6 (2011)
113. Jerison, D., Kenig, C.: The inhomogeneous Dirichlet problem in Lipschitz domains. J. Funct. Anal. **130**, 161–219 (1995)
114. Jörgens, K.: Perturbations of the Dirac operator. In: Proceedings of the Conference on the Theory of Ordinary and Partial Differential Equations, Dundee, Scotland, 1972. Lecture Notes in Mathematics, vol. 280, pp. 87–102. Springer, Berlin (1972)
115. Jost, J.: Partial Differential Equations. Springer, Berlin (2007)
116. Juutinen, P., Linqvist, P., Manfredi, J.J.: The $\infty-$ eigenvalue problem. Arch. Rat. Mech. Anal. **148**, 89–105 (1999)
117. Kadlec, J., Kufner, A.: Characterisation of functions with zero traces by integrals with weight functions. Proc. Am. Math. Soc. **127**, 417–423 (1999)
118. Kajikiya, R.: A priori estimate for the first eigenvalue of the $p-$ Laplacian. Differ. Integral Equ. **28**(4), 1011–1028 (2015)
119. Kato, T.: Perturbation Theory for Linear Operators, 2nd edn. Springer, New York (1980)
120. Kalf, H., Walter, J.: Strongly singular potentials and essential self-adjointness of singular elliptic operators in $C_0^\infty(\mathbb{R}^n \setminus \{0\})$. J. Funct. Anal. **10**, 114–130 (1972)
121. Kalf, H.: Self-adjointness for strongly singular potentials with a $-|x|^2$ fall-off at infinity. Math. Z. **133**, 249–255 (1973)
122. Kalf, H.: A characterization of the Friedrichs extension of Sturm-Liouville operators. J. Lond. Math. Soc. 2(17), 511–521 (1978)
123. Kawohl, B., Lindqvist, P.: Positive eigenfunctions for the $p-$Laplace operator revisited. Analysis (Munich) **26**, 545–550 (2006)

124. Kellogg, O.D.: On the derivatives of harmonic functions on the boundary. Trans. Am. Math. Soc. **33**, 486–510 (1931)
125. Kerman, R., Pick, L.: Optimal Sobolev imbeddings. Forum Math. **18**, 535–570 (2006)
126. Kerman, R., Pick, L.: Explicit formulas for optimal rearrangement-invariant norms in Sobolev imbedding inequalities. Stud. Math. **206**, 97–119 (2011)
127. Kesevan, S.: Topics in Functional Analysis and Applications. Wiley, New Delhi (1989)
128. Kinnunen, J., Martio, O.: Hardy's inequalities for Sobolev functions. Math. Res. Lett. **4**, 489–500 (1997)
129. Klaus, M., Wüst, R.: Characterization and uniqueness of distinguished self-adjoint extensions of Dirac operators. Commun. Math. Phys. **64**(2), 171–176 (1979)
130. Kočubeĭ, A.N.: Extensions of symmetric operators and symmetric binary relations. Math. Notes (1) **17**, 25–28 (1975)
131. König, H.: Eigenvalue Distribution of Compact Operators. Birkhäuser, Basel (1986)
132. König, M.: Über das Verhalten der Lösung des Dirichletproblems am Rand des Gebietes, wenn der Rand zur Classes $C^{2,\alpha}$ gehört. Proc. Roy. Soc. Edinb. **80A**, 163–176 (1978)
133. König, M.: On Juliusz Schauder's paper on linear elliptic differential equations. Topol. Methods Nonlinear Anal. **11**, 351–360 (1998)
134. König, M., On, J.: Schauder's method to solve elliptic differential equations. J. Fixed Point Theory Appl. **9**, 135–196 (2011)
135. Kováčik, O., Rákosník, J.: On spaces $L^{p(x)}(\Omega)$ and $W^{k,p(x)}(\Omega)$. Czech. Math. J. **41**, 592–618 (1991)
136. Krein, M.G.: The theory of self-adjoint extensions of semi-bounded Hermitian transformations and its applications. I. Math. Sbornik **20**, 431–495 (1947)
137. Krein, M.G.: The theory of self-adjoint extensions of semi-bounded Hermitian transformations and its applications. II. Math. Sbornik **21**, 365–404 (1947)
138. Kresin, G., Maz'ya, V.: Maximum Principles and Sharp Constants for Solutions of Elliptic and Parabolic Systems. Mathematical surveys and Monographs, vol. 183. American Mathematical Society, Providence, RI (2012)
139. Kudryavtsev, S.N.: Bernstein widths of the class of functions of finite smoothness. Math. Sb. **190**(4), 63–86 (1999)
140. Kufner, A.: Weighted Sobolev Spaces. Wiley, Chichester (1985)
141. Ladyzhenskaya, O.A., Ural'tseva, N.N.: Linear and Quasilinear Elliptic Equations. Academic Press, New York (1968)
142. Lax, P.D., Milgram, A.N.: Parabolic equations, in Contributions to the theory of partial differential equations. Ann. Math. **33**, 167–190 (1954)
143. Lê, An: Eigenvalue problems for the p–Laplacian. Nonlinear Anal. **64**, 1057–1099 (2006)
144. Lebesgue, H.: Sur le problème de Dirichlet. Comptes Rendus Acad. Sc. Paris **154**, 335 (1912)
145. Levinson, N.: Criteria for the limit-point case for 2nd-order linear differential operators, Časopis Pešt. Math. Fys. **74**, 17–20 (1949)
146. Lidskii, V.B.: Non-selfadjoint operators with a trace (Russian). Dokl. Akad. Nauk **125**, 485–487 (1959)
147. Lieberman, G.M.: Boundary regularity for solutions of degenerate elliptic equations. Nonlinear Anal. TMA **12**, 1203–1219 (1988)
148. Lindqvist, P.: On the equation $\operatorname{div}\left(|\nabla u|^{p-2} \nabla u\right) + \lambda |u|^{p-2} u = 0$. Proc. Am. Math. Soc. **109**(1), 157–164 (1990). Addendum, ibidem **116**(2), 583–584 (1992)
149. Lindqvist, P.: On a nonlinear eigenvalue problem. Ber. Univ. Jyväskylä Math. Inst. **68**, 33–54 (1995)
150. Lindqvist, P.: Notes on the p–Laplace Equation. University Jyväskylä Lecture Notes (2006)
151. Lions, J.L., Magenes, E.: Problemi ai limiti non omogenei (V). Ann. Scuola Norm. Pisa **3**(16), 1–44 (1962)
152. Lions, J.L., Magenes, E.: Non-homogeneous Boundary Value Problems and Applications I. Springer, Berlin (1972)
153. Lyantze, V.E., Storozh, O.G.: Methods of the Theory of Unbounded Operators (Russian). Naukova Dumka, Kiev (1983)

154. Maurin, K.: Methods of Hilbert Spaces. PWN-Polish Scientific Publishers, Warsaw (1972)
155. Melrose, R.: Weyl's conjecture for manifolds with concave boundary. Proceedings of Symposia in Pure Mathematics, vol. 36, pp. 257–274. American Mathematical Society, Providence (1980)
156. Morrey, C.B.: Multiple integrals in the Calculus of Variations. Springer, Berlin (1966)
157. Musielak, J.: Orlicz Spaces and Modular Spaces. Lecture Notes in Mathematics, vol. 1034. Springer, Berlin (1983)
158. Naimark, M.A.: Linear Differential Operators. Part II. Harrap, London (1968)
159. Nardi, G.: Schauder estimate for solutions of Poisson's equation with Neumann boundary condition. L'Enseignement Mathématique **60**(2), 423–437 (2014)
160. Nečas, J.: Les méthodes directes en théorie des équations elliptiques. Masson, Paris. English translation: Direct methods in the theory of elliptic equations, p. 2012. Springer, Berlin (1967)
161. Nguyen, V.K.: Bernstein numbers of embeddings of isotropic and dominating mixed Besov spaces. Math. Nachr. **288**, 1694–1717 (2015)
162. Osękowski, A.: Sharp weak type estimates for Riesz transforms. Monatsh Math. **174**, 305–327 (2014)
163. Ouhabaz, El Maati: Analysis of Heat Equations on Domains. Princeton University Press, Princeton (2005)
164. Pichorides, S.K.: On the best values of the constants in the theorems of M. Riesz, Zygmund and Kolmogorov. Stud. Math. **44**, 165–179 (1972)
165. Pick, L.: Optimality of function spaces in Sobolev embeddings. In: Maz'ya, V. (ed.) Sobolev Spaces in Mathematics I, Sobolev Type Inequalities, pp. 249–280. Springer/Tamara Rozhkovskaya Publisher, Berlin/Novosibirsk (2009)
166. Pietsch, A.: r–nukleare Sobolevsche Einbettungsoperatoren. In: Elliptische Differentialgleichungen II, pp. 203–215. Akademie, Berlin (1971)
167. Pietsch, A.: Approximation numbers of nuclear operators and geometry of Banach spaces. Arch. Math. **57**, 155–168 (1991)
168. Pietsch, A.: History of Banach Spaces and Linear Operators. Birkhäuser, Boston (2007)
169. Pietsch, A., Triebel, H.: Interpolationstheorie für Banachideale von beschränkten linearen Operatoren. Stud. Math. **31**, 95–109 (1968)
170. Pleijel, A.: Remarks on Courant's nodal line theorem. Commun. Pure Appl. Math. **9**, 543–550 (1956)
171. Posilicano, A., Raimondi, L.: Kreĭn's resolvent formula for second order elliptic differential operators. J. Phys. A **42**, 015204 (2009). 11 pp
172. Pucci, P., Serrin, J.B.: The Maximum Principle. Birkhäuser, Basel (2007)
173. Protter, M.H., Weinberger, H.: Maximum Principles in Differential Equations. Prentice-Hall, Englewood Cliffs (1967)
174. Raynal, J.C., Roy, S.M., Singh, V., Martin, A., Stubbe, J.: The "Herbst Hamiltonian" and the mass of boson stars. Phys. Lett. B **320**(1–2), 105–109 (1994)
175. Rellich, F.: Halbbeschränkte gewöhnliche Differentialoperatoren zweiter Ordnung. Math. Ann. **122**, 343–368 (1950/51)
176. Rockafellar, R.T.: Convex Analysis. Princeton University Press, Princeton (1970)
177. Rosenberger, R.: A new characterization of the Friedrichs extension of semi-bounded Sturm-Liouville operators. J. Lond. Math. Soc. **2**(31), 501–510 (1985)
178. Rosenberger, R.: Charakterisierungen der Friedrichsfortsetzung von halbbeschränkten Sturm-Liouville Operatoren, Dissertation, Technische Hochschule Darmstadt (1984)
179. Ruston, A.F.: Fredholm Theory in Banach Spaces. Cambridge University Press, Cambridge (1986)
180. Růžička, M.: Electrorheological Fluids: Modeling and Mathematical Theory. Lecture Notes in Mathematics, vol. 1748. Springer, Berlin (2000)
181. Safarov, Yu., Vassiliev, D.: The Asymptotic Distribution of Eigenvalues of Partial Differential Operators. Translations of Mathematical Monographs, vol. 155. American Mathematical Society, Providence (1997)

182. Schauder, J.: Über lineare elliptische Differentialgleichungen zweiter Ordnung. Math. Zeit. **38**, 257–282 (1934)
183. Schilling, René L.: Measure. Integrals and Martingales. Cambridge University Press, Cambridge (2005)
184. Schechter, M.: General boundary value problems for elliptic differential equations. Commun. Pure Appl. Math. **12**, 457–486 (1959)
185. Schechter, M.: On L^p estimates and regularity. Am. J. Math. **85**, 1–13 (1963)
186. Schechter, M.: Coerciveness in L^p,. Trans. Am. Math. Soc. **107**, 10–29 (1963)
187. Schechter, M.: On the essential spectrum of an arbitrary operator I. J. Math. Anal. Appl. **13**, 205–215 (1966)
188. Schmidt, K.M.: A remark on boundary value problems for the Dirac operator. Q. J. Math. Oxford **2**(46), 509–516 (1995)
189. Schmincke, U.W.: Distinguished self-adjoint extensions of Dirac operators. Math. Z. **129**, 335–349 (1972)
190. Schmüdgen, K.: Unbounded Self-Adjoint Operators in Hilbert Space. Graduate Texts in Mathematics, vol. 265. Springer, Heidelberg (2012)
191. Simader, C.G.: Bemerkungen uber Schrödinger-operatoren mit stark singularen Potentialen. Math. Z. **138**, 53–70 (1974)
192. Simader, C.G., Sohr, H.: The Dirichlet Problem for the Laplacian in Bounded and Unbounded Domains. Pitman Research Notes in Mathematics, vol. 360. Addison Wesley, Harlow (1996)
193. Simon, L.: Schauder estimates by rescaling. Calc. Var. **5**, 391–407 (1997)
194. Stein, E.M.: Singular Integrals and Differentiability Properties of Functions. Princeton University Press, Princeton (1970)
195. Stromberg, K.R.: An Introduction to Classical Real Analysis. Wadsworth, Belmont (1981)
196. Struwe, M.: Variational Methods. Springer, Berlin (1990)
197. Szulkin, A.: Ljusternik-Schnirelmann theory on C^1−manifolds. Ann. Inst. Henri Poincaré **5**, 119–139 (1988)
198. Taylor, A.E.: Introduction to Functional Analysis. Wiley, New York (1958)
199. Thaller, B.: The Dirac Equation. Springer, Berlin (1994)
200. Tidblom, J.: A geometrical version of Hardy's inequality for $\overset{0}{W}{}^{1,p}(\Omega)$, Proc. Am. Math. Soc. **132**, 2265–2271 (2004)
201. Titchmarsh, E.C.: Eigenfunction Expansions Associated with Second-order Differential Equations: Part I. Oxford University Press, Oxford (1962)
202. Tix, C.: Lower bound for the ground state energy of the no-pair Hamiltonian. Phys. Lett. B **405**(3–4), 293–296 (1997)
203. Tix, C.: Self-adjointness and spectral properties of a pseudo-relativistic Hamiltonian due to Brown and Ravenhall, preprint archive mp-arc 97-441 (1997)
204. Triebel, H.: Interpolation Theory. Function Spaces. Differential Operators. North-Holland, Amsterdam (1978)
205. Triebel, H.: Higher Analysis. J. A. Barth, Leipzig (1992)
206. Triebel, H.: Theory of Function Spaces III. Birkhäuser, Basel (2006)
207. Triebel, H.: Nuclear embeddings in function spaces. Math. Nachr. **290**, 3038–3048 (2017)
208. Trudinger, N.S.: On Harnack type inequalities and their applications to quasilinear elliptic equations. Commun. Pure Appl. Math. **20**, 721–747 (1967)
209. Urysohn, P.: Zur ersten Randwertaufgabe der Potentialtheorie. Ein Fall der Unlösbarkeit. Math. Zeit. **23**, 155–158 (1925)
210. Vainerman, L.I.: On extensions of closed operators in Hilbert spaces. Math. Notes **28**, 871–875 (1980)
211. Višik, M.I.: Sur les systèmes fortement elliptiques d'équations différentielles: Math. Sb. **29**, 615–676 (1951)
212. Višik, M.I.: On general boundary problems for elliptic differential equations, Trudy Moscov. Math. Obsc. **1**, 187–246 (1952). English translation in Am. Math. Soc. Tran., Ser. 2 **24**, 107–172 (1963)

213. von Neumann, J.: Allgemeine Eigenwerttheorie Hermitischer Functionaloperatoren. Math. Ann. **102**, 49–131 (1929)
214. Weder, R.A.: Spectral analysis of pseudo-differential operators. J. Funct. Anal. **20**(4), 319–337 (1975)
215. Weyl, H.: Über die Randwertaufgabe der Strahlungstheorie und asymptotische Spektralgesetze. J. Reine Angew. Math. **143**, 177–202 (1913)
216. Wood, I.: Maximal L^p-regularity for the Laplacian on Lipschitz domains. Math. Zeit. **255**, 855–875 (2007)
217. Wüst, R.: A convergence theorem for self-adjoint operators applicable to Dirac operators with cut-off potentials. Math. Z. **131**, 339–349 (1973)
218. Wüst, R.: Distinguished self-adjoint extensions of Dirac operators constructed by means of cut-off potentials. Math. Z. **141**, 93–98 (1975)
219. Wüst, R.: Dirac operators with strongly singular potentials. Math. Z. **152**, 259–271 (1977)
220. Yao, S., Sun, J., Zettl, A.: The Sturm-Liouville Friedrichs extension. Appl. Math. **60**(3), 299–320 (2015)
221. Yaouanc, A.L., Oliver, L., Raynal, J.-C.: The Hamiltonian $(p^2 + m^2)^{1/2} - a/r$ near the critical value $\alpha_c = 2/\pi$. J. Math. Phys. **38**, 3997–4012 (1997)
222. Yosida, K.: Functional Analysis. Springer, Berlin (1965)
223. Zhikov, V.V.: On some variational problems. Russ. J. Math. Phys. **5**, 105–116 (1997)
224. Ziemer, W.P.: Weakly Differentiable Functions. Springer, New York (1989)

Author Index

A
Abels, H., 201
Agmon, S., 91, 194, 196, 204, 210
Allegretto, W., 229
Alonso, A., 137, 140
Ando, T., 156
Arendt, W., 111, 112
Arlinskii, Yu. M., 142, 151, 156, 157, 186
Ashbaugh, M. S., 137, 143, 144, 201

B
Balinsky, A. A., 235, 236, 244, 288, 289, 291–293
Bañuelos, R., 32, 33
Barbatis, G., 235, 244
Batty, C., 112
Behrndt, J., 201
Benedikt, J., 229, 245
Birman, M. Sh., 137, 140
Bourgain, J., 233
Brezis, H., 98, 100, 113
Brothers, J., 105
Brown, B. M., 158, 168, 185, 186
Bruk, V. M., 158

C
Calkin, J. W., 119, 158
Chaudhuri, J., 166
Chernoff, P. R., 289
Cianchi, A., 23
Courant, R., 224
Cuesta, M., 225, 228, 233

D
Davies, E. B., 112, 235, 236
Deimling, K., 220
Denk, R., 112
Derkach, V. A., 158
Diening, L., 268, 269, 272
Douglis, A., 91, 194, 196, 204, 210
Drábek, P., 225, 229, 245
Dunford, N., 1

E
Eckhardt, J., 169, 174
Edmunds, D. E., 12, 14, 18, 20, 21, 23, 38, 51, 97, 99, 105, 115, 118, 119, 151, 154, 164, 178, 184, 185, 228, 236, 247, 249–252, 256, 261, 263–266, 268, 269, 272–274, 279, 280
Enflo, P., 259
Evans, L. C., 24
Evans, W. D., 12, 14, 18, 20, 21, 23, 38, 51, 97, 99, 105, 115, 118, 119, 151, 154, 164, 168, 178, 184–186, 235, 236, 244, 247, 249, 251, 252, 256, 265, 273, 279, 288, 289, 291–293, 296
Everitt, W. N., 166

F
Fabes, E., 210
Fabian, M., 9
Fanciullo, M. S., 218
Faris, W. G., 140
Fazio, G. di, 218
Fichera, G., 7, 63
Figueiredo, D. G. de, 225

Fraenkel, L. E., 41, 55, 63
Franzina, G., 233
Friedlander, F. G., 299
Friedlander, L., 233
Friedman, A., 91, 98, 100
Fukagai, N., 228

G
Gariepy, R. F., 24
Gesztesy, F., 137, 143, 144, 169, 174, 196, 201
Giaquinta, M., 81
Gilbarg, D., 56, 81
Gogatishvili, A., 280
Goldstein, J. A., 106, 110
Gorbachuk, M. L., 158
Gorbachuk, V. I., 158
Gossez, J.-P., 225
Grafakos, L., 24, 27, 28
Grothendieck, A., 259
Grubb, G., 137, 141, 143, 144, 149, 188, 191–194, 196, 197, 200, 201
Gurka, P., 263, 265

H
Habala, P., 9
Hadamard, J., 55
Han, Q., 56, 81
Harjulehto, P., 268, 269, 272
Haroske, D. D., 190, 196, 263
Harris, D. J., 279
Hartman, P., 168
Hästö, P., 268, 269, 272
Hayden, T. L., 7
Helms, L. L., 55
Herbst, I., 289, 290, 292
Hewitt, E., 214
Hieber, M., 112
Hilbert, D., 224
Hinchcliffe, J., 158
Hinz, A., 235
Huang, Y. X., 229

I
Ito, M., 228
Ivrii, V. Ya., 113
Iwaniec, T., 32, 33

J
Jameson, G. J. O., 260

Jaroš, J., 245
Jerison, D., 210
Jörgens, K., 289
Jost, J., 81
Juutinen, P., 228

K
Kadlec, J., 249
Kajikiya, R., 232
Kalf, H., 169, 171, 184, 202
Kato, T., 134, 135, 151, 153, 155
Kawohl, B., 221
Kellogg, O. D., 75
Kenig, C., 210
Kerman, R., 23
Kesevan, S., 110
Kinnunen, J., 249, 252
Klaus, M., 289
Kočubeĭ, A. N., 158
König, H., 258, 259
König, M., 65, 81
Kopalini, T., 280
Kováčik, O., 268, 269, 272
Krein, M. G., 136, 140
Kresin, G., 113
Kufner, A., 249, 253, 256

L
Ladyzhenskaya, O. A., 81
Lamberti, P. D., 233
Langer, M., 201
Lang, J., 228, 261, 263, 265, 266, 268, 269, 272–274, 279
Lax, P. D., 7
Lê, An, 219, 233
Lebesgue, H., 55
Levinson, N., 183
Lewis, R. T., 235, 236, 244, 247
Lidskii, V. B., 259
Lieberman, G. M., 218, 219
Lindqvist, P., 218, 221, 225, 233
Lin, F., 56, 81
Linqvist, P., 221, 228
Lions, J. L., 189, 193, 253
Lyantze, V. E., 158

M
Magenes, E., 189, 193, 253
Malamud, M. M., 158
Manfredi, J. J., 228
Marletta, M., 158

Author Index

Martin, A., 292
Martin, G., 32, 33
Martio, O., 249, 252
Maurin, K., 264
Maz'ya, V., 113
Melrose, R. B., 113
Méndez, O., 210, 268, 272
Micheler, T., 201
Milgram, A., 7
Mitrea, M., 137, 143, 144, 196, 201, 210
Morrey, C. B., 56, 67
Musielak, J., 267

N
Naboko, S., 158
Naimark, M. A., 159, 160, 162, 164, 178, 183
Nardi, G., 81
Narukawa, K., 228
Nečas, J., 12, 14, 24
Nekvinda, A., 249, 250, 252, 269, 273, 274, 279
Nguyen, V. K., 263
Nichols, R., 169, 174
Nirenberg, L., 91, 194, 196, 204, 210
Nisio, K., 156

O
Oliver, L., 292, 294
Osękowski, A., 33
Ôtani, M., 245
Ouhabaz, E. M., 112

P
Pelaut, J., 9
Pichorides, S. K., 28
Pick, L., 23
Pietsch, A., 258–260, 264
Pleijel, A., 233
Posilicano, A., 201
Protter, M. H., 41
Prüss, J., 112
Pucci, P., 63

R
Raimondi, L., 201
Rákosník, J., 268, 269, 272
Raynal, J-C., 292, 294
Rellich, F., 168, 173, 235
Robinson, S. B., 225

Rockafellar, R. T., 213
Rosenberger, R., 169, 179
Roy, S. M., 292
Ruston, A., 258
Růžička, M., 266, 268, 269, 272

S
Safarov, Yu., 113, 265
Santalucia, V. M., 9
Schauder, J., 75
Schechter, M., 194, 196, 204, 289
Schilling, R. L., 1
Schmidt, K. M., 294–296
Schmincke, U. W., 289
Schmüdgen, K., 119, 121, 124, 128, 129, 166, 167
Schwartz, J. T., 1
Serrin, J. B., 63
Shterenberg, R., 137, 143, 144, 201
Simader, C. G., 202, 204, 210, 211
Simon, B., 137, 140
Simon, L., 81
Singh, V., 292
Sohr, H., 204, 210, 211
Stein, E. M., 253
Storozh, O. G., 158
Stromberg, K. R., 1, 214, 221
Struwe, M., 227
Stubbe, J., 292
Sun, J., 176, 178
Szulkin, A., 228

T
Taylor, A. E., 9
Teschl, G., 137, 143, 144, 169, 174, 201
Thaller, B., 288, 296
Tidblom, J., 236, 249
Tix, C., 294
Triebel, H., 10, 189, 190, 196, 250, 253, 256, 263–265, 273
Trudinger, N. S., 56, 81, 218
Tsekanovskii, E., 142

U
Ural'tseva, N. N., 81
Urysohn, P., 55

V
Vainerman, L. I., 158
Vassiliev, D. G., 113, 265

Višik, M. I., 7, 137, 140
Von Neumann, J., 137

W
Walter, J., 169, 184
Wang, G., 32, 33
Weder, R. A., 289
Weinberger, H., 41
Weyl, H., 113
Wood, I., 112, 158, 201
Wüst, R., 289, 290

Y
Yao, S., 176, 178
Yaouanc, A. L., 292, 294
Yosida, K., 9, 107

Z
Zamboni, P., 218
Zettl, A., 176, 178
Zhikov, V. V., 266
Ziemer, W., 24, 105
Zizler, V., 9

Subject Index

Symbols
$(C-)$ subharmonic function, 51
$(C-)$ superhamonic function, 51
C_0- contraction semigroup, 106
C_0- semigroup, 106
J-symmetric operator, 185
m-sectorial operator, 151
$p-$Laplacian, 204, 213
$p-$harmonic function, 216

A
Absolutely continuous norm, 23
Abstract Green's identity, 123
Accretive operator, 151
Adjoint form, 130
Adjoint of an operator, 117
Adjoint pair, 144
Approximation numbers, 258
Approximation property, 259

B
Banach function space, 22
Banach inverse mapping theorem, 5
Banach-Steinhaus theorem, 6
Barrier, 54
Bernstein numbers, 260
Besov space, 263
Boundary space of a linear operator, 125
Boundary triplet, 123
Brown-Ravenhall operator, 287

C
Closable form, 131

Closable operator, 115, 122
Closed form, 131
Closed graph theorem, 5
Closed operator, 115
Closure of a form, 131
Closure of an operator, 115
Coercive operator, 151
Configuration space, 284
Consistent maps, 110
Convex function, 213
Core, 132
Coulomb potential, 287
Courant min-max principle, 101

D
Deficiency, 115
Deficiency index, 117
Dirac matrices, 282
Dirichlet problem, 40, 91
Distribution function, 104
Divergence theorem, 24

E
Ehrling's lemma, 76
Elliptic operator, 57, 83
Essentially self-adjoint operator, 117
Extension of an operator, 115
Exterior ball condition, 55
Extremal m-sectorial extension, 157

F
Faber-Krahn inequality, 104
Field of regularity, 116

Form, 130
Form domain, 134
Fourier transform, 25, 284
Fréchet-differentiable, 96
Fredholm operator, 116
Fredholm-Riesz-Schauder theory, 8
Free Dirac operator, 282
Friedrichs extension, 130
Friedrichs inequality, 16
Fundamental solution of Laplace's equation, 42

G
Gamma field, 128
Gårding's inequality, 86
Gâteaux-differentiable, 96
Gaussian semigroup, 110
Generalised Dirichlet problem, 92
Genus, 227
Graph norm, 115
Green's formula, 24
Green's function, 45
Green's representation formula, 44

H
Harmonic function, 35
Harnack's inequality, 47
Hilbert transform, 26
Hille-Yosida theorem, 109

I
Index, 115
Infinitesimal generator, 106

K
Kato's inequality, 289
Kato, T., 151
Kernel of an operator, 115
Krein extension, 137
Krein-von Neumann extension, 137

L
Lagrange identity, 163
Lax-Milgram lemma, 6
Limit-circle case, 164
Limit-point case, 164
Linear relation, 121
Liouville's theorem, 47
Lizorkin-Triebel spaces, 263

Local barrier, 54
Locally subharmonic function, 52
Lorentz space, 23
Lower bound, 118
Lower semi-bounded, 117
Lower semi-bounded form, 131

M
Maximal function, 250
Maximal operator, 160
Maximal symmetric operator, 118
Mean distance function, 237
Mellin transform, 292
Minimiser, 221
Modulus of continuity, 86
Modulus of convexity, 5
Mollifier, 18
Momentum space, 284
Multi-valued part, 121

N
Newtonian potential, 65
Nodal domains, 21
Non-increasing rearrangement, 104
Non-negative operator, 118
Nuclear map, 257
Nullity, 115
Null space of an operator, 115
Numerical range, 116

O
Open mapping theorem, 5
Outer cone property, 250

P
Paley-Wiener theorem, 299
Pauli matrices, 282
Pauli spinor, 287
Perron's theorem, 53
Picone's identity, 229
Poincaré's inequality, 17
Point spectrum, 8
Poisson equation, 65
Poisson's integral formula, 45
Polarisation identity, 130
Pólya-Szegö principle, 105
Positive operator, 118
Principle part, 83
Principle symbol, 83

Subject Index

Q
Quadratic form, 130
Quasi-convex domains, 201

R
Rayleigh quotient, 100
Rearrangement-invariant space, 22
Regular differential equation, 160
Regular open set, 249
Regular point, 54
Rellich inequality, 235
Resolvent set, 116
Riesz representation theorem, 6
Riesz transform, 26

S
Schatten p-class, 259
Schauder's boundary estimate, 80
Second representation theorem, 135
Sectorial operator, 151
Self-adjoint operator, 117
Semi-Fredholm domain, 116
Semi-Fredholm operator, 115
Sesquilinear form, 130
Singular differential equation, 160
Sobolev space, 14
Spaces with variable exponent, 266
Spectrum, 116
Spherical 2-spinors, 293

Strictly convex function, 213
Strongly elliptic operator, 83
Strongly singular potentials, 202
Strong maximum principle, 39, 61
Subharmonic function, 35
Superharmonic function, 35
Symmetric form, 130
Symmetric operator, 117
Symmetric rearrangement, 104

T
The "soft" and "hard" extensions, 137
Titchmarsh-Weyl function, 165
Trace operator, 24

U
Uniformly convex space, 5
Uniformly elliptic operator, 57, 85
Uniformly strongly elliptic operator, 85
Upper Gaussian estimate, 110

W
Weak L_p solution, 209
Weakly p-harmonic function, 215
Weak maximum principle, 40, 59
Weak solution, 92
Weyl function, 128
Weyl's lemma, 50

Notation Index

Symbols
$G(T)$, 115
$L_2(\Omega, \mathbb{C}^4)$, 3
$\langle \cdot, \cdot \rangle_{\mathbb{C}^4}$, 15
$\mathbb{B}$, 281
$\mathbb{D}$, 281
$\mathbb{D}'$, 281
$\mathbb{H}$, 281
$\mathcal{N}(T)$, 115
$\mathcal{N}_\pm$, 118
$\overset{0}{H}{}^m(\Omega)$, 15
def T, 115
ind T, 115
nul T, 115
$\tilde{\Delta}(T)$, 116
$(\cdot, \cdot)_{m,2,\Omega}$, 15
$(\mathcal{K}, \Gamma_0, \Gamma_1)$, 123
$AC[a, b]$, 160
$B(X)$, 4
$B(X, Y)$, 4
$B_{p,2}^{3/2}$, 201
$B_{p,q}^s$, 263
$C^{k,\lambda}(\overline{\Omega})$, 10
$C^{k,\lambda}(\Omega)$, 10
$C^k(\Omega)$, 9
$C^k(\overline{\Omega})$, 10
$C_0^k(\Omega)$, 9
$E(\mathbf{p})$, 285
$F_{p,q}^s$, 263
G_∞, 122
H, 26
$H^m(\Omega)$, 15
$H^m(\Omega, \mathbb{C}^4)$, 15
$H^s(\partial\Omega)$, 190

$K(X)$, 8
$K(X, Y)$, 8
$L_{1,loc}(a, b)$, 159
$L_{p,loc}(\mathbb{R}^n)$, 204
$L_{p,q}(\Omega)$, 23
$L_p(\Omega)$, 3
$L_p(\partial\Omega)$, 13
M, 269
$M_R u$, 250
$N(\lambda)$, 113
$P(x, y)$, 46
$R(v)$, 100
$R_\mu(A)$, 119
R_j, 26
$T^+(\tau)$, 160
$T_0(G)$, 122
T_F, 137
T_K, 137
$W_{2,loc}^2(\Omega)$, 136
$W_p^\beta(\partial\Omega)$, 24
$W_p^m(\Omega)$, 14
$X \hookrightarrow Y$, 5
X^*, 5
$X_p(\Omega)$, 204
$Z(u)$, 21
Δ, 35
Δ_p, 204, 213
$\Gamma(T)$, 116

$\Lambda(\mathbf{p})$, 285
$\Omega^\star$, 105
$\Omega_{(l,m,s)}$, 293
$\Phi_+(T)$, 116
$\Pi(T)$, 116
$\Sigma(\mu,\theta)$, 151
$\Theta(T)$, 116
α_j, 282
$\delta(x)$, 237
$\delta_{M,p}(x)$, 237
δ_X, 5
$\gamma(A)$, 227
$\gamma(z)$, 128
γ_0, 192
γ_a, 192
γ'_a, 192
$[\cdot]_{k,\lambda,\Omega}$, 10, 11
$\|\cdot\|_{m,p,\Omega}$, 14
$\|\cdot\|_{p,\Omega}$, 3
$\langle\cdot,\cdot\rangle_X$, 4
$\mathbb{F}$, 284
$\mathbb{M}$, 285
$\mathcal{A}(X)$, 258
$\mathcal{A}(X,Y)$, 258
$\mathcal{M}$, 292
$\mathcal{M}(G)$, 121
$\mathcal{M}(\Omega)$, 267
$\mathcal{N}(X,Y)$, 258
$\mathcal{N}(X)$, 258

$\mathcal{N}_\lambda$, 119
$\mathcal{N}_{s,A}(\Omega)$, 191
$\mathcal{P}(\Omega)$, 267
$\mathcal{S}(\mathbb{R}^n)$, 25
$\mathcal{S}'(\mathbb{R}^n)$, 25
$\overset{0}{W}{}_p^m(\Omega)$, 15
$\rho(T)$, 8
$\sigma(T)$, 8
σ_j, 282
$\sigma_p(T)$, 8
$\hat{f}$, 25
$a_k(T)$, 258
$b_k(T)$, 261
e^{tA}, 106
f^*, 22
l_p, 3
$m_i(T)$, 117
$m_\pm(T)$, 117
$p^* = np/(n-p)$, 17
$p_m(x,\xi)$, 83
$t[\cdot,\cdot]$, 130
$t[\cdot]$, 130
u^*, 104
$u^\star$, 104
u_Ω, 17
$\overset{0}{H}{}^m(\Omega,\mathbb{C}^4)$, 15
M(z), 128
supp u, 9

Printed by Printforce, the Netherlands